The Medicine Bow Mining Camps

THE MEDICINE BOW

by

with illustrations by

JELM MOUNTAIN PUBLICATIONS

MINING CAMPS

MEL DUNCAN

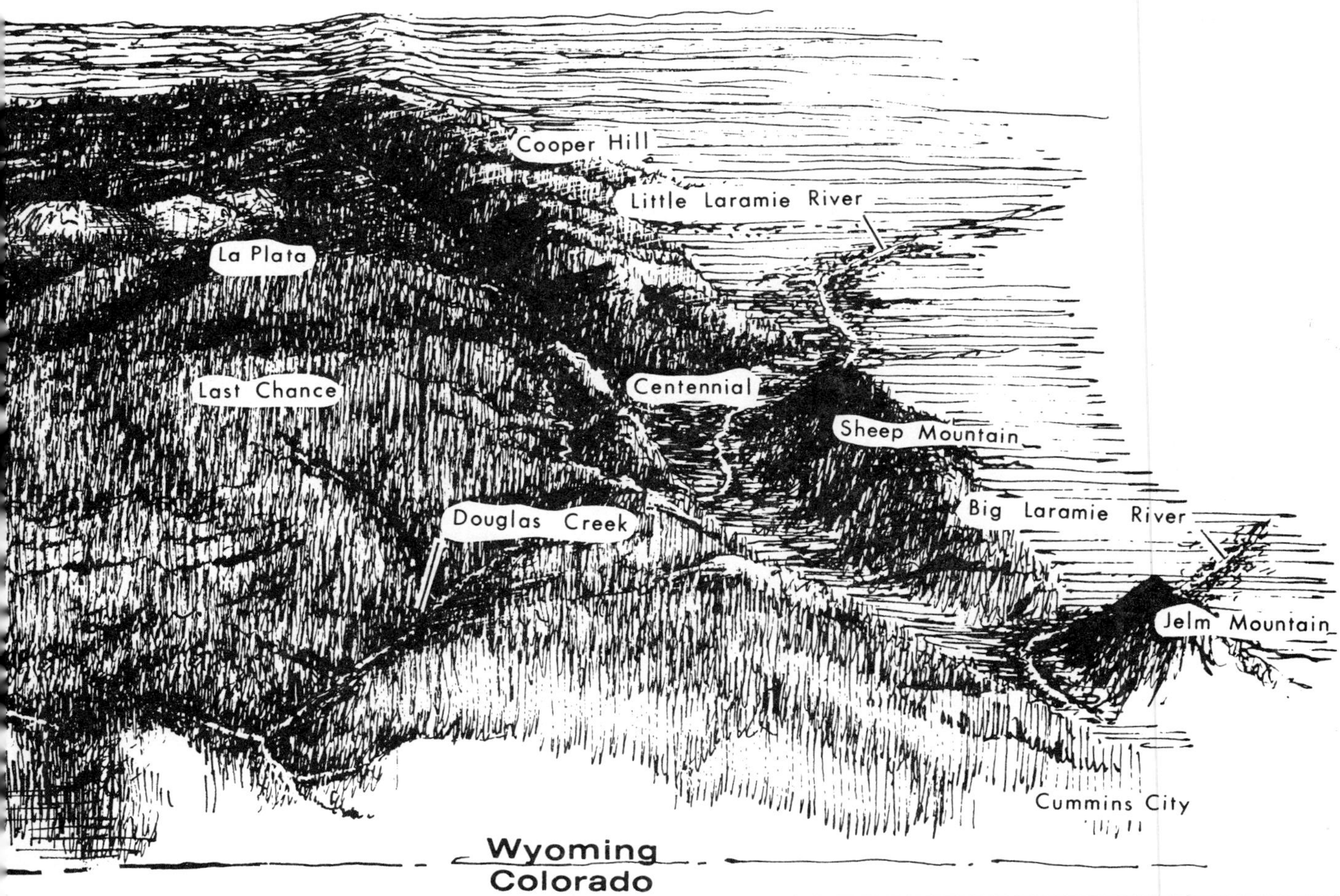

Dan Swift and Joe Barros

LARAMIE, WYOMING 1990

The Medicine Bow Mining Camps

Library of Congress Cataloging-in-Publication Data

Duncan, Mel, 1931 -
The Medicine Bow Mining Camps / by Mel Duncan ; with illustrations by Dan Swift and Joe Barros.
p. cm.
Includes bibliographical references and index.
ISBN 0-936204-90-7 (hbk.) — ISBN 0-936204-89-3 (pbk.)
1. Medicine Bow Mountains (Colo. and Wyo.)—History, Local.
2. Mines and mineral resources—Medicine Bow Mountains (Colo. and Wyo.)—History. I. Title.
F767.M42D86 1991
978.7'86—dc20 90-49405

Jelm Mountain Publications
209 Park Street, Laramie, Wyoming 82070
ISBN 0-936204-89-3 sft. cvr.
ISBN 0-936204-90-7 hrd. cvr.

To my
Wife and Family
Also to Jake, my best friend and constant companion.

ACKNOWLEDGEMENTS

I have accumulated many debts during the gathering of historical data to complete this book. Without the assistance of the custodians of the historical materials I could not have completed the project.

To the keepers of these historical materials I indeed owe a debt of gratitude.

The Wyoming State Archives, Museums and Historical Department, Historical Research and Publications: Rick Ewig, Jean Brainerd for her assistance in my early days of research and the proofreading of the manuscript, Ann Nelson, Roger Joyce and Paula West-Chavoya for her help in accessing historic photos.

Wyoming State Archives, Museums and Historical Department, Archives and Record Management Division: Ida Wozny, Cindy Brown, Paula (Bauman) Taylor, Carl Hallberg, Curtis Greubel and Tonè Mendoza.

The Bureau of Land Management: A special thanks Marty Winemiller of the BLM for her help with the General Land Office material.

The University of Wyoming Librarians: I would especially like to express my gratitude to Emmett Chisum, also to Paula McDougal and Jim Carlson for their help in researching old photographs.

Thanks to Dan Swift for his very appropriate original ink drawings that add so much to the text.

A special thanks to Joe Barros for his portrayal of some of the early mining characters.

Thanks to Graig Marrs for his photo work.

Thanks to Jim Johns for reading the manuscript and providing me with his critique and encouragement.

To Jim Beasley who assembled the book I owe a debt of thanks for a difficult job well done.

Perhaps my greatest debt of gratitude is to those of the past who had the foresight to preserve for the future the documents, records and newspapers that I found indispensable for the type research required for this book.

And finally thanks to all my friends for their encouragement.

CONTENTS

PREFACE

Rising from the Laramie Plains on the east and the Platte River Valley on the west, the Medicine Bow Mountains stretch from Elk Mountain on the north, across the Wyoming and Colorado border, to Cameron Pass on the south. The perennial snow capped peaks spawn a myriad of streams flowing into Douglas Creek and Brush Creek on the west, the Medicine Bow River on the north and the Big and Little Laramie Rivers on the east, all flowing eventually to the North Platte River.

Hardly a square mile of these mountains exist without evidence of mining, whether it be a collapsed tunnel, a pile of tailings, a decaying cabin or merely a ten foot "discovery hole."

Remains of more than half a dozen mining camps, or "cities" as they were frequently referred to, are hidden away in the forest. Who were these people who toiled in the mines, built the cabins and panned the streams in search of precious metals? Why did they not leave some trace or record of their activities? The answer is, of course, they left many traces both written and physical. The county records show thousands of claim filings, The mountains show hundreds of mining scars. All that is required is a gathering of some of these records and a presentation in some semblance of order. Hopefully, this is that presentation.

INTRODUCTION

Works of history are rarely the product of individual effort. The author is, in reality, an editor of the information he has gathered and found appropriate to the subject. However, to present an acceptable product the serious researcher must devote a tremendous amount of time and extend his research to every possible source including individuals, museums, archives, libraries and official depositories. The author then, presents only sufficient original writing to bind the work together. Such is the case with this work of history.

From the beginning I had no intention of skipping over the names and trivial experiences of the miners and prospectors involved. If there is a name attached to the incident described, the name is used. If there is a place involved, the place name is spelled out. Perhaps you will find the name or place of your interest. If names and trivia history bore you, perhaps this book is not for you.

In an attempt to keep the endnotes to a reasonable amount, I have included many references in the text. I have included a few maps to orient those unfamiliar with the Medicine Bow Mountains. I have also included a few prints and drawings for those unfamiliar with the old mining techniques.

This book is intended for both the casual reader and those who wish to delve deeper into the history of the area. Not only the recreational miner but the ghost town enthusiast, the naturalist, the hunter and fisherman and in fact anyone with an interest in the Medicine Bow Mountains should find much of interest. Although the book is an attempt to bring the entire mining era together, each chapter stands alone and can be perused as such. Consequently there may be a small amount of repetition.

The Medicine Bow Mining Camps

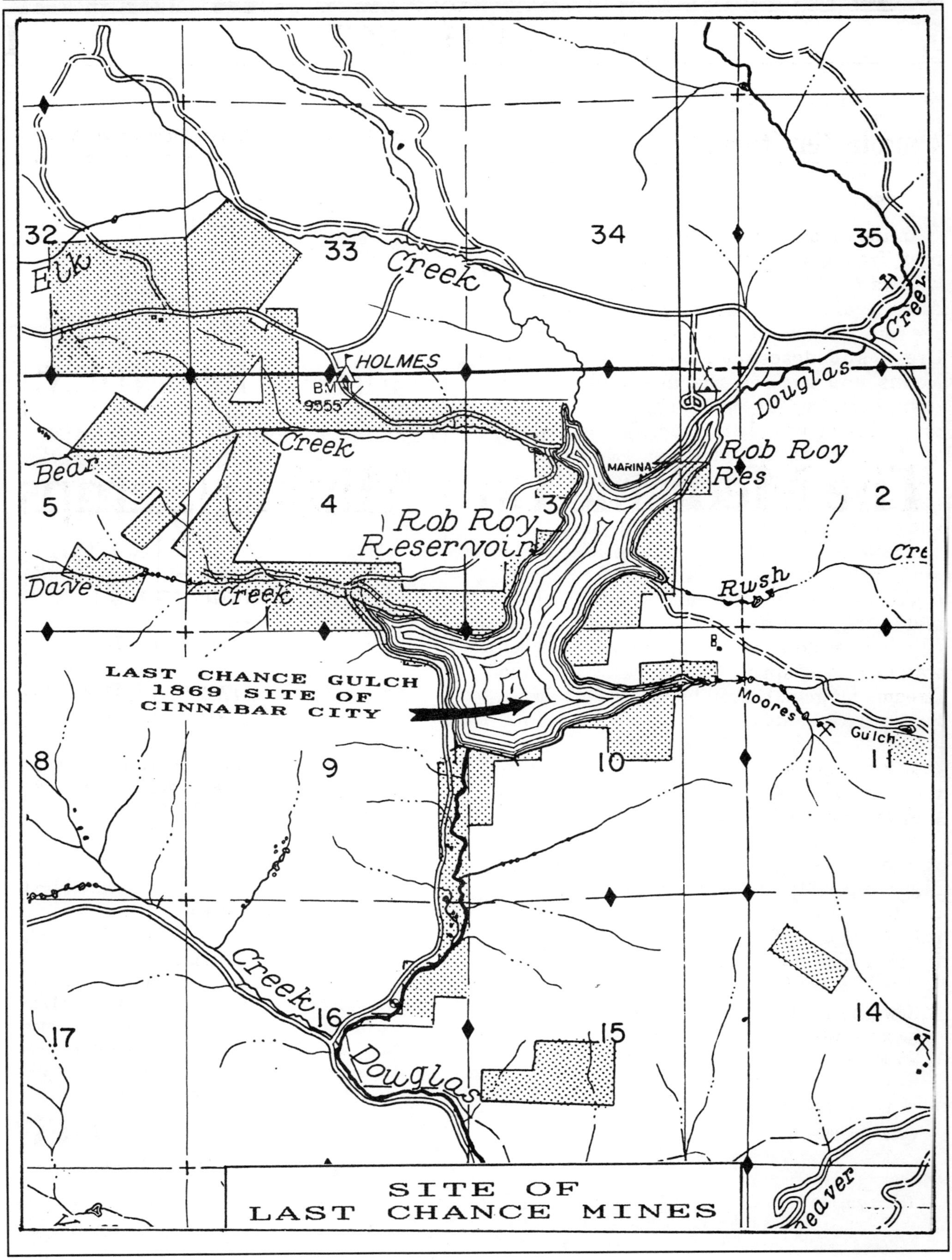
32
33
34
35
Elk
Creek
HOLMES
BM
9555
Bear
Creek
Douglas
Creek
Rob Roy
Res
MARINA
5
4
3
2
Rob Roy
Reservoir
Dave
Creek
Rush
Cre
LAST CHANCE GULCH
1869 SITE OF
CINNABAR CITY
Moores
Gulch
8
9
10
11
Creek
16
Douglas
17
15
14
Beaver
SITE OF
LAST CHANCE MINES

Dakota Territory:

- **1868**

In the area we now know as Wyoming, the first Transcontinental railroad was pushing westward to join its eastbound counterpart. After a midwinter pause west of Cheyenne, the rails crossed the huge new trestle at Dale Creek and descended onto the Laramie Plains and entered the new end-of-track town, Laramie City, on May 4, 1868. The new town, located near Fort Sanders, was already growing beyond the typical end-of-track tent city. General Grenville M. Dodge had chosen the site and the Union Pacific began selling lots even before the rails arrived. Aside from the new settlements along the railroad the territory was largely uninhabited. The arrival of the railroad brought men of every calling and in the mountains to the west of Laramie City, prospectors were soon braving the icy waters of the spring runoff, searching the streams for gold. Driven by the age old desire for the yellow metal, and fueled by the discoveries of the Colorado Cherry Creek prospectors and the Montana gold seekers, these prospectors had every reason to believe they too would be successful.

As early as April, 1868, the territory's pioneer newspaper, the *Frontier Index*, had begun reporting first rumors and then definitive facts of what they called "significant new gold discoveries" in the mountains on the western edge of the Laramie Plains.

April 28, 1868:

New gold mines have just been struck on the Big Laramie River only a half a day's ride from Laramie City, which promise to be as rich if not better than the Sweetwater Mines.

May 19, 1868:

Great excitement. - Prospectors just in from Big and Little Laramie, Twenty-eight miles due west from Laramie City report wonderful discoveries of gold. The results so far, are a long way ahead of anything yet found in the Sweetwater mines. It is confidently believed by old prospectors that there will be fifteen thousand people on the Laramies and in the eastern edge of North Park in less than four weeks. A great many are flocking thither now and numbers on the way.

Friday, May 22, 1868:

INCREASING COMMOTION
GOLD! GOLD! GOLD!
GOLD IN DAKOTA
GOLD ALL AROUND LARAMIE CITY

FRONTIER INDEX.

F. K. Freeman, - - Editor

FRIDAY,.......................MAY 22, 1868.

INCREASING COMMOTION.

GOLD! GOLD! GOLD!

GOLD IN DAKOTA.

GOLD ALL AROUND LARAMIE CITY.

HUNDREDS PREPARING TO DIVE INTO NORTH PARK, AND ON BIG AND LITTLE LARAMIES.

Hundreds are preparing to dive into North Park, and on the Big and Little Laramies.

Just as we told you in the winter about rich mines opening up all around Laramie, all over the mountains and hills which peer up before us on every side. Now we see miners and prospectors making ready to scoop up the precious dust within almost an arms reach of town. Old miners tell us they can average from one to five dollars to the pan in North Park and on the Laramies. From sixty to seventy colors can be gotten up near the heads of these rivers, which rise within a half day's ride of Laramie City and run immediately by the town. Eureka! Eureka!! Eureka!!! Come one, come all; here you will find both gold and silver . . . .

Another article the same day stated:

Laramie City is the great outfitting point for our new

> mines which surround the 'City of the Plains' on all sides. This is to be the golden paradise of the West.
>
> The mining excitement in our midst is becoming intense. Old miners are daily outfitting for the heads of Big and Little Laramies and the eastern edge of North Park - 38 miles due west of Laramie City.

From May 19 to May 22, the estimated distance to the mines had grown from "twenty-eight" to "thirty-eight" miles, the latter a fairly reasonable estimate to the closest of the new discoveries.

May 26, 1868:

> The mines immediately west of Laramie City, are creating great excitement. The pencil whacker of the *Index* will go over to North Park in a few days and stake out a claim or two for 'u-uns and we-uns.' Everybody is getting ready to take a trip to the gold leads.
>
> Mr. Wm. Rollins started over to the new mines - in sight of Laramie City - on Saturday with a party of good old sober sided miners and will commence operations at once.

May 29, 1868:

> In ten or twelve days, the snow in North Park and at the heads of Big and Little Laramie will have so nearly melted away, that miners and prospectors can dive into the new gold fields just at the base of yonder mountain range, only half a days ride from Laramie City. And won't there be a rush for the precious metal? Old miners, who have been over and returned, tell us that these new mineral districts will in less than six weeks surprise the world.

Apparently, the precise location of the new gold field was not yet known to the reporters for the *Index*. While the base of the mountains might be reached in half a day on a good horse, the gold fields were still at least another half a day's ride up into the forest and mountains.

Frontier Index, June 5, 1868:

> Today, parties are fitting out for North Park and Big and Little Laramie, taking supplies to last several weeks. The prospects are said to be splendid.

There appears to be no evidence that the reporters from the *Index* ever joined the parties for their proposed visit to the mining areas, but they did continue to print the news of the mines and the miners.

June 5, 1868:

> Mr. Metcalf and his comrades killed a grizzly bear weighing eleven hundred pounds. The grizzly in question, made a break at the party, the first ball having stung him, but they dodged behind a clump of bushes and kept up the fire until bruin keeled over.

On July 25, 1868, President Andrew Johnson approved the Organic Act of Wyoming, which created the Territory of Wyoming. By that time, the railroad had progressed almost halfway across the new territory. Laramie City, although still under the influence of vigilante law, was struggling to establish a city government and compete for a place in the new territory. With the passing of the railroad, a successful mining region seemed the very thing needed to bring Laramie City to the forefront.

In mid-July the *Frontier Index* moved west to Benton and then to "Greenriver City" where they continued to print mining news, predominately concerning the Sweetwater Mines of South Pass. Their last story about the Medicine Bow Mines was printed September 4, 1868:

> We have just received information of a tremendous rush to Elk Mountain. One man came into Laramie with seven hundred dollars panned out in seven days, with no other implements than pick and shovel. He went in for provisions and was followed back by a large number of persons.

If indeed there was a "tremendous rush," there seems to be a complete lack of evidence in remaining records. The records do, however, present a more complete picture of the mining activities in the central Medicine Bow Mountains.

The principal prospecting and mining was in the central portion of the Medicine Bow Mountains. While some of the mining activity was in the North Park area, the primary discoveries were on a major tributary of the North Platte River, not on the Big or Little Laramie as first reported.

The parties setting out from Laramie City followed a rather indirect route, northwest on the Overland Trail to Mandel's Crossing, then southwest up the Little Laramie to where Middle Fork enters the mountains. Then, by a narrow trail that allowed only pack animals and foot travel, up and over the mountain to what is now known as Douglas Creek.

Here, on a little creek in a small ravine which they named "Moore's Gulch," for the first prospector to stake a claim, Ira B. Moore, they began panning and sluicing the gold-bearing gravel. The first group of miners banded together and took the name "Dakota Mining Company." The main stream, which

Rob Roy Reservoir, 1986, looking West. Moores Gulch in foreground

we now know as Douglas Creek, they named "Antonia River" after finding "Antonia 1830" carved in a tree trunk nearby.[1] The main gulch where they worked they named "Last Chance Gulch."[2] It was located at the present site of Rob Roy Reservoir. Adjoining the Dakota Mining Company's claim John B. Wands staked out a 200 foot claim, which he called the "Morning Star Lode." Adjoining Wand's claim, five men from Laramie City also staked their claim to an additional 1,000 feet of the gulch and named their claim the "Morning Star Lode Extension 2, 3, 4, 5 and 6." [3]

These five men, A.S. Blackbourne, Thomas Feast, John Metcalf, C.W. Hillaker and J.L. Roberts along with Ira Moore, A.O. Libby, Doctor Irwin, N.F. Spicer and Al Cushman of the Dakota Mining Company and about forty others, continued working through the summer of 1868. The miners chose to live in temporary shelters rather than take time from mining to build more permanent log cabins.

During September the claims were visited by Ferdinand Vandiveer Hayden who was on a side trip from his second survey of the west.[4] Joining a large party directed by Generals Gibbon and Potter and in the company of Professor James Hall, New York Paleontologist, Hayden camped with his wagons on the Little Laramie River. He then rode horseback to the "diggings," about ten miles away. The party spent two days in an unsuccessful

Ferdinand V. Hayden on his horse "Patsy"

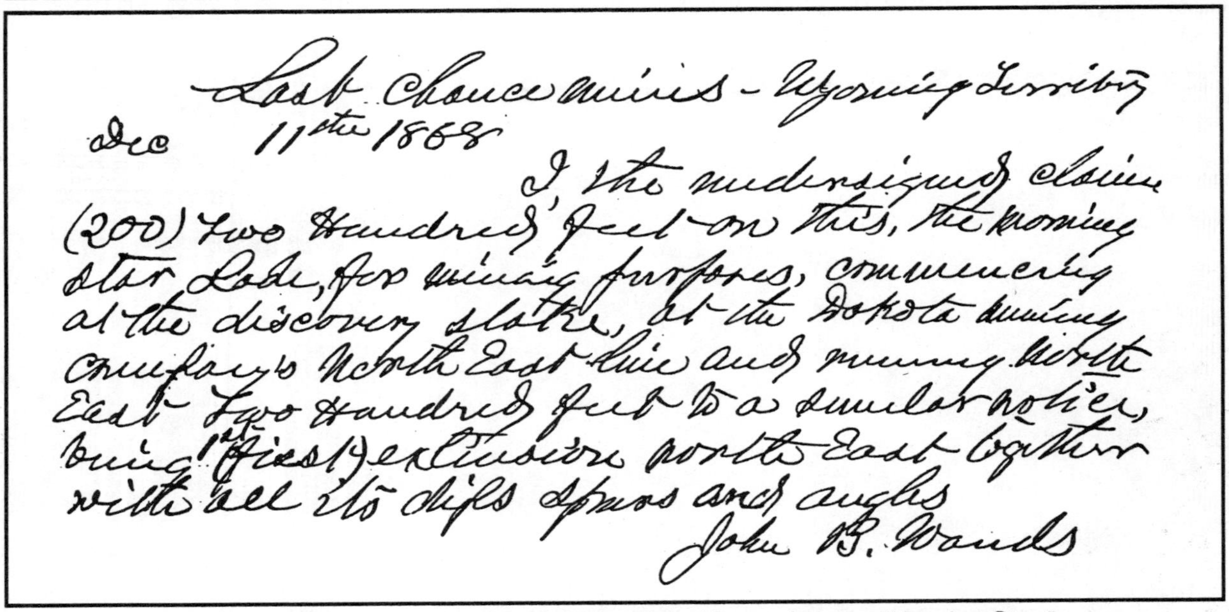

Last Chance mines - Wyoming Territory
Dec 11th 1868
I the undersigned claim (200) two Hundred feet on this, the Morning Star Lode, for mining purposes, commencing at the discovery stake, at the Dakota mining company's North East line and running North East two Hundred feet to a similar notice, being 1st (first) extension North East together with all its dips spurs and angles
John B. Wands

The first recorded claim in the Medicine Bow Mountains, the "Morning Star."

search for the source of the placer golds. In his official report he wrote of the beauty of the area and although not impressed by the value of the mines, noted, "That valuable mines will be found in these mountains at no distant day seems probable." Further, he made the same observation and conclusion that was made by gold seekers both before and after; "The geological evidence is quite conclusive, and the mountains are a continuation northward of the same range in which the rich mines of Colorado are located."[5]

As the summer drew to a close, a lack of water for suitable placer mining and a threat of early snow with no permanent shelters yet constructed, convinced the miners to retire to Laramie City for the winter. The Laramie City to which they returned was not the same tent town they had left. In the first three months of existence, Laramie City had grown to a reported, although doubtful, population of nearly 5,000 people.[6] As the railroad moved westward a major portion of these people followed, but for a short period of time, Laramie City was a booming end-of-track railroad town.

On December 16, 1868, Albany County was established and Laramie City was designated as the county seat. Five days before the forming of the county, the Last Chance miners had begun filing their claims under the Mining Act of July 26, 1866, which provided for claims 200 feet in length for each locator, with an additional claim allowed for the discoverer and a maximum claim of 3,000 feet for an association. The "Morning Star" became the first recorded mining claim in the Medicine Bow Mountains and was entered on page 6 of the new Albany County record book:

> Last Chance Mines - Wyoming Territory
> Dec 11th 1868 –
>
> I the undersigned claim (200) two hundred feet on this, the Morning Star Lode, for mining purposes, commencing at the discovery stake at the Dakota Mining Company's North East line and running North East two hundred feet to a similar notice being 1st (first) extension North East together with all its dips, spurs and angles.
>
> John B. Wands

The next day, additional claims followed:

> Last Chance Mines Wyoming Territory
> Dec 12th 1868 –
>
> We the undersigned claim (five claims) One thousand feet, this the Second extension North East on the Morning Star Lode, commencing at a stake two hundred feet North from the Dakota Mining Company's discovery stake, and running One thousand feet North East to a similar notice together with all its dips, spurs, and angles - the same being Claims Nos. 2, 3, 4, 5 and 6 North East from discovery, claimed and staked this day Dec 12th 1868 –
>
> A.S. Blackbourne
> Thomas Feast
> John Metcalf
> C.W. Hillaker
> J. L. Roberts[7]

Moore and the Dakota Mining Company

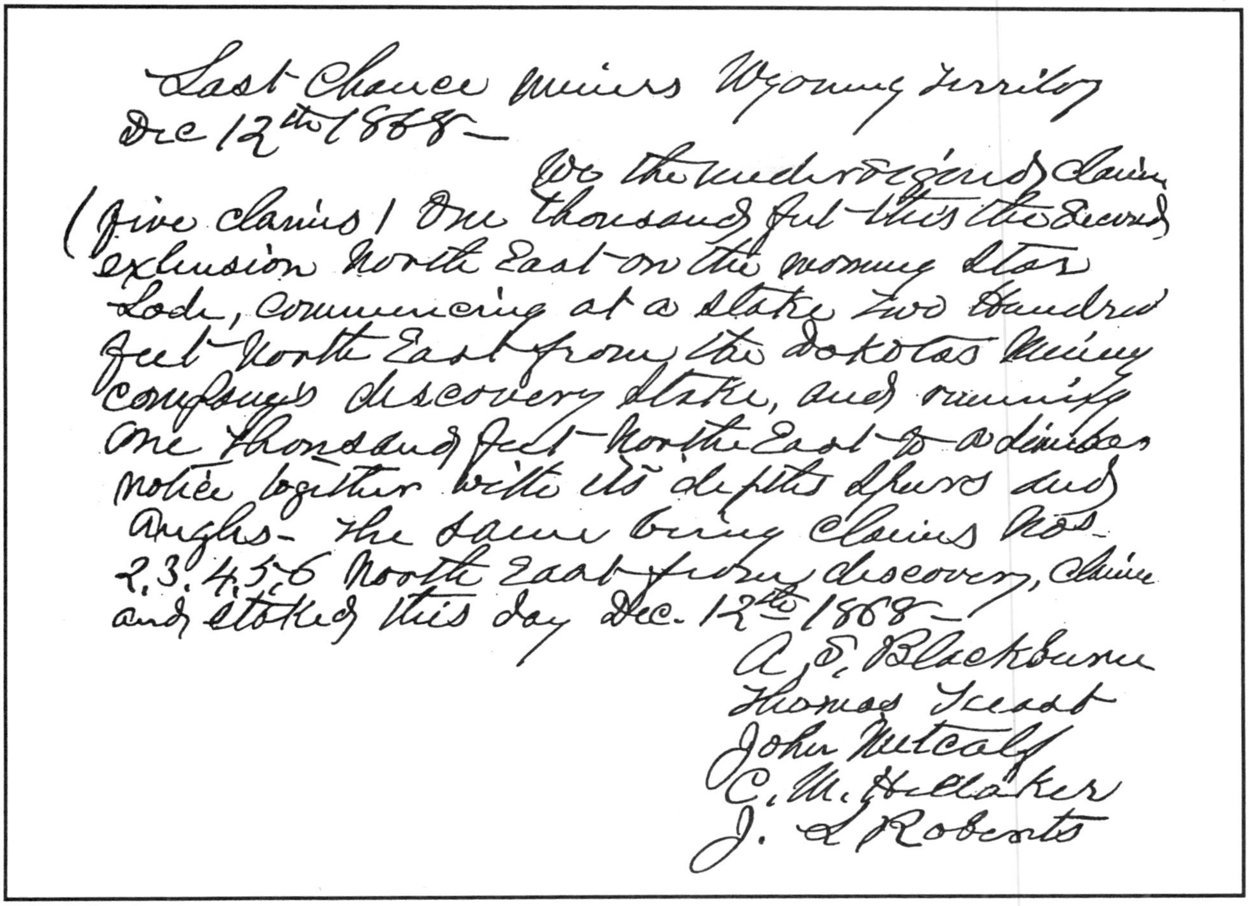
Last Chance Miners Wyoming Territory
Dec 12th 1868—

We the undersigned claim (five claims) One thousand feet this the Second extension North East on the Morning Star Lode, commencing at a stake two Hundred feet North East from the Dakotas Mining Companys discovery stake, and running One thousand feet North East to a similar notice together with its depths spurs and angles— the same being Claims Nos. 2,3,4,5,6 North East from discovery, claimed and staked this day Dec. 12th 1868—

A. S. Blackburn
Thomas [illegible]
John Metcalf
C. M. Hillaker
J. L. Roberts

The "Morning Star" extensions.

failed to file any claims, presumably preferring to hold their claims by markers alone until they could more accurately determine the extent of the placer gold and its origin.

Apparently, the miners had washed out enough gold to carry them through the winter and create a new interest in Laramie City. The gold shown in Laramie City was coarse placer gold, but the Dakota Mining Company members seemed more interested in the gold-bearing quartz ore on the hills above the gulch.

The higher Medicine Bow Mines were deserted for the winter season, although, according to C.W. Shores, about fifty men were employed at Spooner's Mill on Mill Creek, cutting ties and bridge timbers for the Union Pacific Railroad.[8]

- **1869**

In February, 1869, Ira Moore, "one of the largest owners in the Dakota Mining Company," was in Omaha for assays of ore from the Last Chance District and to obtain financial backing for the "Dakota and Omaha" mining companies.[9] Apparently he was only partially successful.

As soon as the snow began melting in the spring of 1869, some of the miners returned to the Last Chance Mines followed by a new group of hopeful prospectors. Following the now familiar route, the miners returned with a revised plan to make the most of the short mining season. Led by Stephen W. Downey, the miners divided into groups of fourteen and began staking claims of 3,000 feet each on the creeks and gulches near their previous year's claims.

The "Reindeer Gold and Silver Lode" on Gold Run Gulch[10] was staked out on May 26, 1869, by fourteen of the miners.[11]

The next claim was the "Humbolt Lode on Last Chance Creek," also filed on by fourteen of the miners, ten of which had their names on the "Reindeer" and four new claimants.[12]

In May, 1869, Territorial Governor, John Campbell, instructed the U.S. Marshal, Church Howe, to conduct a census of the Territory of Wyoming. Howe assigned deputies to each of the four counties, Albany, Carbon, Carter and Laramie.[13]

Stephen W. Downey was designated as Deputy U.S. Marshal for Albany County and completed the census of the mining settlement in the Last Chance Mining District, which he called Cinnabar City. Downey enumerated fifty people in and around Cinnabar City; thirty-four miners, five merchants, four carpenters, three teamsters, two blacksmiths, one saloon keeper and one lawyer, himself. Apparently there were no women in the mining camp. All but D.D. Wolff claimed to be citizens of the United States. Wolff claimed to be a citizen of Colorado. When the first election of the new territory was held on September 2, 1869, the men of Cinnabar City cast seventeen votes for Delegates to Congress.[14]

The miners were almost completely reliant on Laramie City to provide them with tools and provisions. Had they chosen, they had very little opportunity to live off the land. The lakes and streams of the Medicine Bow Mountains were devoid of fish and the forests had only a sparse population of game animals. Elk at that time were primarily a prairie animal and were already following the path of the decreasing bison herds. Antelope were for a time plentiful on the plains but they too were soon to become scarce.

On June 7, 1869, Sam Phillips took twelve dollars in newly washed gold and rode to Laramie City for supplies, apparently the first practical use of the 1869 gold finds. He was accompanied by a man named Williams whose name does not appear on any of the claims but who was apparently placer mining in the area.

On June 17, 1869, another group of fourteen miners claimed another 3,000 feet which they named the "Crescent Lode." Here again, by ten of the previous claimants and four new.[15]

Another 3,000 foot claim was also made on June 17, 1869, by a fourth group of fourteen miners, this one to be known as the "Cosmos Lode." Eight new names appeared on this claim.[16]

During the summer of 1869, under the direction of the Downey Brothers, time out from mining was taken for construction of log cabins and mining facilities.

In a letter to the Laramie City newspaper, Stephen Downey wrote:

> The greatest drawback now is the want of lumber which can only be obtained by the slow process of whipsawing, there being no road to get in with wagons. There is to be a road built into the mines immediately, parties having already undertaken the job.[17]

More than a mile of ditches were constructed for water to use when the spring runoff subsided. One ditch ran from Ruth's Creek[18] to Moore's Gulch, and another from Big Elk Creek[19] to Dave's Creek at Last Chance. (Ditches in the area were visible up to the time of construction of Rob Roy Reservoir).

In later years, several sources reported about $8,000 was taken out of the Last Chance District in 1869 before the miners once again returned to Laramie City to wait out the winter.[20] Laramie City again had changed, the population had dwindled to less than 800 people. (The 1870 census counted 788).

- **1870**

To the miners, 1870 promised to be a great year for mining in the Medicine Bow Mountains.

The *Laramie Daily Sentinel*, May 10, 1870:

> Mining matters are again looming into prominence in our city and vicinity. There was a large number of prospectors and miners on our streets today, purchasing outfits and making preparations for the seasons work. The Bank here purchased a nice lot of coarse gold from the Last Chance miners. We had the pleasure of seeing it and unhesitatingly pronounce it the richest looking gold dust we have ever seen. The prospect is most encouraging for a large yield of gold from the North Park and Last Chance mines the coming summer.

Within a few days the first group of miners departed for the Last Chance Mines. Another smaller group departed for Independence Mountain in the North Park area. Both groups followed the Little Laramie to Middle Fork where the Last Chance miners headed up into the mountains. The North Park miners continued up the valley, over to the Big Laramie River to a portion of the Cherokee Trail and then across to North Park. On May 18 the *Laramie Daily Sentinel* reported the first news of the mining season:

> We have received news from the first party of miners who went into the Last Chance mines last week. They experienced some difficulty in getting there, finding a good deal more snow than was anticipated. They had to shovel through snow from five to fifteen feet

Laramie City, 1870.

deep for a considerable distance, but there were seventeen of the boys, all stout resolute fellows who knew where they were going and what they would find when they got there. Upon arriving at their cabins in the gulch they found them nearly out of sight under the snow. A mountain lion had appropriated one of them to his own use, making his entrance and exit through the chimney. His majesty happened to be out on a prospecting expedition when the boys arrived and took possession, but soon after returned and coolly walking up to the door and looking in, when apparently thinking there was enough company without him, he quietly left, and the boys in consideration of the civil manner in which he had occupied the premises, let him depart in peace. This gang is but the advance guard of hundreds who will go in there to work this spring and summer, and we expect to see a pretty good sized golden stream flowing out from there soon. We shall keep and keep the public informed of their progress.

During the first week in June, A. S. Blackbourne sent a letter to the *Sentinel*, telling them that the three miners in his group were averaging $6.00 a day each, and were expecting soon to double that amount.[21] This was considered very good working wages but hardly a bonanza.

Also during the first week in June, a group of forty miners arrived in Laramie City from the Sweetwater Mines near South Pass. Some of these miners settled in Laramie City and a few headed for the mines in the Medicine Bow Mountains.

By the middle of June, some of the miners had become discouraged and returned to Laramie City. Blackbourne said of these miners, ". . . Not one of them either dug out or washed a single pan of dirt."[22]

By the end of June, the Wyoming Territorial Delegate to Congress, Stephen F. Nuckolls, informed the miners and the *Sentinel* that mail service to the mines could be had if the proper steps were taken, namely a petition and a recommendation for postmaster. Also by the end of June, the president of the Fort Sanders Mining Company, John Metcalf, announced the completion of a water powered arrastra[23] for crushing the ore, using the water power of Douglas Creek.

Mining was beginning to show real progress and hopes ran high at Last Chance Gulch and Independence Mountain. Then in July an event, which was to affect mining for years to come, occurred in North Park. Four of the Independence Mountain miners were killed. The miners had been warned to leave by the Ute Indians led by Chief Colorow. Most of the miners left, but four had chosen to stay at Gilman's camp; Shipman, who was recovering from a broken leg, George Van Dyke, Dutch Jake, and one other, possibly Gilman.

On July 4, 1870, a party of seventeen men from Cheyenne, led by Colonel I. W. French, who had been to the Little Snake mines in search of gold, arrived in the general area where the four men had remained to mine. Investigating a flurry of gunfire nearby, they

Drawing by
Dan Swift

found the Utes and old Colorow shooting now and then at another ridge at what they said were Arapahoes. Colorow asked French for the loan of his guns but French wisely declined, saying he might have to use them himself.

Detouring the Indians, they pushed on to Gilman's camp where they found the body of Shipman, "still in bed, riddled with bullets and with two axes stuck in his head." After burying Shipman, they proceeded down to the Platte River where they found two more bodies, both scalped. Van Dyke's body was not found until later by another party. After burying these bodies, the party continued onward to Laramie City and then Cheyenne. In the meantime, the Indians departed down the Cherokee Trail. Near Grand Encampment Creek, they apparently came upon three trappers, Frank Marion, John (Old Man) Scott and Joe Brum. They were also killed and scalped.

On July 12, 1870 the *Laramie Daily Sentinel* published a short article:

> We learn from good authority that a party of from twelve to fifteen Indians are lurking near Last Chance. It is expected that the miners will have to fight their way home, and that when they leave their property will be destroyed.

When the news of the massacre and the "lurking" Indians reached Laramie, Stephen Downey recognized the potential danger to the unsuspecting miners at Last Chance, which included his father Owen D. and his brother William. Alone, Stephen rode to Last Chance to warn the miners. The miners, unwilling to risk their lives, followed Downey back to Laramie where they spent the next few weeks complaining to the government about the lack of protection.

This incident near Independence Mountain became known as the Shipman Cabin Massacre and a controversy arose over who actually committed the killings, the Utes led by Colorow, the Arapahoes as claimed by Colorow, or the Cheyennes, as reported by the *Cheyenne Daily Leader*.

Pioneer Judge W. L. Kuykendall maintained that the Utes did the killing and then tried to cover up the act with various stories. This was later confirmed. The Judge also believed that had Colorow been punished for the Shipman Cabin incident, the later Meeker Massacre would not have occurred.

It was mid-August, 1870, before the miners returned to their settlement to finish what remained of the mining season.

Some of the prospectors chose to explore possible claims a little closer to Laramie and more distant from the Utes. In June, 1870, Charles Owens claimed 160 acres on the Little Laramie near a small settlement known as Stringtown. The claim was filed as a mining claim but more likely was a claim on the land. On August 24, 1870, John R. Owens staked out a 300-foot lode claim in the Big Laramie Canyon. This quite possibly was the earliest claim in the area which was later known as the Pearson Mining District and Cummins City.[27]

Apparently none of the miners chose to "winter over" in the mountains and all returned to Laramie. As the mining season drew to a close, Colonel Donnellan of the Bank of Laramie bought "several hundred dollars" worth of gold from the miners. The gold was described as ". . . coarse, many nuggets weighing from two to five dollars each." (Years later, Beeler in 1960, wrote that about 4000 cubic yards of gravel were processed in and around Moore's Gulch during the mining seasons of 1869 and 1870, for a yield of 450 ounces or approximately $9,000.)

During the summer and into the fall of 1870, the survey team of Winsor and Cook conducted a survey of the Third and Fourth Standard Parallel North. After being awarded the contract in June, 1870, Nathan P. Cook and Winsor began their survey line westward on the Third Standard Parallel (which runs about four miles north of the present Wyoming and Colorado line) and later returned eastward on the Fourth Standard Parallel, (twenty-four miles north of the Third Standard), completing their survey in late November. Accuracy of these lines was later questioned and they eventually were relocated, however, for several years they served as the only reference in the area.

Drawing by Dan Swift

• 1871

The poor mining season and the Indian depredations of 1870 apparently dampened the enthusiasm of the miners and subdued the 1871 season. A huge forest fire, which burned out of control for much of the summer, further discouraged mining and prospecting.

Unemployment was apparently not a problem. In Laramie City, Ivinson and Company was advertising for one hundred men and fifty teams to work on government contracts at Fort Laramie and Fort Fetterman.

Fifty prospectors were reported working north of Fort Fred Steele. Supplies and transportation were being provided by Laramie City entrepreneurs.

However, little mining was being done in the Medicine Bow Mountains and only one new claim was filed in the Albany County Courthouse, described as being located "North of Moore's Gulch and South of Ruth's Gulch and East of the main creek and known as the Golden Shovel Lode." The 1400 foot claim was filed on June 19, 1871, by John L. Roberts, Thomas Feast, John Crowley, George W. McFadden, Richard DeArmon and Fred Donohugh.[28] (The mining law of 1866 provided for claims 200 feet in length along the vein for each locator, with an additional claim for the discoverer, thus the six claimants were allowed 1400 feet.) It appears these six men were the only miners working on Douglas Creek, and most probably the entire Medicine Bow Range.

Aside from mining, however, timber harvesting was beginning to show some progress. According to the editor of the *Laramie Daily Sentinel*, timber and ties were being harvested all along the east edge of the mountains. Early in May, Charlie Hutton had a log drive down the Big Laramie to Laramie City and Charles Bussard and crew were cutting ties on the north edge of the forest. For a time, it appeared that the mountains were to become a provider of timber, not gold.

During the summer and fall of 1871, 717,000 acres of Albany County were surveyed. The Medicine Bow portion, which included part of Carbon County, was completed by three separate crews, one led by Henry G. Hay, another by John B. Thomas, and the third by W. J. Allason. Hay and Allason surveyed the township lines while Thomas fulfilled his contract for survey of the Tenth Guide Meridian West. The government contract required that guide meridians be monumented every forty chains (1/2 mile). Utilizing the most abundant natural marker available, Thomas marked his survey line almost exclusively on pine trees. As with many of the survey lines in this rugged country, the line was later resurveyed and corrected. These surveys opened the area along the Big and Little Laramie Rivers to homesteading and numerous homesteads were filed. According to the *Laramie Daily Sentinel*, the livestock business was booming.

At Laramie City's Public School House, James W. Marshall, who made the initial discovery at Sutter's Mill, gave a lecture on his fabulous find which precipitated the California gold rush of 1849. Perhaps his lecture provided the incentive for a group of business men who gathered together and made plans to promote and finance more extensive mining in the Medicine Bow Mountains. These six men, L. P. Bradley, H. L. Latham, E. D. Lane, Henry Noyes, A. S. Burt and S. W. Downey formed the Medicine Bow Mining Company and made plans to construct roads, erect mills and generally pursue mining of precious metals.

Apparently none of the miners spent the winter in the mountains. Snow came early in October and continued for a historically cold and harsh winter.

• 1872

Probably the best summation of the 1872 mining is contained in R. W. Raymond's report on mines and mining:

> No new mine districts have acquired prominence in this territory during the year, and there is little to report of the old ones. . . . The close proximity of Utah, with its numerous deposits of rich silver ore, together with the universal habit of miners to stampede to every newly discovered district of note, have depopulated the gold regions of Wyoming.[29]

The scarcity of news articles and the limited number of new filings certainly reinforce this perception.

The Sweetwater Mines were nearly deserted, the Seminoe Mines were being called a fraud and the Medicine Bow Mines had apparently lost much of their attraction. However, some of the miners continued their prospecting, expanding their search for precious metals, whether gold, silver or copper. During the summer of 1872, Daniel Wolff and John and George Renner moved up Douglas Creek and crossed over to the high mountains and a new, yet unnamed mining area. Here they located and claimed numerous new prospects in the area that later became the La Plata Mining District.

Gold production for all of Wyoming Territory for the year ending December 31, 1872, was $11,186.44.

• 1873

From 1868 through most of 1872, the exploration and mining of the Medicine Bow Mountains had been carried out under the Mining Act of July 26, 1866. This act declared the mineral lands of the public domain open to exploration and occupation by citizens of the United States.

It provided for claims 200 feet in length along the vein for each locator, with an additional claim for the discoverer. An association could claim up to 3,000 feet in length with no width specified, only sufficient ground for working the claim.

This act was repealed by the Mining Act of May 10, 1872, which, with some amendments, is still in force today. The act provides in detail for discovery, location, survey and patent of both lode and placer claims and also mill sites. It requires annual labor or assessment work until patent is granted. Tunnel sites for the discovery of lodes are also provided. No limit is placed on the number of claims any individual association or corporation might locate. The act also provides for relocation of abandoned claims. It also states "No lode located after May 10, 1872 can exceed a parallelogram 1,500 feet in length by 600 feet in width."

The act also requires that discovery work must be accomplished in accordance with any state or territorial laws. The general requirement is that the vein or deposit be disclosed to a depth of ten feet or deeper. Hundreds of these ten foot discovery holes are scattered throughout the Medicine Bow Mountains.

As the miners became better versed in mining law, they began to take advantage of the Mining Act of May 10, 1872. In January, 1873, the abandoned "Crescent Lode" of 1869 was relocated and filed on by Blackbourne, Feast and Metcalf. They claimed 1,500 feet by 600 feet, the maximum size allowed by the new law.

The abandoned claim rule was subsequently used extensively by the miners, sometimes several times on the same claim as each group tried their luck where others had failed before them.

The new law offered new incentives to the individual prospector as well as the undeveloped companies. Unfortunately, 1873 brought new financial troubles to the miners as well as the rest of the United States. The Panic of '73 began in the eastern banking world and spread westward across the country. Stock prices and in particular, mining stocks, fell in price when banks began calling in loans and mining investors sold their shares for whatever they could. The developed mines of Colorado were forced to reduce operating costs and new development was practically nil.

The undeveloped Medicine Bow mines however, were less affected by the panic and some exploration and prospecting continued under the new mining law.

Silver was demonetized in 1873 leading some of the miners to believe silver was the metal of the future, and overnight many of the gold mines suddenly became silver mines. The price of silver, however, began to fall from $1.32 in 1872, to $1.15 in 1880, and $.84 in 1889. One by one the mines became gold mines again.

Individual miners merged their finances and labor to form companies. In Laramie City, on September 1, 1873, the New York and Wyoming Placer and Quartz Mining Company was formed and incorporated as the 32nd article of incorporation in the Territory of Wyoming. The capital stock of the company was set at $500,000 and according to the *Laramie Daily Sentinel*, $200,000 was paid in and available for immediate investment. The companies stated object was ". . . to mine for silver, gold and other precious metals in the territories of Wyoming and Colorado and to construct roads and ditches and to acquire all the franchises, privileges, water rights, tunnel rights and all other rights in property which may be acquired under the law. . . ." Their first act was to secure title to a large tract of ground on Last Chance and Park gulches. Their immediate plan was ". . . to build a good road to the mines and put up buildings and a sawmill." Trustees for the company were listed as Joshua Hanna, Thadeus Judson, Richard Wheelock, John Metcalf and Stephen W. Downey. Judson and Wheelock were said to be the "moneyed men and the backbone of the enterprise." Other members of the new company were John Robinson, Albert S. Blackbourne, Robert W. Burkhardt, and Thomas Feast. The *Laramie Daily Sentinel* said of these men, "They are not visionary or extravagant men, but are satisfied that properly worked upon an extensive scale, these mines will furnish employment for thousands of men for years to come." Perhaps visionary meant something other than idealistic and impractical to the writer for the *Sentinel*, for if ever there was a visionary man, at least in mining matters, Stephen W. Downey would qualify. Four of the five years since gold had been discovered at Last Chance, Downey had been in the midst of the mining activity. First as one of the miners, then as a leader amongst the miners, then to invest his time and money in the Medicine Bow Mining Company, the New York and Wyoming Placer and Quartz Mining Company, and later others. Downey was continually at the forefront of mining activity, promoting his visions of a great mining district in the Medicine Bow Mountains.

Downey was a Civil War hero who rose to the rank of Colonel in fourteen months when he was only 22 years old. He was discharged shortly after the battle of Harpers Ferry due to his wounds. He practiced law for a few years in Washington and then moved west to Laramie in 1869 where he continued to prac-

Stephen W. Downey
American Heritage Center, University of Wyoming

tice law, opened a survey office, and became a leading citizen holding many public offices both local and territorial, for the rest of his life. Sire of a large family, his name is perpetuated in the history of Laramie. No matter how involved he became in his other activities, Downey always found time and money to pour into the Medicine Bow mines. His mining interest, however, was not restricted to the Medicine Bow mines. In 1873, his interest in mining led him into the "Black Hills" east of Laramie where with his associates, Metcalf, Feast and Burkhardt, he prospected the eastern slope. This was probably some of the earliest prospecting in that region which was later to become the Silver Crown Mining District. At one time, he also held extensive claims in the Salt Creek oil fields which, if held, would have eventually made him a wealthy man.

In August 1873, R. W. Burkhardt surveyed the Last Chance and Park Gulch placer mines. Burkhardt filed the original survey copy in Downey and Grants survey office and a copy in the county courthouse.

The winter of 1873-74 was considered mild and prospectors and miners continued to work east of Laramie but apparently not in the Medicine Bow Mountains.

• 1874

By 1874, the miners seemed to grow weary of the mediocre results of their labors at Last Chance and began to spread their explorations to the hills and mountains surrounding Laramie. Already they had moved into the high Snowy Range. In early 1874, new claims were filed in the "Black Hills" east of Laramie. On the Laramie Plains near Cooper Lake, N. K. Boswell and J.J. Kelly were prospecting for precious metals and minerals. On the Big Laramie River, "Father" Thomas D. Pearson, "Judge" George Ritter and John LaFeavre were staking out new claims in the newly organized Pearson Mining District. On Mill Creek a major district was anticipated and Judge Melville Brown, Ora Haley, Stephen Downey and a group of miners and business men made claim to extensive placer grounds.

In Laramie the Mill Creek and Douglas Creek group of miners and backers formed another new company, the Home Mining Company. At a meeting in Judge Melville Brown's office, Walter Sinclair was elected Secretary and Ora Haley, Treasurer. Stephen Downey and his brother William were named trustees as apparently were all the remainder of the group, N. C. Worth, William Lewis, James Lemmon, N. Colisey, John Robinson and the organizer and president, Judge Brown. The company planned to begin work immediately on Douglas Creek, to develop large scale placer mining operations.

About the same time, the New York and Wyoming Placer Mining Company renewed their operation and Thadeus Judson, president and principal financier of the company, visited Laramie City and then the mines of Last Chance.

The Home Mining Company spent the entire mining season opening new claims, putting in a tailrace and a bedrock flume. It was October before they were ready to operate, but due to decreasing water supply and the onset of winter, they were unable to pursue their placer mining to any satisfaction.

On June 25, 1874, the Postmaster General of the United States notified George Ritter

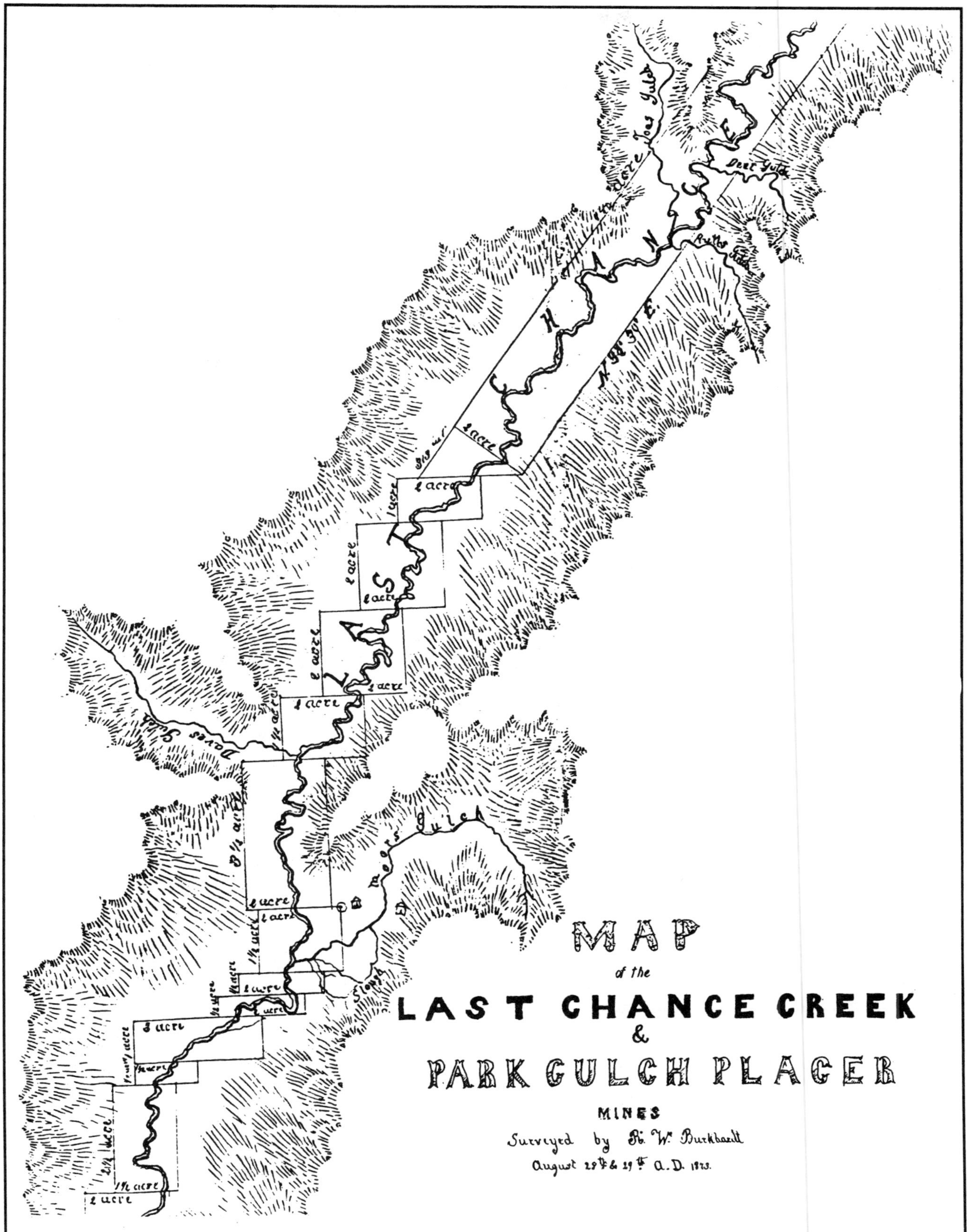
LAST CHANCE
MAP
of the
LAST CHANCE CREEK
&
PARK GULCH PLACER
MINES
Surveyed by R. W. Burkhardt
August 28th & 29th A.D. 1873.

that the Last Chance post office had been approved and he was appointed postmaster. Aside from his mining activities, which had moved from Last Chance to the Big Laramie, Ritter was serving as Justice of the Peace, Probate Judge and County Treasurer for Albany County. He also owned and operated Ritter's Restaurant in Laramie and had served on the Third Territorial Legislative Assembly. Although Ritter was no longer living and mining at Last Chance, he accepted the appointment with the intention of passing the duties on to another. Official mail service from Laramie City began immediately and was to continue, weather and road permitting, for years to come.

In spite of the years of extensive prospecting and mining in the Medicine Bow Mountains, the fact remained that through 1874 not one significant discovery had been made. Through 1874, somewhere around one hundred men had filed on somewhere around twenty claims in the Last Chance District. Total production was around $20,000 (1,000 ounces) of placer gold for almost four years of work. Not one stamp mill had been constructed, and only one primitive arrastra had been set up to crush the ore from the little developed lode mines. The prospectors were still working the placers with pick and shovel, gold pan and sluice box. Although many of the prospectors were washing out enough gold to earn a meager living, the fact was, they could probably have earned as much working in a coal mine, cutting ties, or working in the growing cattle industry. Still, many stayed on, perhaps lured by the beauty and seclusion of the mountains and streams, but more likely by the hope of that yet undiscovered bonanza.

By the end of 1874, little remained of the Last Chance District except a few cabins and the post office. The post office continued in operation for several years, George Ritter passing on his duties to William H. Lewis on July 29, 1874, Lewis giving up to James Beagle on December 12, 1876, Beagle to William Bordway a short time later, on February 1, 1877, and finally to Fletcher Dunham who took over the duties on November 6, 1877. Dunham was the last Last Chance postmaster and subsequently the first Douglass City postmaster when the office was moved to the new location.

By the end of 1874 the first gold rush to the Medicine Bow Mountains had more or less run its course. A few miners and prospectors lingered in the mountains but the intensity of the interest was gone. To be sure, this interest would intensify time and time again, but for the time it had faded. Although those who appeared on the first census were indeed true Wyoming pioneers, many would vanish from the territory, apparently seeking their fortunes elsewhere. However, some of the prospectors stayed on, continuing their quest for valuable metals.

The Downeys continued their interest in the Medicine Bow mines and would continue to be the leading force behind the mines. James Vine established a business in Laramie but would return to dabble in the mines from time to time and apparently grubstake others. Albert S. Blackbourne would return time and again to the mountains for years to come, working at times as an accomplished carpenter. Pete Kennesaw, who was on the 1869 census, stayed in the area and would be heard from again. John Walbright also on the early census would also reappear on the mining scene. The next discovery in the Medicine Bow Mountains and resultant gold rush were only a short time away.

Last Chance faded into virtual non-existence but the discoveries made in the area generated an interest and launched a widening search for precious metals that continues to this day. The greater portion of the original Last Chance discoveries is now covered by the waters of Rob Roy Reservoir.

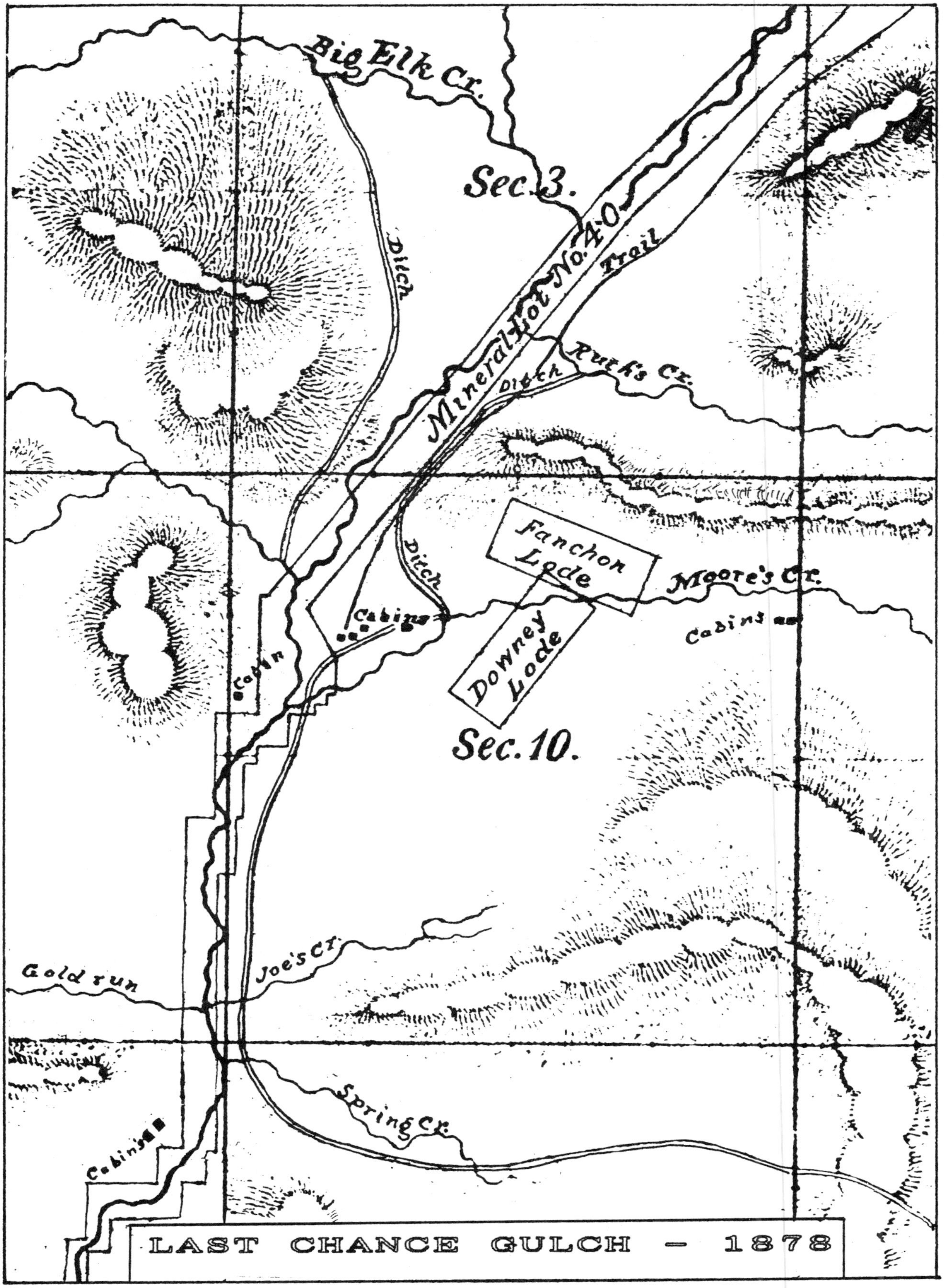

LAST CHANCE GULCH – 1878

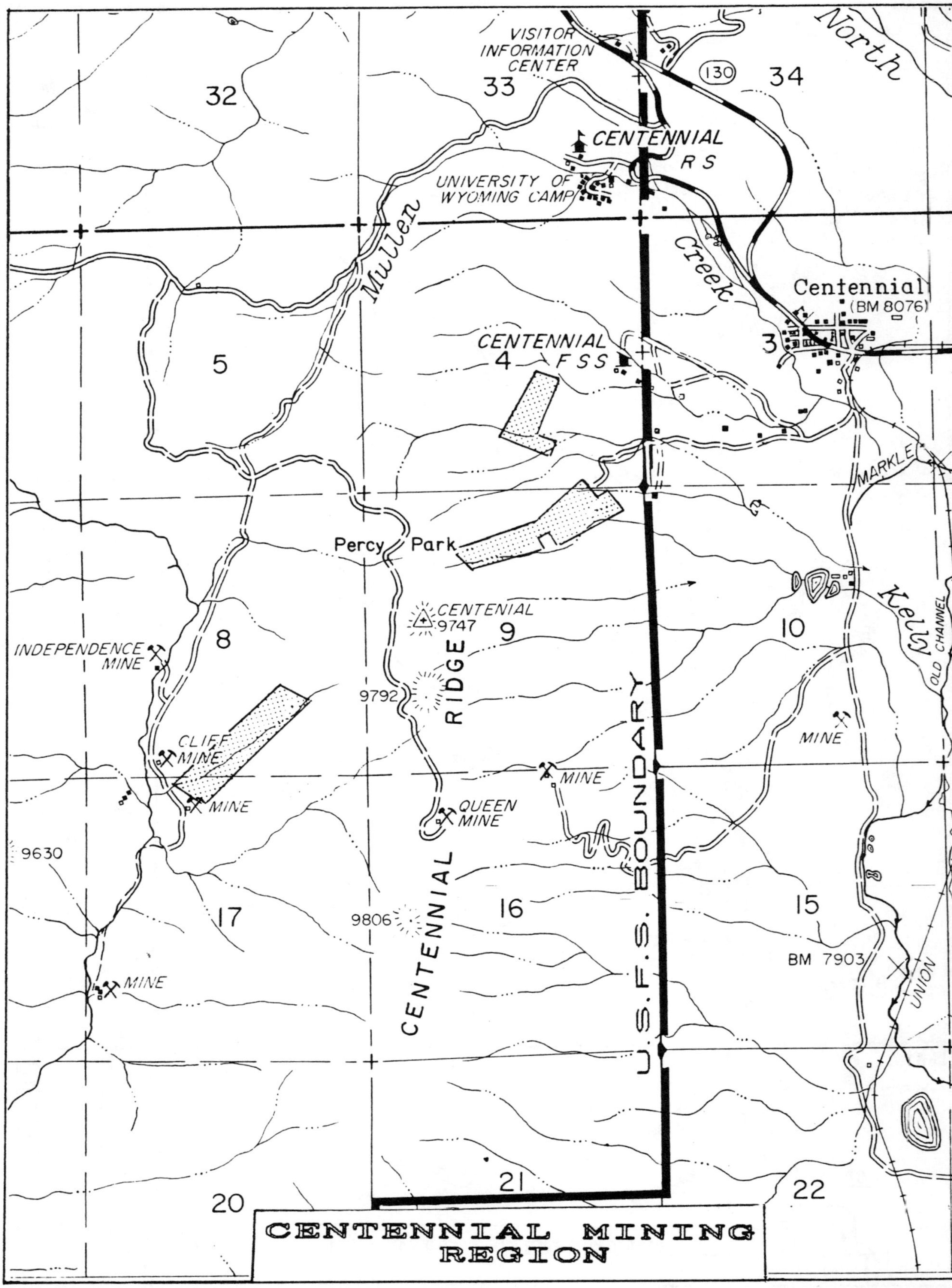

CENTENNIAL MINING REGION

Although the Last Chance placers failed to produce the great quantities of gold originally forecast, prospecting in the Medicine Bow Mountains continued into the cold winter months of 1874. Cold and deepening snow forced the prospectors out of the high country to the lower ridges that form the foothills of the range.

On December 22, 1874, the *Laramie Daily Sentinel* reported a new discovery:

> A party of hunters a few days ago discovered a quartz ledge upon the Little Laramie which upon examination carries a large amount of free gold. These gentlemen Messrs. Lambing, Fowler and Skidmore have gone to work to develop the ledge. Today they went out with a team and will sink the shaft and bring a load of the rock into town and send it to Colorado to be crushed.
>
> We have seen quite a number of pieces of the rock and every piece is spotted with gold. The lode is four or five feet wide with well defined wall rock. These men are experienced miners and know what they are about, and we think they have got hold of a good thing.

"Hunters" they may have been, but for several preceding months this same group had been prospecting along the Big Laramie and Little Laramie Rivers and had staked out a claim in the Big Laramie Valley.[1]

A week later the *Sentinel* of December 28, carried another article, reporting on the progress made at the new discovery:

> We mentioned a few days ago the discovery of a promising quartz lode on the headwaters of the Little Laramie some twenty-five miles west of this city. A party went out there last week with means to open it and see how it would prospect. They have returned to town and brought in quite a quantity of the rock. In nearly every piece of it free gold is visible and it is estimated by experienced miners, that it will yield from $150 to $200 per ton. The party dug a hole a few feet deep on the lode and report it to be a strong well defined lode, about seven feet between wall rocks. Some of the rock has been sent to Denver and other places for assay and we shall hear further of it in a few days.

• 1875

Without waiting for the assay results to return, the prospectors, with the help of William Downey as surveyor, staked out a claim 600 by 1500 feet "...on the mountain between the North Fork and the Middle Fork of the Little Laramie River in what is supposed to be Albany County ...." (The Albany and Carbon County line was, at that time, east of the present county line. Centennial Ridge, as well as most of the other Medicine Bow mining areas were actually located in Carbon County). Taking no chances, the original locators filed their claim in the Albany County Courthouse in Laramie City and also the Carbon County Courthouse in Rawlins on January 4, 1875. The "original" locators now numbered four; Isaac Lambing, W.G. Fowler, Jacob Durst and Ira W. Skidmore, who apparently in honor of the impending first 100 year celebration of the United States, named their new claim; "We claim the said gold bearing lode or vein under the name of the Centennial."[2] (Years later it was Lambing who was given the actual credit for naming the mine.)[3]

> **W. O. Downey,**
> **Surveyor & Mining Engineer.**
>
> **MINING CLAIMS SURVEYED**
> **AND**
> **PATENTS SECURED.**
> **Land Office & Mining Bureau,**
> **in William's Block.**
> Laramie City, Wyoming Territory.

Early in February the assay results were returned and the *Laramie Daily Sentinel* of February 4, 1875, reported:

> The first assay of ore from the Centennial Lode, near the little Laramie, surprised everybody: It showed $965 of gold to the ton, while rock recently taken out at a depth of twelve feet is still richer and would not

Centennial Mining Claim

To all to whom this notice shall come Know ye that We the undersigned citizens citizens of the United States and over the age of twenty one years do hereby locate a mining claim of fifteen hundred linear feet along the course of this vein or lode by six hundred feet in width. Beginning at the centre of a Shaft on said vein or lode and running thence N. 47° 30' West along said vein or lode 1000 feet to a Stake in a monument of Stone and designated as the beginning point running thence South 47° 30' West 300 feet thence running S 43° 30' East 1500 feet thence running N 47° 30' East 600 feet thence running N 43° 30' West 1500 feet then running S 47° 30' West 300 feet to the place of beginning

Said Lode is Situated in the Mountains between the North fork and the Middle fork of the Little Laramie River in what is supposed to Albany County and in the Territory of Wyoming The Ore of this Lode is Gold bearing We claim the said Gold bearing Lode or Vein under the name of the Centennial This claim is made under the act of Congress May 10th A.D. 1872 and according to a survey made by William O. Downey Surveyor on the first day of January A.D. 1875

J. P. Lumbing
W. G. Fowler
Jacob Durst
Ira V. Skidmore

Territory of Wyoming }
County of Carbon } ss

This Instrument was filed for record July 1st 1875 at 10 oclock A.M. and recorded July 3rd 1875

V. L. Smith Recorder

Territory of Wyoming }
County of Albany }

This Instrument was filed for record at 3½ Oclock P.M. January 4 1875 and duly recorded in Book A of Mining Claims page 107

J. W. Meldrum Register

(Book B 21)

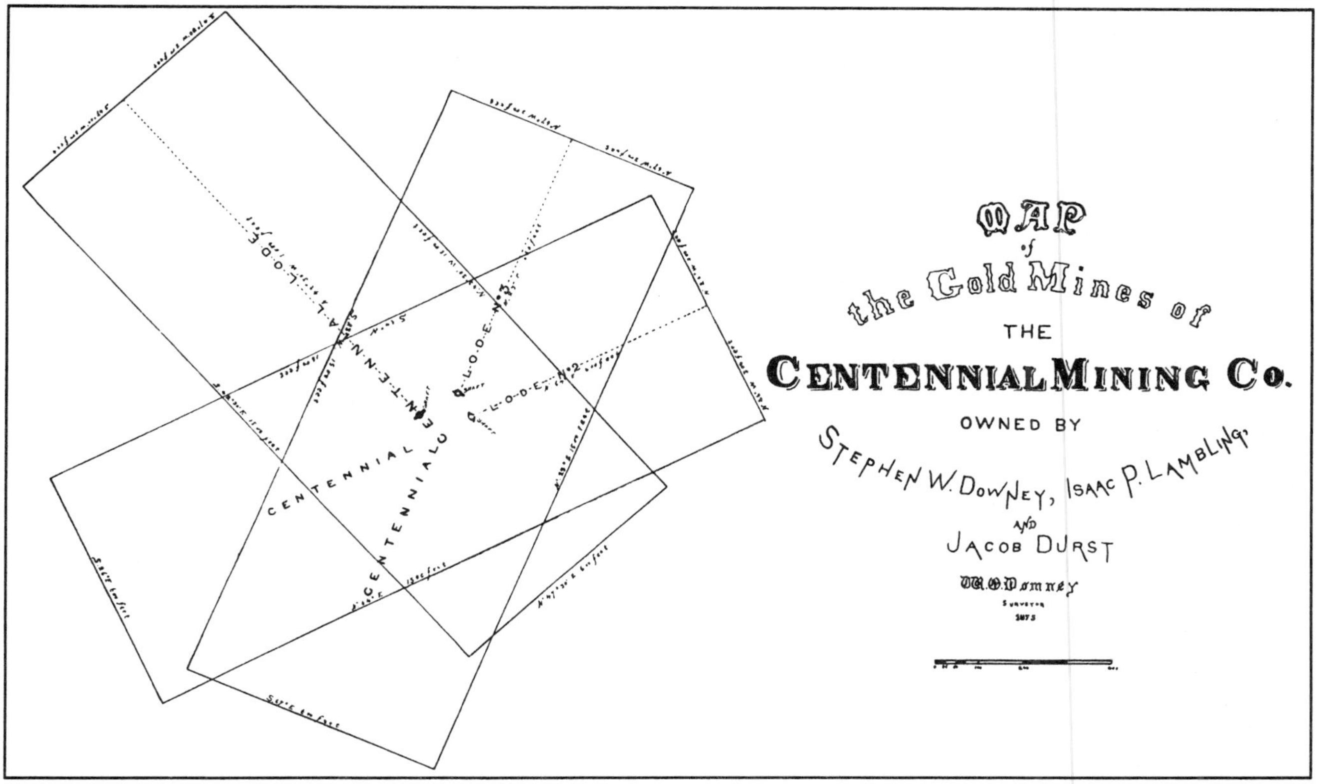

surprise competent judges should it run $2,000.00 to the ton. It would not in fact surprise anyone now should the "Centennial" prove to be the richest auriferous quartz in the country equaling in value, the "great bonanza" on the Comstock Lode in Nevada.

The Centennial Mine quickly became the focal point of mining in the Medicine Bow Mountains. Here was an area which had easy access from Laramie City, an area that could be worked year around, and an area that had assay reports better than any of the surrounding territorial mines.

J. H. Triggs in his 1875 *History and Directory of Laramie City* expressed the hopes of the city for extensive mineral development of the area:

> Standing at Laramie City as we look to the west we behold the Medicine Bow Range. Then just beyond and in full view, rises up that grand old leviathan, the Snowy Range, the Great Mastodon of the Western Hemisphere.
>
> Locked up in these everlasting hills are countless stores of wealth waiting only for the hand of energy to be applied. Nature has entrusted us to the key. What if it requires almost an iron grasp to turn it? What if our first excavations hardly seem to repay our efforts? We will slacken not our grasp. The same great mineral belt, which lies along the eastern slope of the Rocky Mountains extending throughout the entire territories of Colorado and New Mexico from which Colorado alone has extracted her millions of dollars in gold and silver, extends also through our territory and in close proximity to Laramie City. . . . Within the Medicine Bow Range to the west of us, . . . are found ledges of gold, silver, copper and rich placer mines . . . .
>
> During the past winter . . . a district has been discovered bearing lodes of wonderfully rich free gold quartz, . . . the first discovery made in this district was the Centennial Lode . . . and thus far the Centennial Lode promises to be an immense fortune.

The new prospect at Centennial seemed to spur the Medicine Bow miners on to explore a wider area of the Medicine Bow Mountains. Expeditions left Laramie City almost weekly and reported new finds on the Big and Little Laramie Rivers, Sheep Mountain, Brush Creek and on Douglas Creek where great hope of producing mines still existed.

Whether the original Centennial locators were employed by Stephen W. Downey when they made their discovery or whether they were independent prospectors who later came under his influence is not clear. What is evident, is that by early 1875, the new discovery was under Downey's control. At the Centennial Mine, under his direction, a careful search was conducted to determine the direction and extent of the gold bearing vein.

Many of the previously discouraged miners

who had returned to Laramie City became convinced that this was indeed a new and valuable discovery and began a small gold rush to Centennial Ridge. New claims quickly followed the original Centennial. The First West Extension of the Centennial was filed by J. E. Bush, William Leonard, Silas Durst and Jonnas Wedges.[4] The Second West Extension was filed by the surveyor who had laid out the original claim, William O. Downey, along with S. B. Walters and John H. Martin.[5] The First East Extension, filed at the same time was claimed by George Dean, Christopher McPherson and Henry Lewis.[6] A few days later, the Second Southeast Extension, the Third Northwest Extension, Centennial Lode #2, Centennial #3 and more in a profusion of filings that adjoined and overlapped the original claim on all sides.

In Laramie City, the Centennial Mining Company was formed by Lambing, Durst and Downey. A ten stamp mill to process the gold bearing quartz was ordered and near the mine, construction was started on a building to house the mill and proposed blacksmith shop. On July 20, 1875, the machinery arrived in Laramie City and by August 3, when the *Sentinel* correspondent visited the mine the machinery was in place awaiting only the completion of the tramway to begin operation. The tramway, which was 425 feet in length, was required to descend 260 feet to the mill. It was put into operation on September 1, 1875. The correspondent for the *Sentinel* noted: "The mill is of a good kind for crushing the rock and saving the gold but is of limited capacity, we judge, though the builders guarantee it to run through from five to eight tons per day.[7] Similar criticism of the mill continued throughout the boom. Mortimer N. Grant ran the stamp mill and did the amalgamating until November, 1876, when he was sent to California to examine the mines there and the methods used in working them.

On July 23, 1875, a post office was opened and Isaac Lambing, one of the original discoverers, was designated as the first postmaster.[8] The new post office became a mid-point stop on the Laramie City to Last Chance postal route which was operated by James Ingersoll. Lambing estimated the post office served 500 people, 100 in the town and the remainder in the surrounding hills.

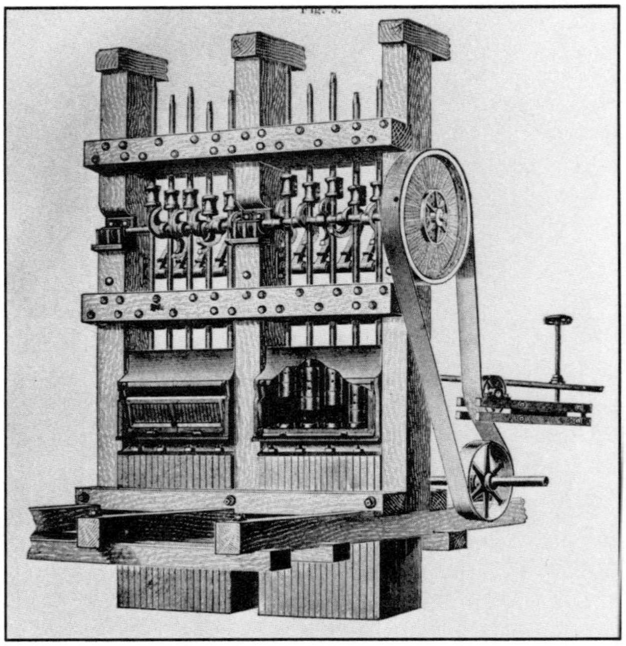

Ten Stamp Mill.
Courtesy of Bud Owens.

When the Albany County Commissioners met on July 30, 1875, the Centennial Precinct was designated and the "house" at the Centennial Mine was selected as the polling place. I. P. Lambing, William Waters and J.M. Brown were named as election judges. An election was held on September 7, 1875 and although actually located in Carbon County, Centennial's twenty-one votes were counted in Albany County and helped to elect the Albany County Legislators.[9] (The following year, and through 1886, Centennial's vote was counted in Carbon County).

Although Centennial Ridge as well as most of the claims were in Carbon County, most of the claims continued to be filed in both the Albany County Courthouse in Laramie City and the Carbon County Courthouse in Rawlins. Those claims filed in Albany County usually contained the phrase, "supposed to be in Albany County." In November 1875, the miners, through the *Laramie Daily Sentinel*, petitioned the state legislature to change the western boundary of Albany County so as to include the Centennial Mining District within the county. If the legislature should fail to act, the miners threatened to outvote the rest of Carbon County and have the county seat moved to Centennial.[10] At the time, the miners were unsuccessful and almost eleven years would pass before the county line was actually moved.

As the summer mining season progressed, prospecting continued and additional claims were made up and down Centennial Ridge. Two thousand feet north of the first discovery the miners struck what they thought was part of the original vein. Two miles to the south, several new discoveries were made, the most notable being the "Blind Lode" claimed by William "Billy" Waters and John H. Martin.[11] Telling of his discovery and early work on the mine, Waters said, "It rained nearly all the time, night and day, for the first three days, and we worked in the rain all day and then wrapped our blankets around us and stood up under a tree all night."[12]

Near the "Blind Lode," Benjamin C. Dutton and J. W. Collins staked out their "Union" and "Eagle" claims, each the maximum size of 600 by 1,500 feet.[13]

Some of the prospectors apparently had some difficulty understanding the limitations of the mining law. On August 18, 1875, A. H. Walker claimed the "Twin Spring Gulch," adjoining Waters' and Martin's claim, with the simple statement, ". . . running down the creek southwest to or beyond the Middle Fork of the Little Laramie River 200 feet or whatever the law calls for." No other description was specified.[14]

Across the valley on Sheep Mountain, two new lode claims were made. The first, called the "Agnes" was discovered by "Captain" James Lemmon and claimed by "General" Noah Worth, George Pickett, P. G. Murphy, John Riffil and Lemmon. It was described as being ". . . on the summit of Sheep Mountain, 12,000 feet above the level of the sea."[15] (Most recent maps indicate the maximum elevation of Sheep Mountain as 9,583 feet). Adjoining the "Agnes" was the "Clara." Shortly after discovery the two mines were combined and became known as the "Consolidated Agnes" whose owners claimed 3,000 feet along the vein, which was not in strict accordance with the law, but apparently no one complained. Two extensions were also located next to the "Consolidated Agnes," one by "Colonel" Dana and Professor Thompson and the other by Frank Priest and son. Great hopes were placed on these Sheep Mountain claims but the Centennial Mines continued to receive the greater attention and more important, the money and equipment to carry on mining in an extensive way.

Work on the first Centennial shaft followed the quartz vein into the ridge, stockpiling the ore. Upon completion of the tramway, the ore was carried down to the mill for processing. Crushing continued into the fall and winter and the first "cleanup" was delayed until almost Christmas, 1875.

• 1876

In January, 1876, the product of the mill was taken to the mint at Denver for final processing. The resulting gold brick weighed almost 90 ounces and was valued at about $1,800.00, a nice return, depending, of course, on the amount of time expended and ore processed. At the same time, some of the quartz ore was taken to Denver for assay. The results were startling, even if this was selected ore, which no doubt it was. The official assay result was "2,263.02 ounces of gold per ton for a value of $46,780.34 and 209.98 ounces of silver per ton for a value of $270.49 for the assayers total of $47,051.83 per ton."[16] The assayers arithmetic may not have been perfect, but considerable faith was placed on these assays and if a large quantity of this grade of ore could be processed, this truly would be one of the richest mines in the country. The report of the assay appeared not only in the regional newspapers, but also on the front page of the *New York Times*.[17]

James H. Hayford travelled to Denver to observe the assay and the smelting of the ore from the first mill cleanup. Hayford, editor of the *Laramie Sentinel* from 1869 to 1895, was a man of strong convictions and seldom failed to express his opinions through the columns of his newspaper, whether on politics, morality, or economics. As a promoter of Laramie City and the surrounding area, he was often at odds with the Cheyenne press about which city was the fastest growing and which city offered the best opportunities for new business.

The gold brick was first shown in Denver but received little comment from the Denver press. However, the Laramie City newspapers did somewhat better:

> While the train stopped in Cheyenne we had a chance

James H. Hayford
Wyoming State Museum

to stick this eight pound gold brick under the nose of several of our Cheyenne friends. They are surprised and make up a face like a boy swallowing bitter medicine - he hates it but must take it. Nicholls got the brick in his hand, ran up to the wondering crowd, and cried out: 'Look at that, gentlemen! That is from the Centennial Black Hills Mine!' [18]

The *Daily Leader* of Cheyenne responded in their next days' column:

> Hayford was in town Saturday, and displayed a few specimens said to be taken from the Centennial Lode. If the owners of the Centennial had selected a more reliable messenger, they would have done better. As it is everybody in this section is in doubt, and no one knows whether the ore exhibited came from the Centennial or the Comstock Lode. Thus Hayford's reputation for stretching the truth sticks to him wherever he goes.

However, the same paper, the same day, also printed a short article:

> Some of the rich ore recently brought in from the Centennial Mine in Albany County has been deposited with the Territorial Librarian Judge Slaughter. It is the richest looking ore ever produced from a mine in this Territory. If there is a good quantity of it in the Centennial Mine, the owners will have a bonanza indeed.[19]

Encouraged by the output of the Centennial Mine, Billy Waters sold his town property in January and invested in a five-stamp mill for his mine, however, it was mid-summer before he began operation.

Continued gold output from the Centennial Mine was anticipated but was not realized. The spring which supplied water for the mill slowed to a trickle then ceased flowing altogether. Mid-winter was no time to consider digging a ditch or moving the mill down to the river and so production from the mill ceased. Stockpiling of the ore, however, continued.

On February 17, 1876, the Centennial Mining Company incorporated. This was the 60th article of incorporation in the 6 year old Wyoming Territory. The amount of capital stock was fixed at $500,000, divided into five thousand shares of one hundred dollars each.[20] Downey, Rumsey, Lambing and Durst, all of Laramie City, scraped up what they could, but by necessity turned to "Eastern Capitalists" for the majority of their financing. George A. Davis and Edwin H. Taft, of Providence, Rhode Island, and Charles A. Avery, of New York, provided the majority of money to finance the company.

With the onset of the spring of 1876, the Centennial mill was again in operation and

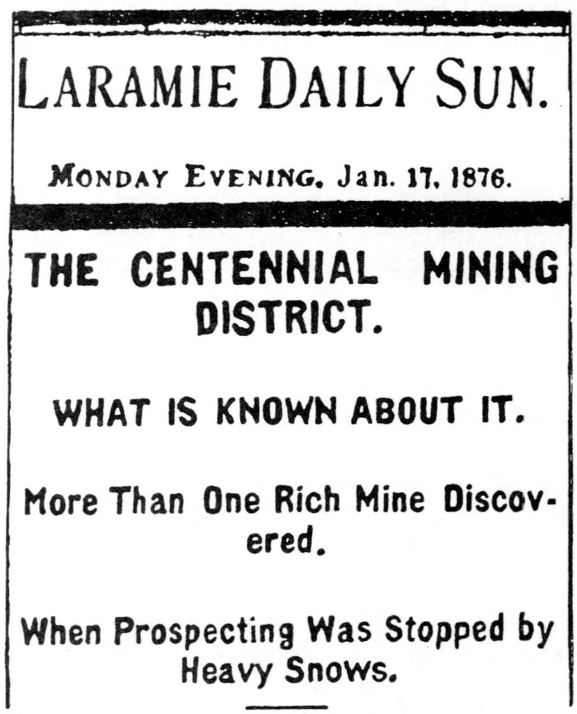

LARAMIE DAILY SUN.

MONDAY EVENING, Jan. 17, 1876.

THE CENTENNIAL MINING DISTRICT.

WHAT IS KNOWN ABOUT IT.

More Than One Rich Mine Discovered.

When Prospecting Was Stopped by Heavy Snows.

crushing seven to eight tons of ore per day. During May, the Centennial Gold Mining Company built offices and also a residence for the superintendent. Little and Coolbroth built a hotel about the same time. Peter Holt opened the Centennial Dining Hall and prepared to receive day boarders. Business was apparently good for he soon took on partners, Mr. and Mrs. Ulen.

Also in May, J. W. Collins temporarily gave up his mining long enough to establish a stage line between Laramie City and Centennial to carry freight, mail and passengers "at reasonable rates." The stage left Laramie City on Monday, Wednesday and Friday, returning on alternate days. Tickets were available at C. S. Dunbar's Market.

By June, 1876, the Centennial Mining Company, under the direction of Stephen W. Downey, was planning a new and improved stamp mill which would do custom crushing for the small mines of Centennial Mountain.

Business must have been good, for in July, Thomas Bird established a second stage line, advertising his stage would leave Laramie City on Sunday, Tuesday and Thursday and return on alternate days. He also advertised that he would carry freight, passengers and express as well as the U.S. Mail. Tickets for his line were sold at the bank, Laramie City's one and only, operated by Edward Ivinson. Through the Centennial boom and bust, a stage line was to continue to serve Centennial under various owners and operators until 1907 when the long awaited railroad finally arrived in Centennial.

CENTENNIAL

Stage Line.

THOMAS BIRD, Prop'r.

Leaves Laramie Sunday. Tuesday and Thursday returning alternate days. Carries

Passengers, Freight and Express

to and from the

Centennial Mining District.

Carries the United States Mail.

☞TICKETS FOR SALE AT THE BANK.

jun28

During the summer of 1876, the town was laid out, cabins built and in addition to the mill, a boarding house, blacksmith shop, superintendent's house and stage station became part of the new camp.

When a special correspondent of the *Chicago Times* visited the mining camp in August, he said the ten-stamp mill under the supervision of Lambing was running day and night. He also said the camp had a superintendent's house, a boarding house, a blacksmith's shop, a store, a stable and a dozen log cabins. Lambing told the correspondent that there were 500 men in the surrounding mountains prospecting and mining and the number was increasing daily.[21]

The Centennial mill was processing seven to eight tons of ore and producing $150 to $200 of gold per day and on August 4, 1876, Lambing took a $6,000 gold brick to Laramie City. George H. Stayer, an "Eastern capitalist" who was also a shareholder, stated that in his opinion, ". . . the mine was worth a modest ten million dollars." [22]

In August, 1876, another company, the Carbon Gold and Silver Mining Company, was formed in the Centennial District. Officers were Isaac Ludlum, George H. Hildreth and J. A. Roden. Capital stock was set at $5,000,000 divided into 500,000 shares at ten dollars a share.[23] Apparently the company was not incorporated and offered no stock for sale to the general public.

The Centennial Gold Mining Company, first under the supervision of Isaac Lambing and then N.M. Ambrose, expanded their search for precious metals into the mountains east of North Park. The expedition, financed by the company and led by Mortimer Grant and Nathaniel K. Boswell, acquired by discovery and location, seven lodes and 160 acres of placer ground in the Colorado district.

Although no new or notable discoveries were made along Centennial Ridge, the mining camp endured and in the election of November 7, 1876, Centennial Precinct cast sixty-three votes, the fourth largest precinct of Carbon County.[24]

On December 12, 1876, Lambing trans-

Mortimer N. Grant
Wyoming State Museum

ferred his postmaster duties to Thomas Markle.

Prospecting continued in the Centennial District and another company, the St. Cloud Mining Company, was formed. The Centennial Gold Mining Company, by virtue of its members became one third owner in the new company and its mines, the "St. Cloud," "Hidden Treasure," "Mountain Gorge," and "I.X.L." Under the supervision of A. B. Campbell, numerous shafts and tunnels were sunk on these mines with no reported important discoveries.

• 1877

In January, 1877, a fatal accident occurred at a mine about one-half mile south of the Centennial Mine. Thomas Hodgson was killed in a snow slide caused by explosives used in excavating a shaft. John Kelly, George B. Davis and Hodgson were partners working the claim and had set the blast to go off as they cleared the entrance and moved to the side. When the blast went off, the five or six feet of accumulated snow began to slide, one hundred feet above them and five or six hundred feet along the slope. All three of the miners were carried down the mountain and into the trees. Kelly escaped unhurt, Davis with minor injuries, but Hodgson was killed when he struck a rock or tree. The body was wrapped in blankets and buried in the snow awaiting instructions for funeral arrangements.[25]

Work continued on the Centennial Mine, into the summer of 1877. The first shaft followed the rich vein of quartz into the ridge where apparently it faulted, leaving the miners with a solid wall of red granite confronting them. Another shaft was started close by and the debris was dumped into the first shaft. This second shaft was sixty feet deep, at an angle of 30 degrees where it also encountered the red granite wall against which the quartz vein terminated.

A third shaft was begun under the supervision of N.M. Ambrose. This shaft, however, accidentally intercepted the old number one shaft filled with debris. Work on the shaft became impossible without pumping air and

N. K. Boswell
University of Wyoming, American Heritage Center

finding no evidence of the gold bearing quartz vein, work was discontinued at the end of August, 1877.

The Medicine Bow mines finally had a significant discovery, finally had a stamp mill, finally enough money to pursue mining in a reasonable fashion, but to no avail. The discovery, while extremely rich was also extremely limited. The stamp mill and machinery were useless without nearby ore to work.

Some six to eight tons of ore left in and about the Centennial shafts which together with what was taken from the other nearby company mines amounted to 35 tons of ore. When processed through the mill, this ore produced 37 ounces of gold for a value of $740.00. Expenditures for the Centennial Gold Mining Company amounted to $3,567.53 which not only depleted the company funds but required a draft upon Downey's private account for the amount of $1,121.30 against which Downey was to receive the value of thirty-seven ounces of gold.[26] The failure of the Centennial Mine was Stephen Downey's greatest disappointment and years later was said to have contributed to his death.

While the shutdown of the Centennial was a great disappointment to those working the mine, the general attitude was that of optimism and belief that it was only a matter of time until the lost vein or another of equal value would be found. However, the population of the little mining camp began to dwindle and when the election of 1877 was held, the Centennial District cast only 26 votes. Centennial hotel operator, J. W. Coolbroth garnered 22 of these votes in his unsuccessful bid for the State House of Representatives.[27]

The Billy Waters stamp mill was moved to Douglas Creek in December, 1877.

• 1878

Other mines in the Centennial District continued development work but apparently without much success. The "Virginia" owned and operated by French and Simmonds, the "Bullion Lode," the "Golden Eagle" and the "Terrible" were all referred to as "notably promising."

At the Michigan Mine on Centennial Ridge, another fatal accident occurred on April 21, 1878. "Captain" James Lemmon and John French were working the Michigan Mine and had set an explosive charge which damaged some timbers at the head of the shaft. Against the advice of Lemmon, French ventured into the shaft. A rock weighing an estimated five tons fell on French, killing him instantly, "... severing his head from his body and literally crushing him to jelly." Word was sent to nearby Centennial, recruiting volunteers to extricate the body. Two hours were required to remove the remains which were taken to Laramie City for burial. French, 65 years of age, left a wife, five sons and three daughters, all in Michigan except one son, Lucius, who lived in Laramie City and also worked the Medicine Bow mines. Joseph S. Simmonds, his son-in-law, handled the arrangements in Laramie City. French was the original discoverer of the Michigan Mine which he had named for his home state.[28]

> **FATAL ACCIDENT.**
>
> **John French Killed in a Mine by a Falling Rock.**
>
> This afternoon at 2 o'clock Messrs. Nelson Burchard, Joseph S. Simonds and Lucins French, a son of deceased, arrived in town with the body of John French, the original discoverer and part owner of the Michigan mine, Centennial district, who was killed at 3 o'clock yesterday afternoon under the following circumstances, as detailed to us by Mr. Simonds, who is a son-in-law of deceased:

From June 5, to June 15, 1878, William Downey completed a government survey of the township where the Centennial mines were located, (T15N, R78W). In his field notes Downey described the eastern one-half of the township as very good grazing area but not fit for cultivation. He also noted that "... quite a number of ranchers had located stock ranches along the streams and their flocks and herds of cattle are to be seen in every direction." Downey also noted the following settlements had been made: James Anderson, Section 11; Thomas Markle, Section 10; Perry Bickford,

Section 10; J. M. Coolbroth, Section 4; S. W. Downey, Section 4; William Harris, Section 11; Peter Christianson, Section 22; George Garratt, Section 23; and Charles H. Bussard, Section 36. He made no note of buildings, mines, or settlements in Section 3, the present site of the town of Centennial. (The early mining camp was clustered around the Centennial Millsite in Section 4, about 1/2 mile from the present town of Centennial.)

His notes continued, "The western half of the township is very rough and unfit for cultivation or grazing purposes. Many lodes or veins of gold bearing quartz transverse the axis of the mountain. Numerous prospect holes have been dug upon the sides and top of Centennial Mountain in search of gold bearing veins. The Centennial tunnels and shafts and the Michigan tunnel are the most extensive workings on the east side of the mountain. The Centennial Gold Mining Company shaft, tunnel, quartz mill, mining store and other buildings are situated on the West 1/2 of the S.E. 1/4 of Section 4. These cost the company about $18,000. The Centennial Mine has produced about $25,000 of gold bullion. At present work is suspended." [29]

The principal participants in the Centennial gold rush dispersed into the surrounding mountains, hoping to discover another bonanza that would equal the first Centennial. Stephen Downey, although deeply disappointed in his Centennial enterprise, directed his efforts to the Keystone Mine near the old Last Chance diggings. Coolbroth moved his hotel business to the new camp on the Big Laramie that was to become Cummins City.

Centennial's boom was over, but unlike the other Medicine Bow mining camps both before and after, Centennial lived on. When the 1878 election for a Delegate to Congress was held, Centennial cast 38 of it's 40 votes for Stephen W. Downey who went on to serve in the Forty-sixth Congress.

• 1879

During the summer of 1879, N.M. Ambrose, the former supervisor of the Centennial Mine, returned to Centennial to inspect the discovery of Reuben Ferguson who reported he had found the lost Centennial vein. However, Ambrose regarded it as only a spur of the old vein and no new gold production was reported. The mill at the time was reported to be in good working order and ready to operate at an hours notice.[30]

In the 1879 election, Centennial cast only 23 votes.[31]

• 1880

According to the U.S. census, taken in June, 1880, only sixty-one people remained in the entire Centennial District. Serryn Bruyn was keeping the saloon, Sanford Rusk was the butcher, Joe McMann a carpenter, and most of the others were listed as farmers, ranchers or stockraisers and their families. Only three claimed to be miners, Baker, Kelly and Schafer.

An additional nineteen people were enumerated in the vicinity of Sheep Mountain. Of these, only one, Peter P. Rynders, listed his occupation as mining. Apparently the serious miners and prospectors had left Centennial for more promising ground. The gold rush to the Black Hills of South Dakota, which began in 1874 and lasted through 1879, drew some of the Centennial miners. Another minor rush to the Big Horn Mountains began about the same time and attracted a few miners. Others returned to city life and less precarious trades, and some returned to the surrounding mountains in hopes of finding a rich new lode which could match the original Centennial.

Of the sixty-one people counted during the census, twenty-eight did their civic duty and voted during the election of 1880. Of these, twenty votes were cast for one of their own, Franklin O. Sawin, surveyor and prospector of the area mines, in his unsuccessful bid for a seat in the House of Representatives.[32]

The Centennial stamp mill was moved to Cummins City in August, 1880.

• Later

During his tenure in Washington D.C., Stephen Downey enlisted the financial aid of nine Washington D.C. investors who apparently offered to support his search for the lost Centennial vein. Downey was again unsuccessful in his new search.[33]

By 1882, the vote had dwindled to fourteen and by 1884 only seven votes were cast in the Centennial Precinct.[34]

In 1882, Thomas Markle transferred his

The Utopia Mining and Milling Co.,

Robt. W. Innes,
Treas. and Gen'l Mgr.,
Grand Rapids, Mich.

MINES:
Centennial Mining District,
Centennial, Albany County, Wyoming.

Sept 25 1899

Wyoming State Museum

postmaster duties to Frank L. Dustin and returned to ranching. Lemmon, Davis, Grant, P.G. Murphy, Worth, Sawin and others were to appear again in the various Medicine Bow mining camps.

In 1886, the Ninth Wyoming Legislative Assembly changed the western boundary of Albany County thereby placing Centennial in Albany County where, according to the miners, they had belonged all the time. The seventeen votes cast by the Centennial precinct in 1886 were counted in Carbon County but by 1888 the votes (again seventeen votes) were counted in Albany County.

The 1890 census is missing but the 1900 census listed 110 people in the Centennial District, mostly ranchers.

In January, 1893, the Centennial Mine was reclaimed and renamed the "Colombia" and work was again pursued by Rhoades and Custer. Again, no significant amount of gold was recovered.

The Billy Waters Mine set up a two stamp mill in 1894. Although located some distance from the mine some 100 tons of ore was run through the mill producing a modest amount of gold. The Billy Waters Mine was later relocated as the "Free Gold." In later years (1925) it was said that the "Centennial," the "Utopia" and the "Billy Waters Mine" were the only mines which had produced gold.

By 1901, Centennial was becoming known as a resort area rather than a mining camp. The "new" town was surveyed and platted, by Fred Miller, in Section 3, below the original

Headframe of the Queen Mine located on Centennial Ridge - 1988.

Tailings of the original Centennial Mine .

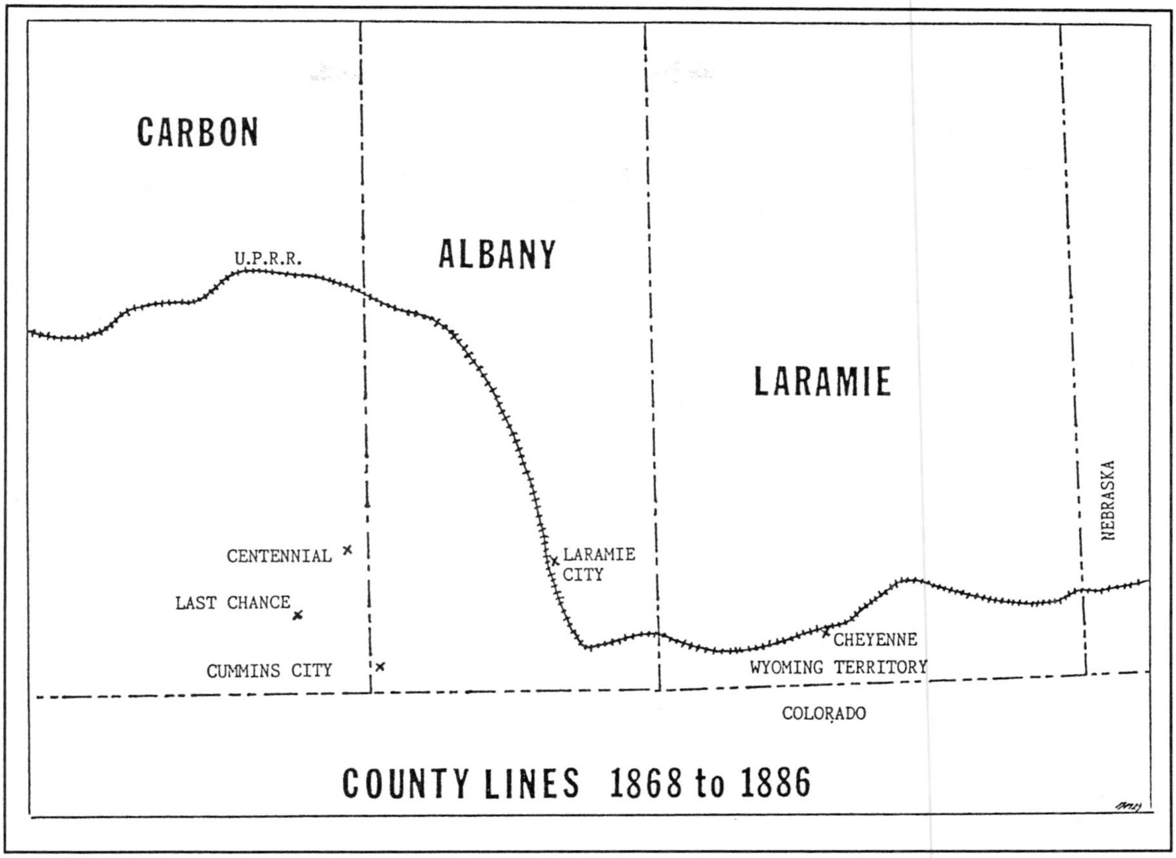
CARBON
ALBANY
LARAMIE
U.P.R.R.
NEBRASKA
CENTENNIAL
LAST CHANCE
CUMMINS CITY
LARAMIE CITY
CHEYENNE
WYOMING TERRITORY
COLORADO
COUNTY LINES 1868 to 1886

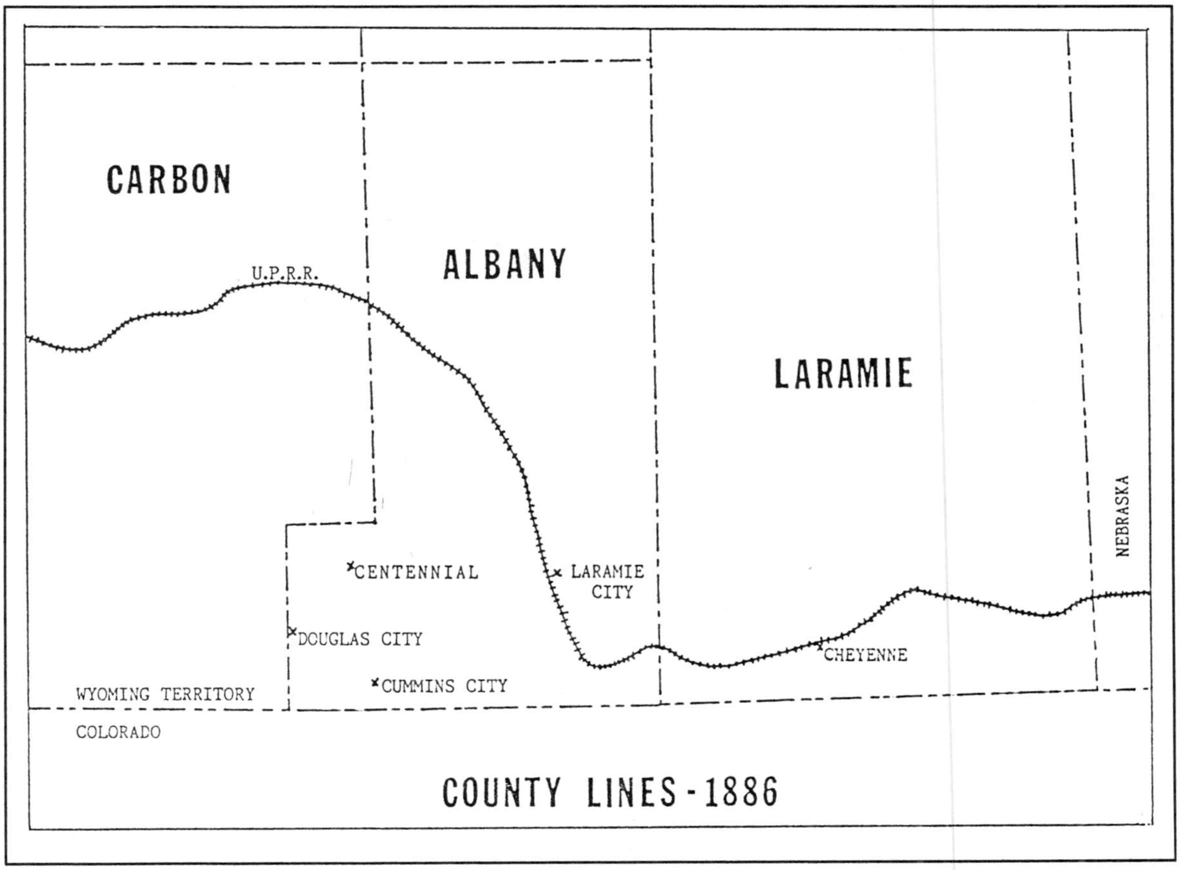
CARBON
ALBANY
LARAMIE
U.P.R.R.
NEBRASKA
CENTENNIAL
LARAMIE CITY
DOUGLAS CITY
CHEYENNE
CUMMINS CITY
WYOMING TERRITORY
COLORADO
COUNTY LINES - 1886

mining camp where fifty-six blocks and over 1,000 lots were laid out.

The long awaited Laramie Hahns Peak and Pacific Railroad arrived in Centennial on June 17, 1907, and signaled the last of the horse drawn stage. The last stage from Laramie arrived in Centennial on August 8, 1907.

In 1908, the Commercial Gold Mining Company supplanted the Utopia company and under the direction of Bernard Holtum sold stocks and engaged in a limited attemp to re-open the mines. Again, with almost predictable results, the mines failed to produce.

Throughout the years prospectors have returned to Centennial Ridge time and time again, still seeking the lost Centennial vein or a bonanza of their own.

In 1920, the WYOCOLO Gold Mining Company was formed, working the mines on the west slope of Centennial Ridge, notably the "Cliff," the "Independence" and the "Columbine." Although an extensive amount of work was accomplished, again the return was not sufficient to sustain a company. The remains of this enterprise are visible to this day on the western slope of Centennial Ridge along the Middle Fork of the Little Laramie River.

Even now, Centennial Ridge draws prospectors who no longer need a large quantity strike to make their day. With gold selling for 20 to 30 times the pre-1932 price, even a small discovery can reap rewarding profits.

Centennial stands out as the one district that, for a short time anyway, was able to produce a nominal amount of gold for a reasonable amount of labor and turn a profit. From the one mine, Centennial produced approximately one-fourth of all the gold produced in the hundreds of mines in the Medicine Bow Mountains.

Centennial Mountain remains a veritable museum of early Medicine Bow mining activities. The structures of 1876 and 1877 were reported to have been "removed" in 1907 and the remaining buildings are difficult, if not impossible, to accurately date. However, the tailings of the original Centennial Mine and Mill are visible to this day.

Centennial Mill steam engine.

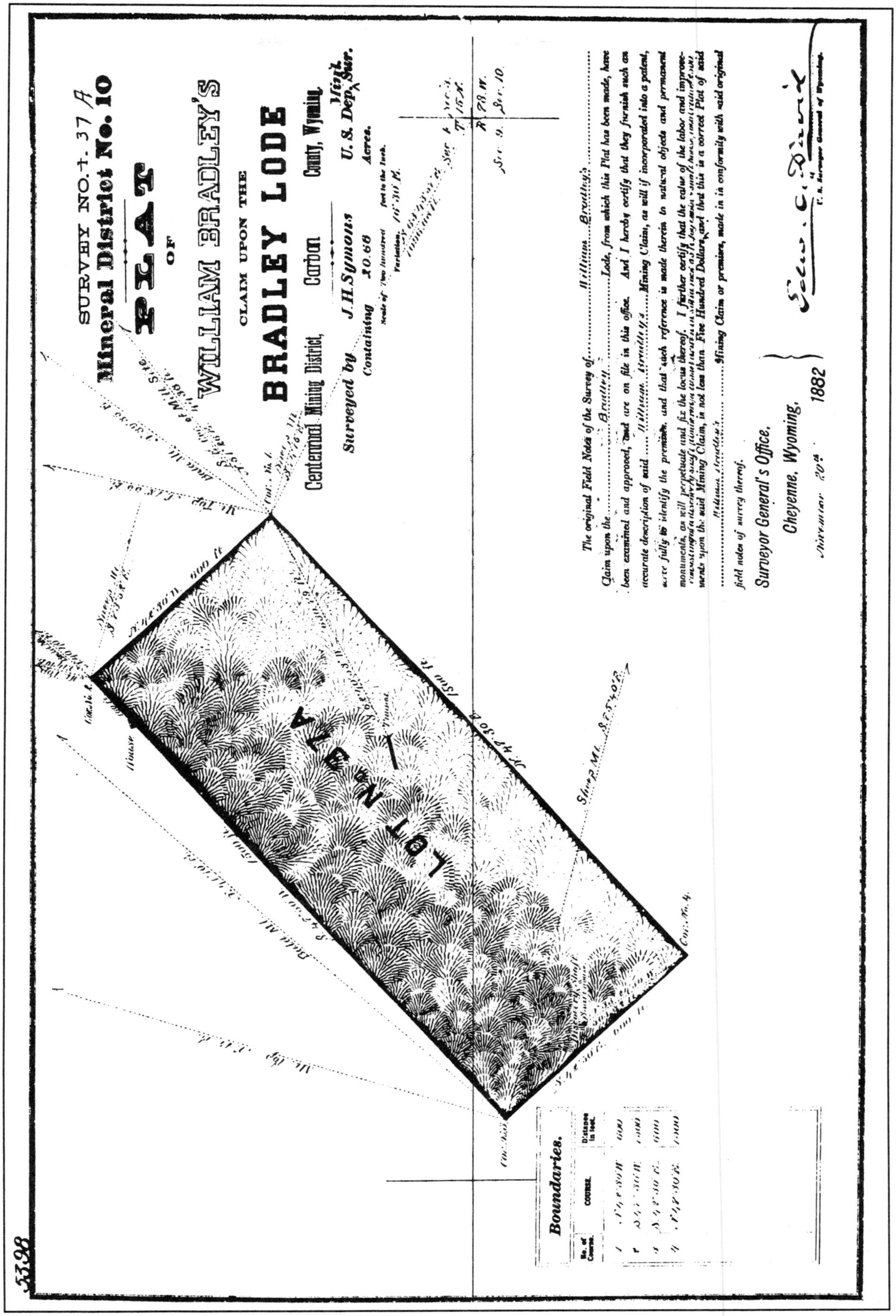
5398
SURVEY NO. 37 A
Mineral District No. 10
PLAT
OF
WILLIAM BRADLEY'S
CLAIM UPON THE
BRADLEY LODE
Centennial Mining District, Carbon County, Wyoming.
Surveyed by J.H. Symons U.S. Dep. Mini. Sur.
Containing 20.66 Acres.
Scale of Two hundred feet to the Inch.
Variation. 16°30' E.
LOT No 37A
Tunnel.
House
Discovery Shaft
Cor. No. 1
Cor. No. 2
Cor. No. 3
Cor. No. 4
N. 47°30' E. 1500 ft.
Sec. 4 | Sec. 3.
T. 15 N.
R. 78 W.
Sec. 9. | Sec. 10.
Boundaries.
No. of Course. | COURSE. | Distance in feet.
The original Field Notes of the Survey of William Bradley's Claim upon the Bradley Lode, from which this Plat has been made, have been examined and approved, and are on file in this office. And I hereby certify that they furnish such an accurate description of said William Bradley's Mining Claim, as will if incorporated into a patent, serve fully to identify the premises and that such reference is made therein to natural objects and permanent monuments, as will perpetuate and fix the locus thereof. I further certify that the value of the labor and improvements upon the said Mining Claim, is not less than Five Hundred Dollars, and that this is a correct Plat of said William Bradley's Mining Claim or premises, made in in conformity with said original field notes of survey thereof.
Surveyor General's Office,
Cheyenne, Wyoming,
November 20th 1882
U. S. Surveyor General of Wyoming.

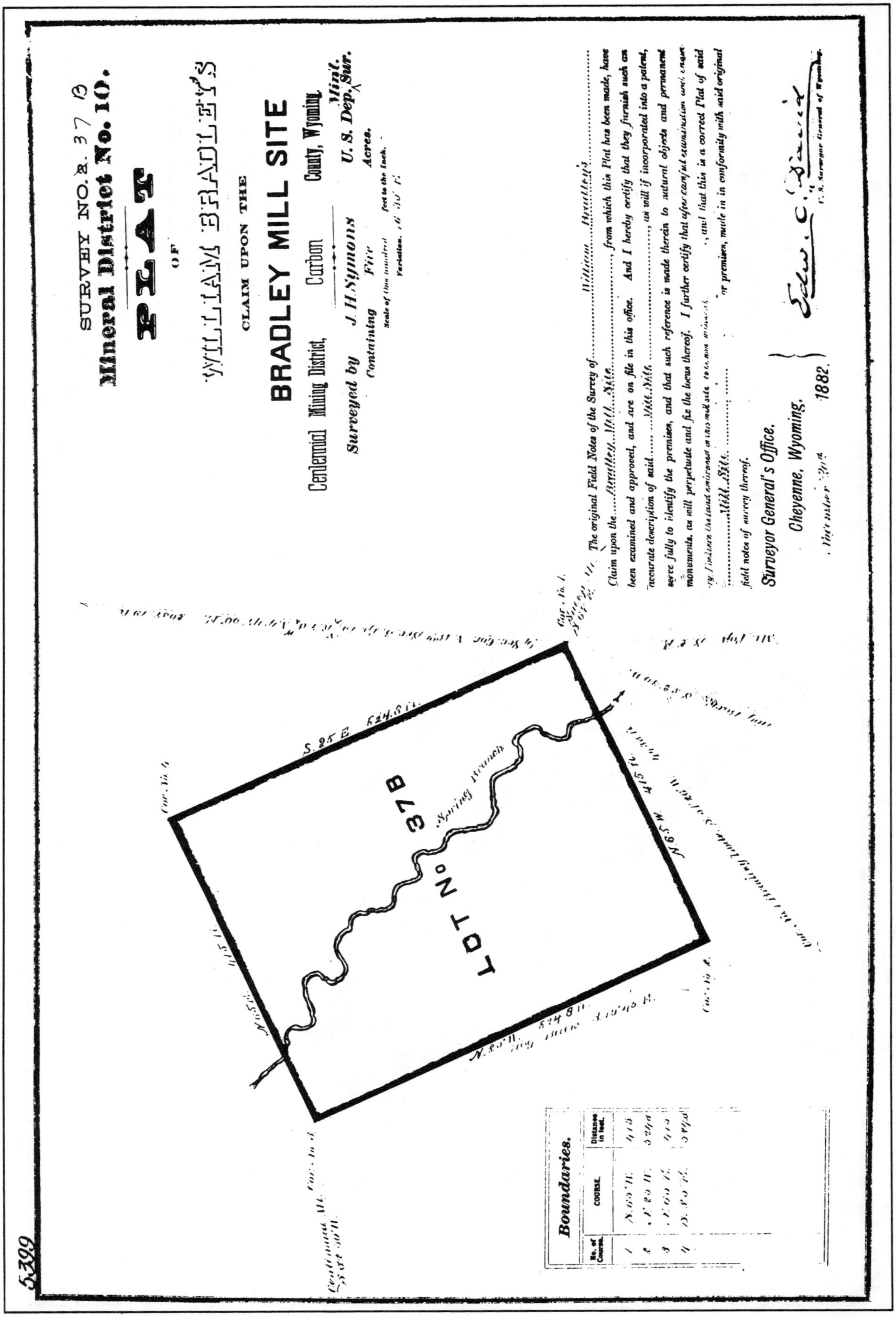
SURVEY NO. 37 B
Mineral District No. 10.
PLAT
OF
WILLIAM BRADLEY'S
CLAIM UPON THE
BRADLEY MILL SITE
Centennial Mining District, Carbon County, Wyoming,
Surveyed by J.H.Symons U.S. Dep. Minl. Sur.
Containing Five Acres.
Surveyor General's Office.
Cheyenne, Wyoming,
1882.
LOT No 37 B
Spring Branch
S. 25 E. 524.8 ft.
N. 65 W. 415 ft.
N. 25 W. 524.8 ft.
Boundaries.
5399

Know all men by these presents That I the undersigned that is to say Fred A Miller under the laws of the State of Wyoming in reference to Townsites and Additions to be platted, the sole owner and proprietor of the lands shown on this plat and in this dedication does hereby lay out a Town to be known as Centennial, in lots and blocks being numbered in rotation from No. 1 to No. 56 all of which Blocks lying as described herein, as follows Being the Northeast quarter of Sec. 3. Township 15 N. Range 78 West of the Sixth Principal Meridian, situate in Albany County, State of Wyoming, and having caused the same to be surveyed, and staked out into lots, Blocks, streets and alleys, and a plat thereof to be made with reference to known monuments, accurately describing all the subdivisions of such tract, numbering the same by progressive numbers and giving the dimension length and breadth and the course of the streets therein.

Does hereby state and make known that the said subdivision of the within described Blocks aforesaid as the same appears on the accompanying map, is with the free consent and in accordance with the desire of the said Fred A Miller the owner and proprietor aforesaid, and the streets and alleys are hereby dedicated perpetually to the public use.

In Witness Whereof the party to this dedication has hereunto set his hand and seal this twelth day of August A.D. 1901

Fred A Miller

State of Wyoming } ss
County of Albany }

Be it remembered that on this twelth day of August A.D. 1901 before me, a Notary Public, within and for the County of Albany in the State of Wyoming personally appeared Fred A. Miller who is personally known to me to be the identical person whose name ~~whose name~~ is subscribed to the foregoing instrument and accompanying plat of Centennial in the County of Albany in the State of Wyoming, and the dedication of the same as set forth and described in the foregoing instrument of writing as his free and voluntary act and deed

In witness whereof I have hereunto set my hand and official seal

This twelth day of August A.D. 1901

Notary Seal

C S Greenbaum
Notary Public.

No. 3445 2½

Filed for record at 4:00 p.m. Aug. 12, 1901.

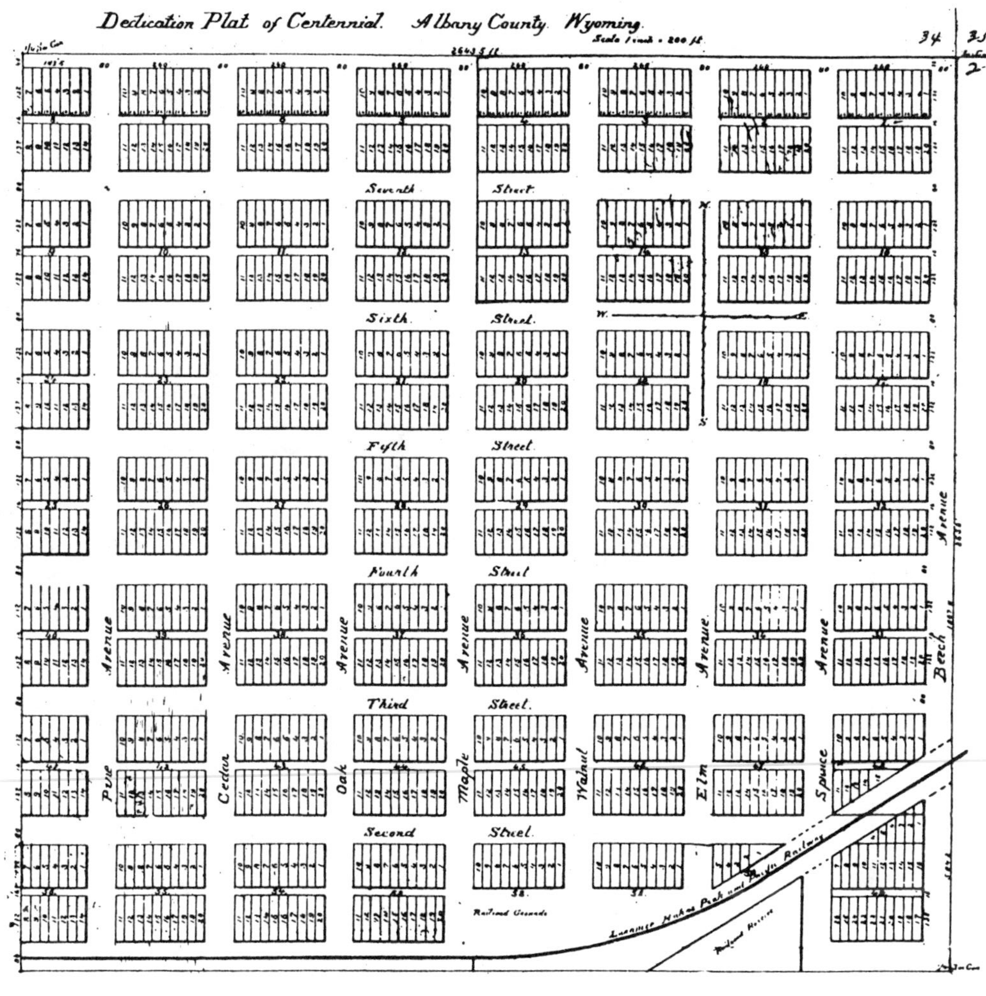

PROPOSED PLAT
OF
CENTENNIAL
1901

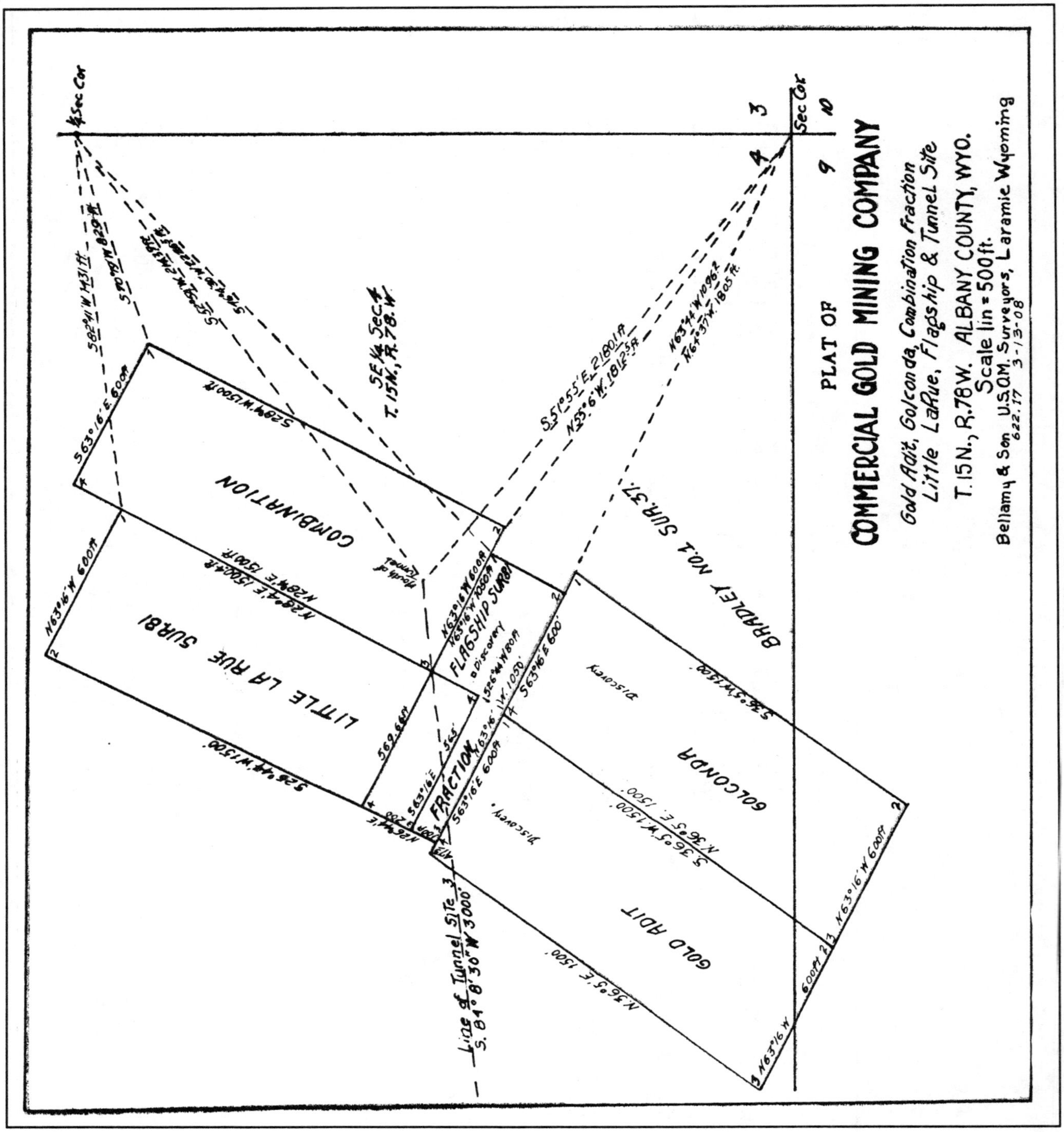
PLAT OF
COMMERCIAL GOLD MINING COMPANY
Gold Adit, Golconda, Combination Fraction
Little LaRue, Flagship & Tunnel Site
T.15N., R.78W. ALBANY COUNTY, WYO.
Scale 1in=500ft.
Bellamy & Son U.S.D.M. Surveyors, Laramie Wyoming
COMBINATION
LITTLE LA RUE SUR.81
FLAGSHIP SUR.
FRACTION
GOLCONDA
GOLD ADIT
BRADLEY NO.1 SUR.37.
SE¼ Sec.4
T.15N., R.78.W.
Line of Tunnel Site
S.84°8'30"W 3000'

MINERAL LOCATIONS ; 1909

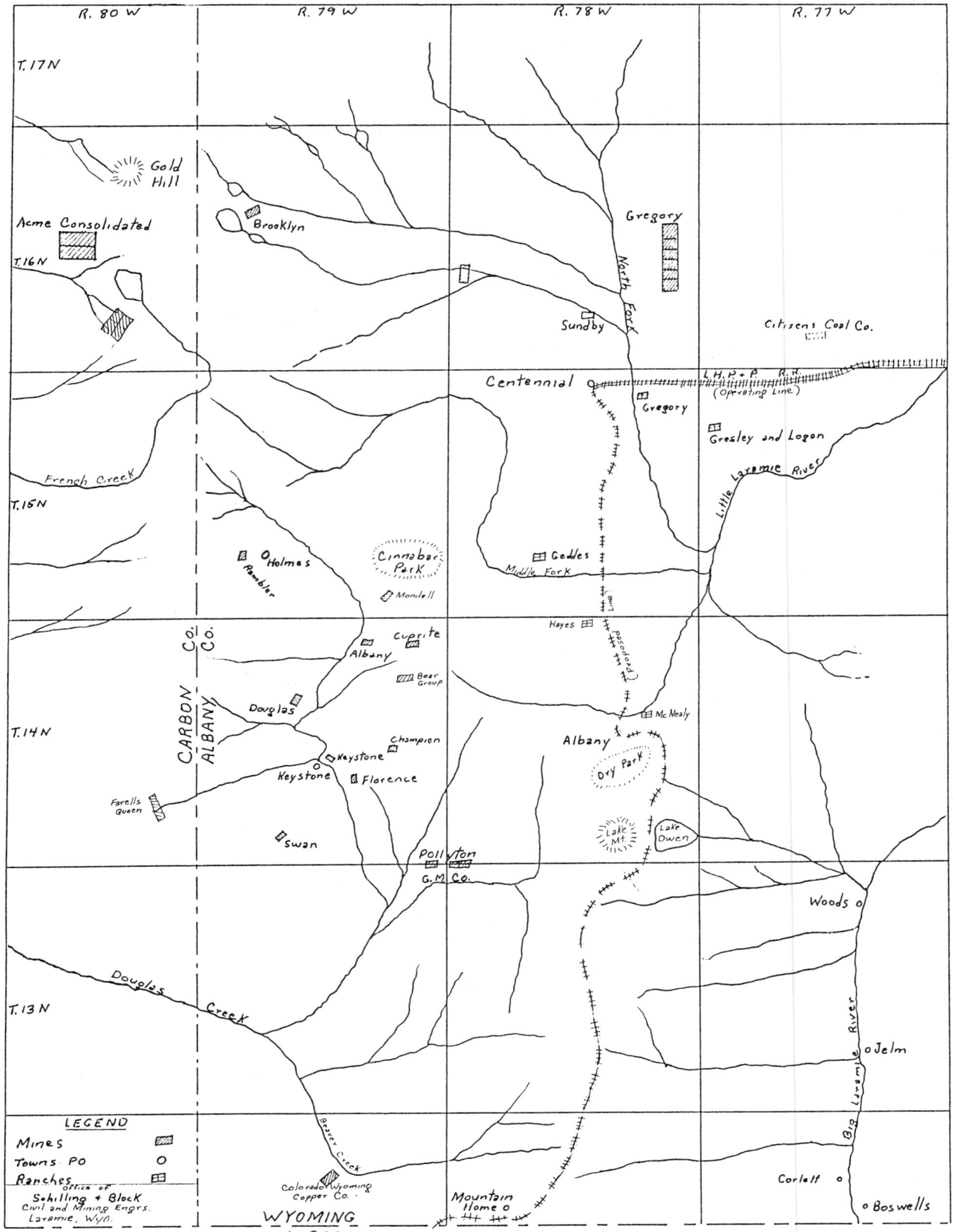

From the Centennial Post
December 18, 1909

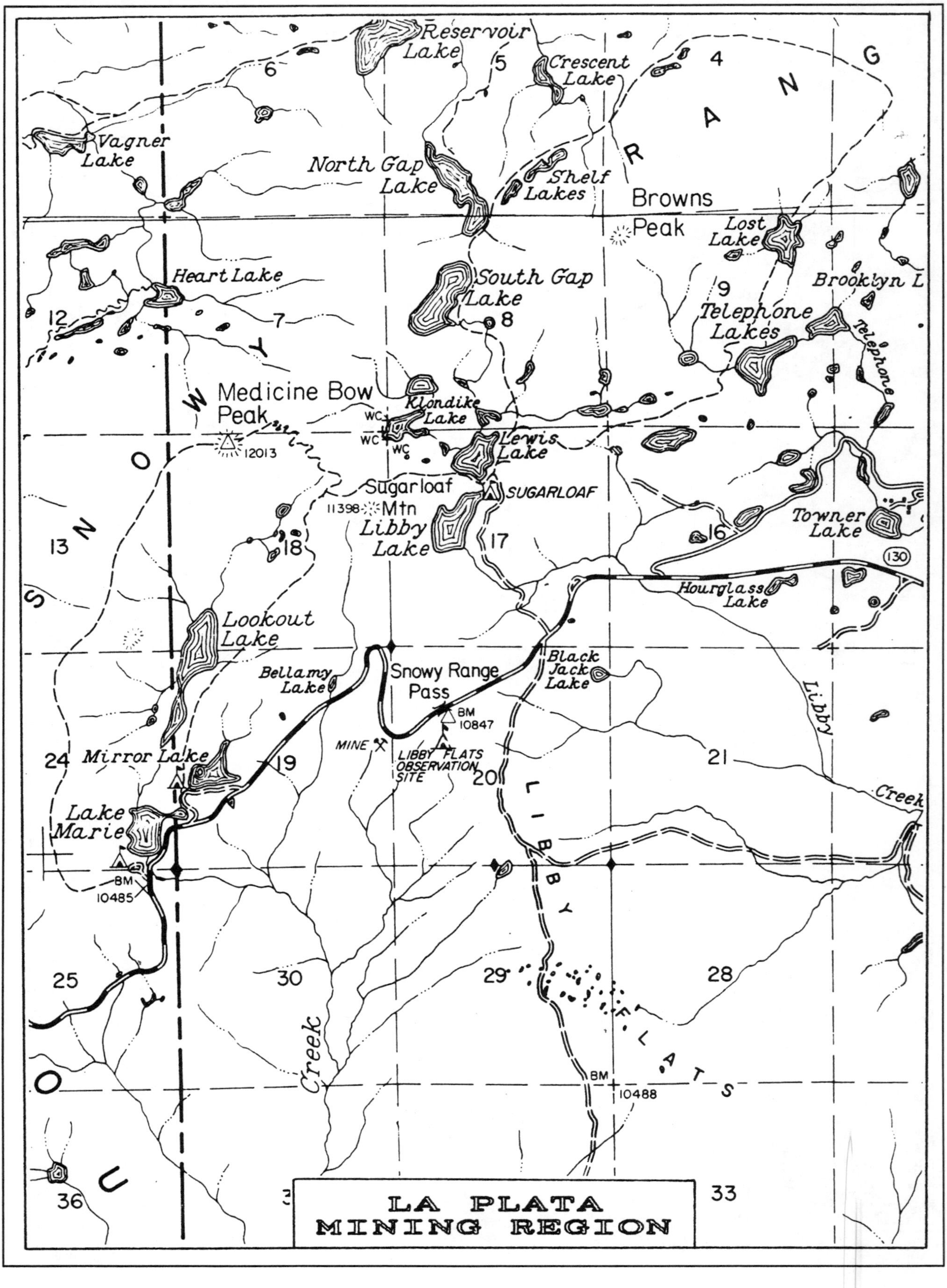

Reservoir Lake
Crescent Lake
Vagner Lake
North Gap Lake
Shelf Lakes
Browns Peak
Lost Lake
Heart Lake
South Gap Lake
Brooklyn L
Telephone Lakes
Telephone
Medicine Bow Peak
12013
Klondike Lake
Lewis Lake
Sugarloaf Mtn
11398
SUGARLOAF
Libby Lake
Towner Lake
130
Hourglass Lake
Lookout Lake
Bellamy Lake
Snowy Range Pass
BM 10847
Black Jack Lake
Libby Creek
MINE
LIBBY FLATS OBSERVATION SITE
Mirror Lake
Lake Marie
BM 10485
Creek
BM 10488
LIBBY FLATS
SNOWY RANGE
1 4 5 6 7 8 9 12 13 16 17 18 19 20 21 24 25 28 29 30 33 36
LA PLATA MINING REGION

Even before the Centennial discovery, the prospectors on Douglas Creek had begun to expand their search for gold. As the placers of Last Chance began to show less and less profit, several of the miners moved further up the stream in hopes of finding the source of the fine gold of the Douglas Placers that assayed well but produced little.

During the summer of 1872, Daniel D. Wolff, John and George Renner moved up Douglas Creek and crossed over to the high mountains and a new, yet unnamed mining area. Leaving the timbered areas, they began prospecting among the alpine-like lakes, rocks and streams of the high Medicine Bow Mountains which they called, as we do today, the Snowy Range. Here during the summer of 1872 they located and claimed the "... Independence Lode on Bear Mountain in the Medicine Bow Range at the junction of the Limestone and Porphyry formation." The recently enacted Mining Law of 1872 placed no limit on the number of lode claims an individual or corporation might locate. Wolff, apparently well aware of the new law, proceeded to make several new claims. On August 1, 1872, Wolff claimed the "... Cloud Cap Lode, running through the central mountain of the Medicine Bow Range directly west from the Little Haystack Mountain." About the same time, Wolff was joined by Franklin O. Sawin and together they claimed the "... Alpha Lode, situated between the Big Haystack and Hayrick Mountains in the Medicine Bow Range." Wolff and Sawin also claimed the "... Albany County Lode, situated East from the face of the Central Pyramid Mountain in the Medicine Bow Range." [1] Wolff then claimed the "Maid of Laramie Lode" adjoining his previous "Cloud Cap Lode" and then the "... Silver Wave Lode, commencing at the southern end of the Northern Pyramid Mountain and extending along the eastern face of the Pyramid Mountains in the Medicine Bow Range." He then claimed the "Lady of the Lake Lode ... on the east side of Virginia Lake and directly East of the Northern Pyramid Mountains."[2]

Although actually located in Carbon County, these claims were filed in the Albany County Courthouse on August 1, 1872. Most of the claims were for silver although gold, copper, lead and cobalt were also listed. The claims were made under the Mining Law of 1872 and were all the maximum size allowed, 600 by 1500 feet.

Before the close of the mining season, Wolff claimed the "... Silver Queen Lode, located on Bear Mountain at the junction of limestone and porphyry formation in the Medicine Bow Mountains" near the "Independence Lode." [3]

Sawin acquired another partner, David Rice, and together they claimed the "Crystal Lode, ... on the north side of Timberline Mountain near its base in the Medicine Bow Mountains" and the "Smiling George Lode on the west side of Timberline Mountain."

This small group of miners was probably the first in this area and they applied geographic names which seemed appropriate, at least to them. Haystack, Hayrick, Pyramid and the other names used by the miners certainly describe the area but apparently these names were of short duration and appeared only on the claim descriptions filed at the county courthouse.

Apparently some silver but little or no gold was produced from this group of mines. According to the Bureau of Statistics, Treasury Department, Wyoming Territory gold production for 1872 totaled only $11,186.44. This figure accounted only for the gold which passed through the United States assay offices within the territory and did not account for any gold carried out of the territory for assay or sale. The bulk of this reported gold production most likely came from the Sweetwater mines and very little from the Medicine Bow mines. Noticeable, however, was the Wyoming Territory silver production, which was zero, while Nebraska, which had no silver mines, totaled $342,463 from the Omaha reduction works. This product was undoubtedly from shipments along the Union Pacific and a portion could have easily come from the Wyoming Territory and the Medicine Bow mines.

Courtesy of Bud Owens

• 1877

Work on the mines apparently continued but there was little news until 1877. Then in the rugged boulder strewn fields of the high Medicine Bow Mountains, new interest developed near the 1872 claims of Daniel Wolff, Frank Sawin and other early prospectors. Sawin had remained in the area, but apparently Wolff had not. Sawin had, during this time, established a claim on the "Telephone Lode" which he believed had a strong showing of silver in the limestone deposits.

A meeting was held at the Sawin's cabin on October 1, 1877, and the bylaws of the newly named La Plata Mining District were drawn up and boundaries established; "Commencing at the Falls on Libby Creek, running thence up Libby Creek, to the extreme south fork of Libby Creek, thence up said south fork to the head of Couger [sic] Creek, thence down Couger Creek to French Creek, thence across French Creek due west to the summit of Snowy Range, thence due north eight (8) miles thence east 8 miles, thence in a southerly direction to the place of beginning." The document was signed by C. M. Gregory, President; C. T. Justus, Secretary; and F. O. Sawin, Recorder.[4] Although the forming of the new district was encouraging to the prospectors, the records for the remainder of 1877 reveal little activity.

• 1878

In 1878, the Douglas Creek mining was drawing the greatest interest. The La Plata District, however, was becoming known, not only for it's promise of silver, but also it's high mountains and pristine beauty.

By August, 1878, there were about thirty claims located including the "North American" and "Telephone" which were the only ones showing development and they were down only 40 feet.[5] Although both mines were showing indications of good paying ore, both were plagued by water in their shafts. A hand pump was being used in an attempt to control the water. C. M. Gregory was superintendent of the North American Mine. "Captain" James Lemmon was superintendent of the Telephone which was owned by Frank Sawin, Noah C. Worth and Melville C. Brown. According to "W.F.C.," correspondent for the *Laramie Sentinel*, there were three good log cabins, three tents and several "wickie ups" in the camp. He also said, "There are about twenty-five men, and one colored boy, in camp."[6] As of mid-July, 1878, the nearest wagon road was eight miles from the La Plata mines. The miners, feeling the need for a road into the district, held a meeting on August 7, and organized a road building group. "Major" William O. Downey was appointed chairman and Gregory, Fales and Moore were designated as surveyors. The surveyors reported back on August 8, and after a short meeting at the North American cabin, chose the route via Nash's Fork and Libby Creek. Wasting no time, the entire camp turned out to do the road work under the direction of N. T. Webber.[7] Webber, who had a sawmill near Centennial, was planning another mill within four miles of the mines and was interested in improving the road for timber hauling.

In the Fall, 1878, the La Plata District came to the forefront for a short time, not only for its mining activity, but for the interest and resulting controversy in climbing the highest peak in the Medicine Bow Mountains. It began quietly enough with the announcement that "Captain" James Lemmon and his daughter, Marion Forbes, had left Laramie City for the summit of Snowy Range. They planned to be gone for two or three weeks.

A few days later, another party also made a trip to the Snowy Range. The correspondent, signing himself "Observer" sent the details to the *Laramie Daily Sentinel*:

> Our party consisted of Mrs. Dr. Baker, Mrs. Captain Kelly, and Miss Vernie Markle, with Messrs. Fales and Kelly and your correspondent as escort.

DAILY SENTINEL

LARAMIE CITY. WYOMING.

J. H. HAYFORD, : : EDITOR.

Wednesday Evening August 11

ABOVE THE CLOUDS

Three Ladies Ascend the Highest Point of the Snowy Range and There Plant the Stars and Stripes

The start was made from Centennial valley at 6:30 o'clock in the morning. Our route lay along Libby Creek on a very easy grade for four miles, thence over the divide to Nash's Fork, thence through the Big Park on to La Plata City, arriving there at ll o'clock.

"La Plata City," at that time consisted of probably not more than a dozen log buildings, only three of which were dwelling cabins, scattered about the mines. The correspondent identifying himself as "Observer" continued with a description of the area and the ascent of the mountain.

La Plata City is situated on a beautiful stretch of level country interspersed by the most beautiful of lakes. Our party was agreeably surprised at the exceptionally fine scenery, which would have amply repaid us for our trip of twelve miles.

Everyone in camp, lounging around the camp fire, spinning yarns about old time mines, miners and mining, was entirely taken by surprise to see a posse of ladies and gents riding past. Then the news that the first ladies ever on the Snowy Range had arrived in camp spread like wildfire. Within ten minutes all the residents of the District turned out en masse, and congratulated the ladies on their appearance on the range and their fortitude in enduring the trip.

On the range we were joined by Messrs. Higgens, McGrath and Suprise, residents of the place, when we marched on and mounted our banner (the flag of the Union) on the top of the North American shaft house, when we returned to the Telephone cabin for a few minutes rest, it being proposed by Messrs. Higgens and McGrath that the ladies ascend the pinnacle of the Snowy Range, Cloud Cap mountain. Our party again mounted their respective horses (or mules as the case may be) to complete our journey to the summit. Our route lay parallel with the main range, in a southwesterly direction, over the most beautiful country beheld. Our party was unanimous in voting that it was the most beautiful of mountain scenery. It would be impossible for me to describe it and do it justice - it must be seen to be enjoyed. A smart gallup of three miles over a level country, brought us to the range proper, then a climbing of 400 feet brought us to a beautiful terrace on the one side, and a cliff of 300 feet high, almost perpendicular, on the other. Here we were compelled to leave our horses and continue on foot, over masses of loose bowlders [sic] to the highest peak on the Snowy Range, arriving at the summit at 2:30 o'clock, p.m. Preparations were immediately made for hoisting the Stars and Stripes. Everything being in readiness, Miss Vernie Markle, assisted by Mrs. Baker and Mrs. Kelly, at the sound of the signal gun, pulled the cord and unfurled our glorious old flag, over 12,000 feet above sea level. After firing thirteen guns (giant powder), the "Star Spangled Banner" was rendered by the party; then, in a neat speech, Mr. McGrath thanked the ladies, in behalf of the residents of La Plata, for the honor they (the ladies) conferred on them by coming in their midst, also complimenting them on their energy and fortitude in undertaking such a fatiguing and tiresome journey.

Mrs. Kelly responded on behalf of the ladies, thanking the gentlemen of La Plata for the kindly greeting they gave the ladies on their arrival at the range, more especially their escort, Messrs. McGrath, Fales and Suprise. Then our party turned faces toward camp, for we were all exceedingly hungry, having no lunch along.

After regaining our horses, we passed through a dense forest of spruce pine. We suddenly came upon a beautiful lake, which was promptly named the Ladies' Lake, in honor of the brave ladies who were the first to visit the Snowy Range, also to unfurl the flag of our country on its highest peak. We then followed the south fork of French Creek; then across the divide to the north fork of Libby Creek; then a gallop across some level parks, where one of our party - don't like to mention names - while being a little too attentive to the lady riding beside his side, suddenly lost his balance and slipped off his horse, measuring the green sward with his length. We arrived at La Plata at 7 p.m., where a bountiful supply of eatables was spread for our nearly famished tourists. After doing the supper ample justice, we again mounted our chargers, and, with a hearty farewell from all, turned towards home, beholding one of the most delightful of moonlight scenes, and arriving at Markle's ranch at about ll o'clock p.m.

Too much cannot be said for the ladies, especially Mrs. Baker, to complete the trip without apparent fatigue, when stalwart men acknowledged being tired.

> We promise to write again on some future occasion, I will close,
>
> OBSERVER[8]

THE LA PLATA DISTRICT.

An Interesting Letter from an Occasional Correspondent

LA PLATA, August 12, 1878.

EDITOR SENTINEL:

Seemingly, the first ascent of Cloud Cap, (now known as Medicine Bow Peak) was history. However, a few days later, another correspondent, signing himself "JUSTICE," also wrote the *Sentinel*, disputing the claim. His claim was that Mrs. Forbes had reached the true summit of Cloud Cap in the company of Frank Sawin on the same day and had seen no other climbers. Mrs. Forbes, he said, had traveled to the La Plata camp by wagon and made her assault on the mountain directly from the camp, indicating she had probably ascended a different part of the mountain than the Baker expedition.[9] Careful examination of the two expeditions indicates the Baker party most probably ascended the southwest extremity of the mountain which indeed would be almost 300 feet lower than the highest portion which would be considerably closer to the La Plata site. The three mile ride, the terrace, the cliffs, all closely describe an ascent from the southwest end of the mountain. The lake the Baker party named Ladies' Lake could be any one of several in the region, most likely, what is now known as Silver Lake. Even today it is difficult to visually determine the highest point on the mountain but to Mrs. Forbes most likely belongs the honor of the first female to ascend Cloud Cap Peak or as we now call it, Medicine Bow Peak.

On August 2, Joseph A. Nock filed his claim on the "Chrysopolus Lode."[10]

August 17, 1878, the *Laramie Weekly Sentinel*:

> La Plata miners under the direction of N.T. Webber have constructed a wagon road from the Centennial to the La Plata mines. May now be reached by any sort of vehicle from a top buggy to a seven ton freight wagon.

During August, Judge Melville C. Brown visited his holdings in the La Plata District. On his return he reported his observations to the *Laramie Daily Sentinel*, assuring them he was ". . . not a bit excited or at all sanguine, but as a matter of fact, there is enough silver out there in the La Plata District to last the world a million years for all purposes, even to the building railroads with it."[11]

Brown, who was 40 years old in 1878, had gone to the California gold fields when he was 18, where he was engaged in mining and merchandising. In 1863-1864, he was engaged in mining and politics in the Boise, Idaho, area. He spent some time in the Montana gold fields and then moved to Cheyenne in 1867, where he established a law office. A short time later (1868), he moved to Laramie City where he set up his law office. In 1871 he was elected as a member of the Wyoming House of Representatives and in 1889, president of the Wyoming Constitutional Convention. He was twice mayor of Laramie City and served as prosecuting attorney for Albany County and several other United States and Wyoming offices. Through all his public service, Brown always found time to dally in the mining business, often grubstaking prospectors in his favorite areas and frequently prospecting on his own. Little wonder that after his eloquent statement on the value of the mines, the La Plata miners named the second most predominant mountain in the area, Brown's Peak.

On July 13, 1878, James T. McGrath (locally known as Jim McGraw) and John Higgens located the "Mayflower Lode." This location was filed on November 14, 1878.[12]

On August 5, 1878, L. L. Thompson and A. L. Suprise located the "Monticello." On this same date, Louis Suprise, Andrew Suprise, A. L. Suprise and L. L. Thompson located the "Ida Lode." Two days later, Louis Suprise, Julia Suprise, A. L. Suprise and A. L. Thompson located the "Croesus." These claims were not recorded in the county records until May 24, 1880.[13]

The third week in September, 1878, a rather prestigious group visited some of the Medicine Bow Mines. The group included Governor Hoyt; Professor James B. Baylor, of the United States Coast and Magnetic Survey; Professor J. G. Murphy, Territorial Assayer;

"Colonel" Stephen W. Downey; Dr. James H. Hayford of the *Sentinel*; and "General" Noah C. Worth. After a visit to the Douglas Creek mines, the party spent one day at La Plata. On his return to Laramie City, Hayford wrote that the La Plata mines were "... *supposed* to contain *silver*, at the rate of from 100 to 1,000 ounces per ton. But he would be a bold man who would risk his reputation (if he had any reputation) by venturing an opinion as to is richness." [14] Professor Murphy declined to comment on the ore.

On October 17, 1878, R. H. Fiddick filed a claim on the "Bismark." [15]

On October 19, 1878, L. A. Moore, claimed the "Bonanza" and several other prospects near the North American Mine. In an effort to raise enough money to develop the claims he then sold a one-fourth interest to George Fales for a reported $500.[16]

Water continued to be a problem and in November, 1878, a steam pump was obtained for the North American mine. F. O. Sawin informed the *Sentinel* that four mines planned to work through the winter.[17]

Sawin continued to be the primary source of information on the La Plata mines and on November 23, reported the Galena Mine had struck "good pay ore the full width of the shaft and is taking out the auriferous (gold bearing ore) at a rapid rate." [18] He also said the North American had just completed getting in their winter supply of wood and provisions, apparently just in time, for the *Sentinel* reported the first major snowstorm of the season had hit the region on November 19. Sawin traveled down the mountain by "snowshoes" (Nordic Skis) to Centennial where after a day or two of collecting supplies, he planned to return to La Plata for the winter.[19]

• 1879

Although mining was being carried out in much of the Medicine Bow Mountains, there was still no suitable route across the mountains. On February 25, 1879, thirty-seven leading residents of Warm Springs (Saratoga) petitioned the Carbon County Commissioners to survey a practical route from the mouth of Brush Creek to La Plata and Douglas Creek with the view of having a county road established.[20] Apparently only a preliminary survey was completed at the time and several years would pass before an established route was developed and then without the county's assistance.

Melville C. Brown
Wyoming State Museum

Early in the spring of 1879, J. B. Adams filed on an extension of the "North American." [21]

On April 1, 1879, H. H. Suprise claimed the "Tenderfoot" in the La Plata District.

On May 16, 1879, the La Plata post office was designated and Frank O. Sawin became the first postmaster.[23]

During the 1879 mining season the first recorded exchange of mining property in the La Plata District took place with the Quit Claim Deed of the "Chrysopolas Lode" by Joseph Nock to Richard Barret and William F. Crummer of Galena, Illinois.[24]

• 1880

On April 1, 1880, another exchange took place when James T. McGrath sold the old "Mayflower" claim which had been relocated as the "Crown Point," to A. J. Crawford, for a reported $500.00. [25]

Several years passed with very little activity reported in the La Plata District and the post office was discontinued January 6, 1881.[26]

During the summer of 1881, a new district was organized between La Plata and Centennial. Formed on the promise of silver, the district was named Silver Hill and the camp Silver City. It was located near the confluence of Silver Run, Gold Run and Libby Creeks, (near the present site of the Snowy Range Lodge). Alex H. McCuen (McCune), J. H. Kelly, William Bradley, T. J. Harvey, F. L. Dustin and several other prospectors made numerous claims on what they believed to be good silver deposits.[27] The district never developed beyond the exploratory stage but as late as 1903, claims were still being made in the Silver Hill District.

Frank O. Sawin failed to keep up with the assessment work on the Telephone Mine and in 1885, P. A. Steinhoff relocated the mine, using the abandoned claim act, under the name of the "New Telephone." [28] Around 1888, Sawin left the area to prospect elsewhere and later was to appear in the Gold Hill District.

Most of the others faded into obscurity. One of the few to stay was William "Billie" Class who stayed on to mine his more than twenty claims. Class Lake was named after him. Although predominately active in the Douglas Creek District, W. H. Lewis was also a miner in the area and Lewis Lake was named after him.

La Plata was quiet for some years, but again came to the forefront with the claims of 1891 and 1892. The *Laramie Republican* September 3, 1891, reported twenty-five to thirty miners in the old La Plata camp and that the "New York" and "Brooklyn" mines were receiving the most work. The road from Centennial was said to be in deplorable condition and the citizens of Laramie were urged to join in on the road work. Early in October, B. W. Towner and John McFayden filed on two new claims, the "Nina May" and the "Aztec."[29] Still, no paying mines could be developed. B. W. Towner moved on to try again at Cooper Hill but left behind his name on Towner Lake.

In 1891, following the Gold Hill boom, there was again a surge in claim filings, and around 1892, William "Billie" Class made numerous claims around the region now known as the Sugar Loaf Recreation Area. Class filed on the "Blackstar" on September 12, 1892.[30] Again, no mine produced enough ore to ship.

The La Plata post office was again authorized on October 18, 1892. Harry S. Hall was the designated postmaster but apparently the post office was never in regular operation.[31] La Plata had its own polling place at the Telephone Mine cabin but was considered part of the Centennial Precinct. In 1892, eighteen votes were cast.

LA PLATA MINING DISTRICT

The Region Changing From a Silver To Copper and Gold Deposits—The Coming Hahn's Peak Railroad.

When the township was resurveyed in 1898, United States Deputy Surveyor, William M. Gilcrest noted, "Much prospecting for gold has been done in this township but only one mine is being worked now."[33]

From 1901 to 1904, more claims were made on the headwaters of South French Creek, several by John H. Morris. In 1904, with the anticipation of the arrival of the Laramie, Hahns Peak and Pacific Railway, the miners of La Plata declared the region was changing from silver to copper and gold.

In 1905, Morris joined with William H. Peabody and John Crooks to form and incorporate the French Creek Mining Company.[34] Although the company set their capital stock at $200,000, little if any stock was sold and they were apparently operating on a shoestring. Morris was traveling to and from Centennial for supplies, on foot, a distance of more than sixteen miles. On one such trip, in June, 1906, he met his friend Billie Class and elected to return to the mountains with him. Traveling on the Centennial to La Plata road they returned together as far as the La Plata camp. Here, they parted company, Class going to his cabin near what is now called Class Lake, and Morris continuing over the divide to the headwaters of South French Creek and his cabin. Four days later, Class returned to Centennial but not without first making a good neighbor call on Morris, a side trip of several miles on foot. Class found the cabin locked and

Billie Class Stamp Mill.

Red Mask Cabin.

although a little suspicious, continued on his way. On previous occasions, people had forced entry and found Morris asleep and the old gentleman had become highly indignant. On one occasion, he was found completely helpless from being poisoned by canned goods, but nevertheless, indignant. Some two weeks later, Morris' good friend, Freeman Bower stopped by the cabin on his way to Gold Hill. He too found the door locked and so continued on his way. On his return trip, Bower was accompanied by William Cole also of Gold Hill, and they again visited the cabin. On closer investigation, the door was found to be bolted from the inside.

Remains of Red Mask Claims, c. 1935

After a short consultation, they decided to break down the door. According to the *Centennial Post*, "One of the most grewsome sights the boys had ever witnessed confronted them, as Morris was found dead in bed in an advanced stage of decomposition." After another quick consultation, the two continued on to Centennial from where Coroner Robinson of Laramie was summoned.

Robinson left Laramie the same evening, spent the night at the Buckeye Ranch, and after a short stop at Centennial, left on horseback in the company of Cole, G. A. Campbell and Ed Hall for the Morris Cabin.

After reaching the cabin they found it advisable to bury the remains near the cabin, as they considered it impossible to transport the body back to Centennial over the rough trail. The coroner determined death was due to heart disease and that he had died the same night he arrived at his cabin with his pack of supplies which was found untouched in the cabin.[35] Morris was 57 at the time of his death. He left a widow and six children. His claims and shares in the company were deemed worthless by the probate judgement.

In 1912 and 1913 several "new strikes" were made in the district, still known as La Plata. Although great expectations were again expressed, again nothing developed. Prospectors returned to the La Plata time and again, but always with the same result, good assays could be obtained from selected ore but a paying mine could not be developed.

The French Creek Mining Company was still in existence in 1922 when Billie Class, apparently doing assessment work for the company filed his proof of labor record on the

Remains of Red Mask Claims, c. 1955

Red Mask remains, 1990.

mines in South French Creek Valley (Section 30, T16N, R79W).

Little evidence remains of the old La Plata District, much of which lies in a prime recreation area. A few discovery holes which were not reclaimed by the Forest Service, and the remnants of a few cabins are all that remain of the once bustling little district. Only the decayed logs of John Morris' cabin remain but the pile of rocks marking his grave are still visible. The remnants of Billie Class' stamp mill can still be seen between Lewis Lake and Class Lake. The picturesque but deteriorating ruins near the summit of Snowy Range Highway (U.S. 130) date back to the Red Mask claims of James McLachlin made in 1924.[36]

Even to this day, an occasional prospector can be seen, retracing the ground of the prospectors of more than a hundred years ago.

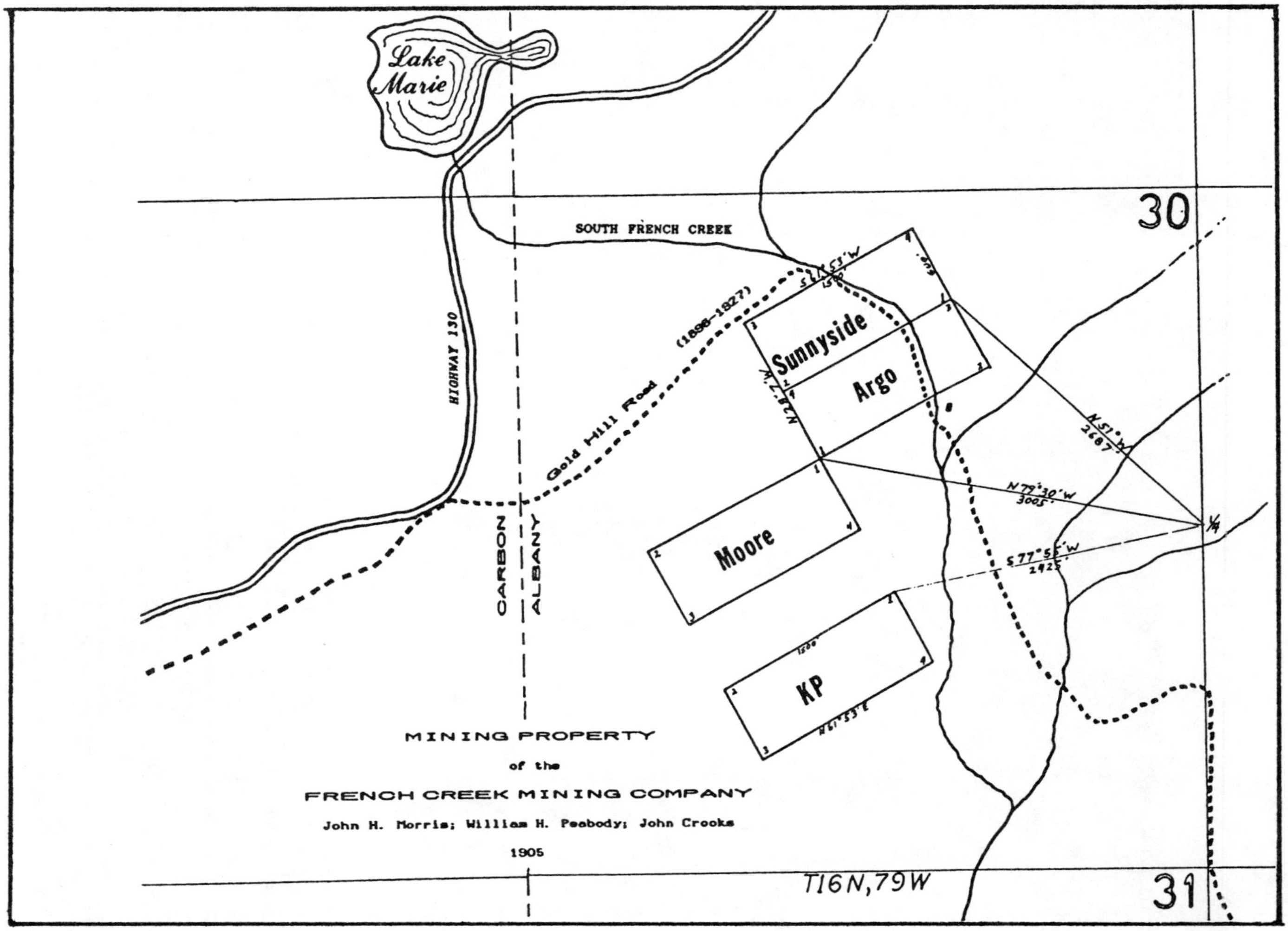

John Morris grave site, 1990

John Morris cabin site, 1990.

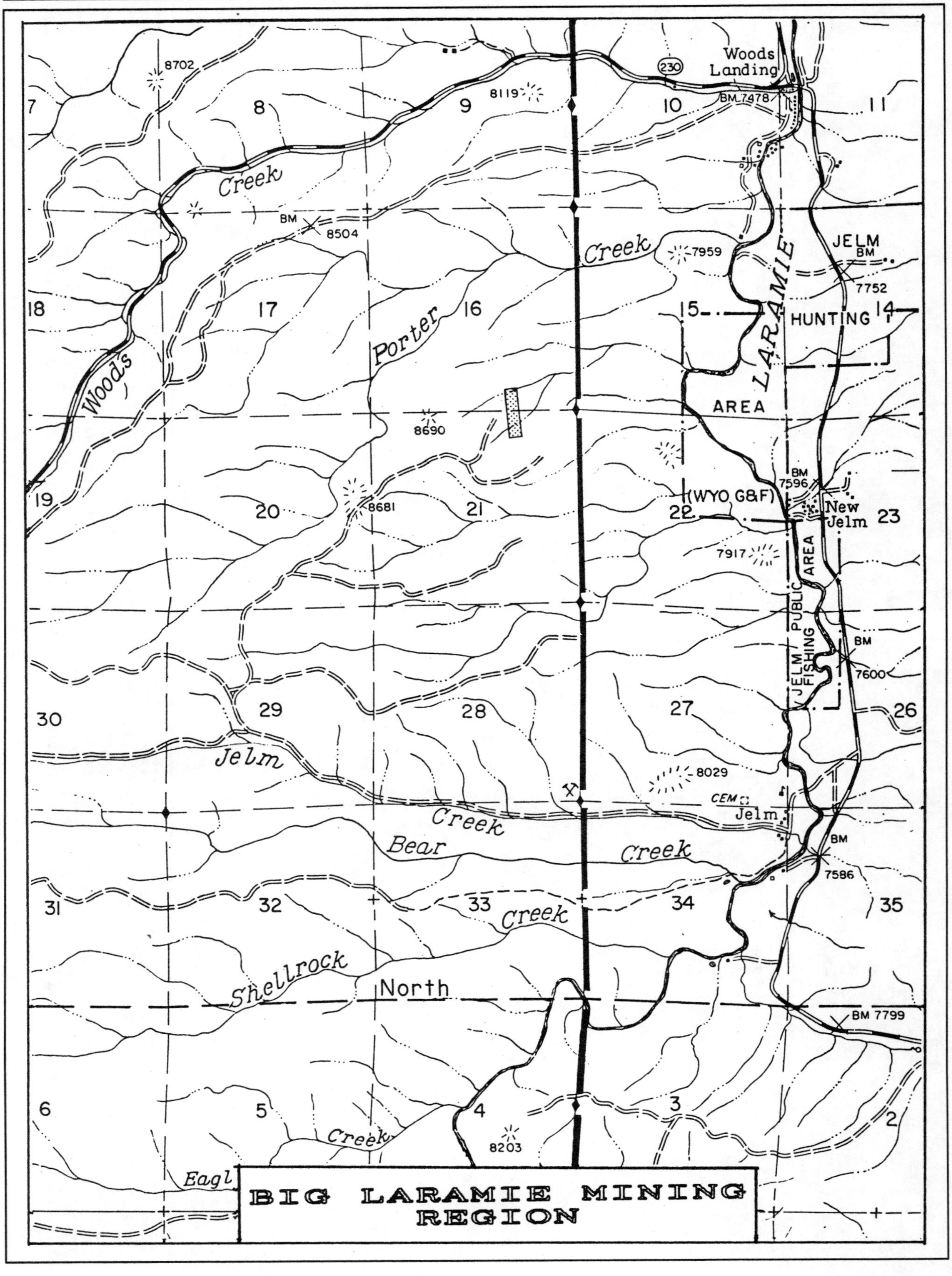
BIG LARAMIE MINING REGION
Woods Landing
230
BM 7478
Woods Creek
Porter Creek
LARAMIE
JELM
BM 7752
HUNTING AREA
(WYO G&F)
New Jelm
JELM FISHING PUBLIC AREA
BM 7596
BM 7600
BM 7586
BM 7799
BM 8504
8702
8119
7959
8690
8681
7917
8029
8203
CEM
Jelm
Jelm Creek
Bear Creek
Creek
Shellrock Creek
North
Creek
Eagl
7
8
9
10
11
18
17
16
15
14
19
20
21
22
23
30
29
28
27
26
31
32
33
34
35
6
5
4
3
2

IV THE BIG LARAMIE MINES IV

In the valley of the Big Laramie River, between the Medicine Bow Range and what is now known as Jelm Mountain, prospecting began during the early period of mineral exploration of the Medicine Bow Mountains, gained momentum after the Centennial excitement, peaked in 1881 and then died off before the mid 1880s.

The first settlers of this valley were not miners but tie hacks, cutting and floating ties down the river to Laramie City to be sold to the railroad, an industry that was to continue for years after the 1880s mining boom. Charlie Hutton had a tie drive in 1871 and James J. McGreavey[1] established a tie camp during the 1870s. McGreavey himself eventually made several mining claims along with his partners Dennis McMannus and John Sparks. McMannus and McGreavey remained to see the mining boom but Sparks left for Nevada where he became quite successful and eventually governor. Sparks had come to Wyoming Territory with a Texas cattle drive in 1873 and made a small fortune selling public lands to eager ranchers and miners. Popularly known as "Honest John," Nevada historians denote him as anything but honest.

As early as 1872, McGreavey's camp was designated a precinct polling place and in that year sixty votes were cast for territorial delegates to Congress.[2] During the mid-1870s McGreavey's camp became a stopover point on the North Park Trail and was often used as a reference point to help locate mining claims in the valley. Later the camp became the site of Cummins City and eventually Jelm City.

The Big Laramie mining era began in 1874 when Thomas D. Pearson, George Ritter and John LaFevre located a placer claim at the mouth of Bear Creek on the Big Laramie River, from which, using only a small warped and cracked sluice, they initially washed two ounces of gold, creating enough excitement to warrant organizing and naming their little area the Pearson Mining District.[3] Aside from a vaguely defined mining claim in the Big Laramie Canyon by John R. Owens, on August 24, 1870,[4] the 1874 claim of Pearson, Ritter and LaFevre appears to be the first recorded on the Big Laramie.[5] During October, 1874, a group of "prospectors," Isaac P. Lambing, W. G. Fowler, Ira W. Skidmore, Jacob Durst and Stephen W. Downey, located the Lambing Lode on the Big Laramie River in the newly designated Pearson Mining District.[6] (Notably, this same group, a few months later claimed the Centennial Lode that led to the organization of the Centennial Mining District.)

As the population of the valley began to increase, McGreavey's tie camp remained the social and political center for the valley and in 1874 the precinct cast twenty-eight votes.[7] The designated polling place was Pearson's cabin. Pearson, Tom Waldron and Gurney Canard were the appointed election judges.

In May, 1875, Pearson claimed 160 acres for "farming purposes . . . on the East side of the Big Laramie River near the mouth of a gulch running out of the pass through Gillum Mountain . . . adjoining the Pearson Mining District." [8]

In April, 1875, Isaac Lambing, Walter S. "Buck" Bramel and Walter Sinclair located a copper bearing claim one-fourth mile east of the McGreavey tie camp which they named the "Poor Man" and an adjoining claim aptly named the "Parallel."[9] The claims failed to produce and by July, Lambing had returned to Centennial to become the first postmaster.

In August, Pearson and Gurney Canard began hydraulic mining their placer claim. Canard's wife was living at the claim, probably the first woman settler in the valley. According to Canard, his wife had panned out twenty dollars worth of gold dust "just for amusement," during her leisure hours.[10]

The fall of 1875 brought a lull in mining activity in the Medicine Bow Mountains and only the Centennial Mines seemed to show any progress. According to the *Laramie Sentinel*, the winter of 1875-1876 was long and severe. It appears that many of the miners experienced hard times and turned to other means to earn their livelihood.

The *1875 Laramie City Directory* listed

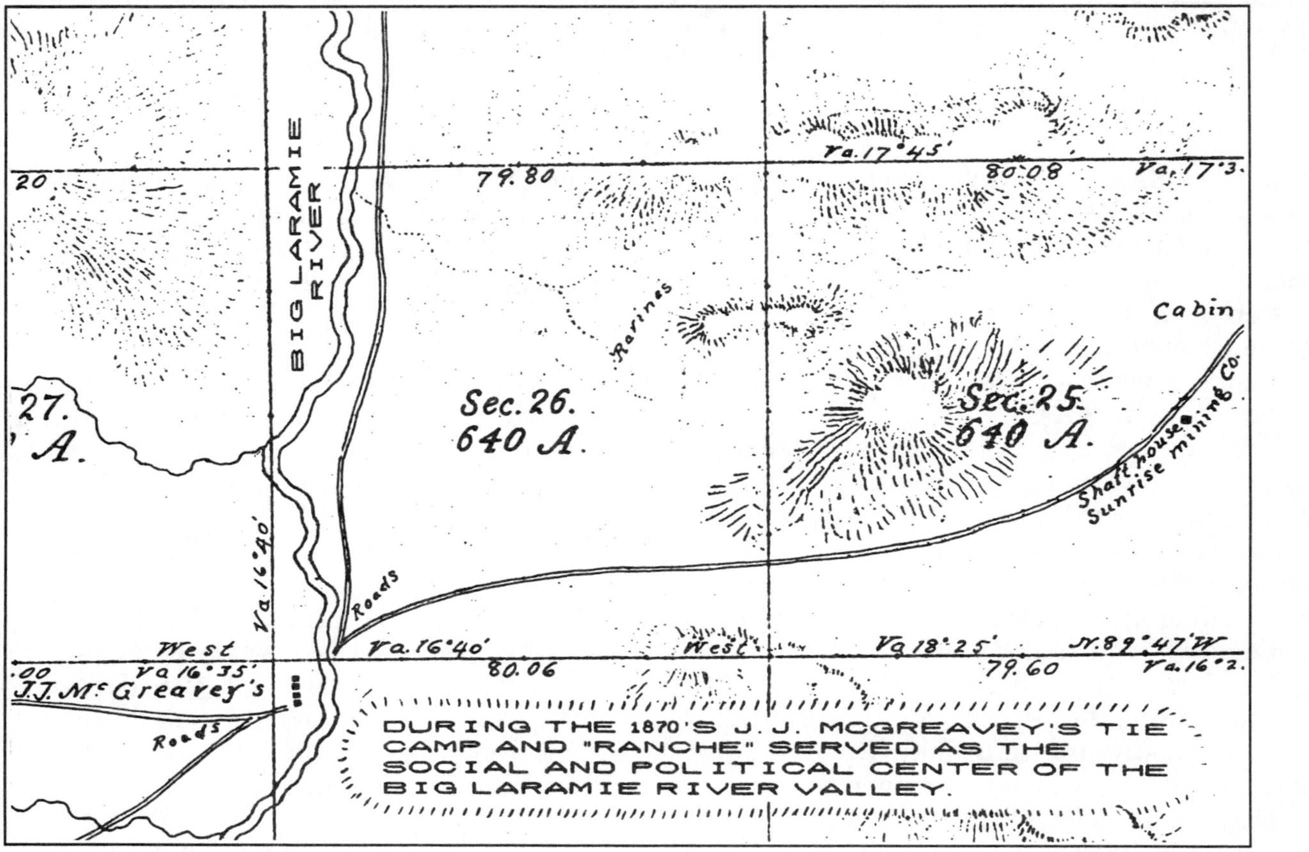

DURING THE 1870'S J. J. MCGREAVEY'S TIE CAMP AND "RANCHE" SERVED AS THE SOCIAL AND POLITICAL CENTER OF THE BIG LARAMIE RIVER VALLEY.

several of the former miners. George McFadden, who had been mining on Douglas Creek, was listed as the owner and operator of a photography and artist salon. McFadden was to return to prospecting and mining time and time again during the next twenty-five years.

John Metcalf and Tom Feast, two of the Medicine Bow miners, started a building and contracting business. However, they too were later to return to the mountains and take part in the forming of mining companies.

Melville C. Brown, who still maintained his interest in the mines, had opened his own real estate and law office.

William Downey and Mortimer Grant had a surveying business. However, both would return to the mountains time and again to pursue their mining interests.

James Vine, one of the earlier claimants in the Last Chance District was dealing in furniture.

George Ritter, a former Last Chance miner, Justice of the Peace, postmaster and most recently one of the Big Laramie miners, had been elected Albany County Treasurer. Ritter apparently found his needs desperate enough to warrant absconding with somewhere between $5,000 and $16,000 of the county's funds. He was subsequently apprehended and after two trials became a model prisoner at the Territorial Prison located just west of Laramie City.

In 1876, the Centennial Mine still captured the greatest interest of the miners and investors, but in the valley of the Big Laramie another group of prospectors was busy staking out a new group of claims.

On May 6, 1876, John LaFevre, J. J. McGreavey and Dennis McMannus claimed the Gold Hill Lode, located "... two miles west of the Big Laramie River and near and north of the McGreavey Tie Road."[11] (Although not notably rich, the Gold Hill claim precipitated the ensuing gold rush which followed some four years later). On July 4, 1876, McMannus claimed the "Far West ... situated on what is known as Ring Mountain about two miles east of the Big Laramie River at the place where McGreavey's tie camp is situated and about one half mile south from Gunn Springs."[12]

In August, 1876, a group of twelve men located and claimed the "Elkhorn Ledge ... situated about two miles east of the Big

Laramie River at the place where McGreavey's tie camp is situated and between Gilem and Ring Mountains." [13]

Gilem Mountain is no doubt an early spelling variation of what is now known as Jelm Mountain. Variously spelled Gilem, Gillem, Gilham, Gillum, Jelmn and finally Jelm, the name has been linked to a tie hack, E. C. Gillem who reportedly worked in the area prior to 1875. (Prior to this time the mountain had been called Mount Agassiz after the great naturalist, Louis Agassiz who accompanied Hayden on his exploration of the west.)[14]

Ring Mountain has retained the name that McMannus placed on it when he originally found the circular rock formation on its west side.

As 1876 neared an end, LaFevre and McMannus located the "Great Western" about two miles east of the Big Laramie River and about one half mile south of Gunn Springs, apparently adjoining their "Far West" claim of July 4.[15] About the same time, W. S. "Buck" Bramel made claim to the ". . . Western Lode, situated on the southwest slope of Gilham Mountain about one mile from the east bank of the Big Laramie River," which he filed on December 21, 1876.[16] (The "Western" along with the "Gold Hill" was to figure prominently in the gold boom which followed).

In December, 1876, J. Ward Boyles, manager of the Sunrise Mine, sent some specimen ore to Denver for assay. The results reported by the Office of Leopold Balback, were $1,104.48 of silver and $558.09 of gold per ton.[17] These assays were richer than any since the Centennial announcement of the previous year. Announcements of this nature had in the past, sparked at least a modest gold rush, but for whatever reason, this announcement failed to spark any notable new activity regarding the Big Laramie Mines.

In 1876, McGreavey's voting precinct was changed to Dawson's, located at the Dawson Brothers tie camp. Only sixteen votes were recorded in 1876 and twenty in 1878.[18]

From 1877 through 1879 there was very little new mining activity along the Big Laramie. In 1877, the Mutual Mining Company was formed and in April, the company claimed a five acre tract for their mill site located near the Sunrise Mine on the south slope of Jelm Mountain.[19] The "Sunrise" and Pearson's mine were apparently doing only a modest amount of work. Occasionally an ore specimen was brought in to Laramie City from the Pearson Mine and speculation was that Pearson, Blackbourne and Miller had found the mother lode.

The 1877 shutdown of the promising Centennial Mine dampened the spirits of the already discouraged miners. However, some prospecting and mining continued. In January, 1878, Rudolph Kleeman, superintendent of the Sunrise Mine, contracted with Reese Davis to sink a shaft on the mine to 150 feet in an effort to prove the worth of the mine.

In April, 1878, a fatal accident occurred at the mine located in the pass between Jelm and Ring Mountains. Kleeman, Davis and John LaFevre went to the mine to inspect a promis-

A TERRIBLE DEATH.

ANOTHER FATAL ACCIDENT IN A MINE.

Rudolph Kleeman Falls Down the Shaft of the Sunrise.

A Sheer Plunge of One Hundred and Forty Feet.

ing vein at the 150 foot level. Davis went down the shaft first, using a horse powered whim[20] which lowered one ore bucket while raising the other. Davis prepared a landing spot for the inexperienced Kleeman and signaled him to descend. Kleeman came down in one of the ore buckets and when he arrived at the landing, Davis noticed he was "trembling considerably." After completing his inspection, Kleeman again climbed into the bucket for the trip to the top. Kleeman had a stiff leg and apparently could not ride in the bucket like the more agile miners. As the bucket neared the top, Kleeman cried out that he was about to fall. LaFevre called back telling him to hang on as he was almost out, but Kleeman fell from the bucket, striking the other bucket as

HORSE WHIM.

he fell, landing in three feet of water at the bottom of the shaft. Davis was able to drag him from the water into the drift. Kleeman lived for about twenty minutes before dying from his injuries which proved to be two broken legs, a fractured skull and internal injuries.[21] Burial was in Laramie on April 21, 1878. Kleeman left a widow and six children. Of note, during probate, the appraisers of his estate, who were also Big Laramie miners, deemed his interest in the Sunrise Mine worthless.[22]

In 1878, William Downey surveyed the township around McGreavey's where most of the mines were located. His general description of the area said it was unfit for cultivation but was very good grazing land and was well watered by the Big Laramie River. He noted that gold bearing quartz had been discovered in the east half of Section 25, which he marked on his map as the Sunrise Mining Company. He also noted the settlements that had been made: J.J. McGreavey had improvements consisting of a house and outbuildings on the NW 1/4 of Section 35; and G.A. Burg had improvements consisting of a house and stable in the NW 1/4 of Section 2.[23]

In December, 1878, McGreavey sold his "ranche" and tie camp to John J. Strode for a reported consideration of $500.00.[24] Seven months later, on July 8, 1879, Strode sold the old McGreavey ranch and tie camp, consisting of 160 acres, to Frank L. Olmstead for a reported $450.00.[25]

During the late 1870s, along the eastern edge of North Park, several mining camps were established. Nearly all the supplies and people moving into the area went by way of Laramie City. McGreavey's became a stopping place for those going to and from North Park. Mail service by way of McGreavey's was established into the park with Sam Hanworth as carrier. "Snowshoes" (Nordic skis) were sometimes required on the mountain trail but service continued even to the half dozen or so of the more courageous miners who remained at the mining camps after the Meeker Massacre of September, 1879.

In May, 1879, the Cache La Poudre and North Park Toll Road Company was incorporated, in Colorado, to build a wagon road along the Cache La Poudre River, past Chambers Lake, over Cameron Pass into North Park. It was completed as far as Chambers

Lake by 1880 and into the North Park mining camps in 1881. However, the road from Laramie City by way of McGreavey's remained the preferred route during construction and even after the toll road through Cameron Pass was completed.

Promotion of the North Park road was an item of considerable interest to the citizens and businessmen of Laramie City as well as the Big Laramie miners and ranchers. Many of the Albany County Commissioners' meetings were spent discussing the various proposals for the route to North Park. At first the route following the river was considered too difficult and the route through the pass between Jelm and Ring Mountains (Sunrise Pass) was proposed as the public road. Although the river canyon route required three bridges, this road eventually won the greatest support, due at least partially, to the $50.00 the river route petitioners placed on deposit for surveying and marking of the road.[26]

• 1880

In the Spring of 1880 the Big Laramie miners were joined by John Cummins. Cummins, fresh from the North Park mines, proceeded to promote the development of the mining camp and on April 5, 1880, he located and claimed his first Wyoming claim, the "Highland Chief Lode ... situated on the southwest slope of Gilham mountain, one mile east of the Big Laramie in Albany County." The claim was 600 by 1,500 feet.[27]

On April 10, 1880, just south of McGreavey's old camp, Gust A. Burg claimed 20 acres (600 by 1,500 feet) On "... the East 1/2 of the SE 1/4 of the NE 1/4 of Section 34 T13N, R77W in Albany County." The claim, surveyed by William Downey was one of the few accurately and legally located.[28]

During April, a new arrival at the new mining camp began providing information on the mining activity along the Big Laramie River to the *Laramie Sentinel.* Traveling from Laramie City on April 15, he reported the wagon road was in good shape. Since his arrival in camp, he said, it had snowed every day, but fifteen to twenty prospectors were hard at work putting in as many hours as daylight and weather would permit. He also reported Bramel's "Western Lode" as the most worked, and said that John Cummins, P. T. Smith, Hamlin Clark and J. P. Landt, under the firm name of Cummins and Company had purchased five sixths of the claim. In his correspondence he also remarked that although he was only a "tenderfoot" he could see free gold as well as anybody and that the future of the camp was promising.[29]

Ore bucket, Encampment Museum.

Cummins and his partners, Smith, Clark and Landt were all from the Denver area and had joined together in an apparent effort to make their fortune buying and selling mining property rather than holding and working the claims. Cummins had been working in North Park where he had established a small mining camp. He had moved into the Big Laramie camp after selling his North Park claims and raising an alarm of the Indian activities in the region. An alarm which apparently did little to scare off the North Park miners but did have the desired result of attracting more prospectors to the new camp in Wyoming Territory.

On April 16, 1880, the entire population of the mining camp held a meeting, presided over by P. T. Smith, and resolved to permanently name the area as the Bramel Mining District. Jerome Cross was elected recorder and district boundaries were set: Fifteen miles on each side, running to the Colorado state line and encompassing some 144,000 acres. It was also decided that lode claims should be restricted to 300 feet in width rather than the maximum 600 feet allowed by the 1872 mining law. A petition for the establishment of a post office was circulated and unanimously signed. George W. Moore, the camp's first businessman, was recommended as postmaster. It was further resolved that the camp would henceforth be known as Cummins City.[30]

On April 24, 1880, Cummins and P.T. Smith were in Laramie City and were interviewed by the *Sentinel.* They reported that they had received certification of assays from "ten or a dozen" samples. The *Sentinel* noted, "Though the assays are official and undoubtedly correct, they are so large we are almost afraid to publish them, running from $36,792.00 up to $72,000.00 per ton in gold and being rich in silver besides."[31] What they did not say was that only Cummins knew the exact origin of the ore samples. Assays are useful only to those who know the location and extent of the ore.

On May 1, 1880, Cummins bought McGreavey's old tie camp property from Olmstead for a reported amount of $515 and began making his plans for the new "city."[32] Before long the Laramie City and Cheyenne newspapers were referring to the mining camp as Cummins City.

Also, on May 1, Brophy and Mills were in Laramie City loading lumber to build a 24 by 60 foot hotel. They hired several carpenters and headed for Cummins City with a contract to open for guests in five days.[33] The road to Cummins City and North Park was reported open and a large number of miners were in Laramie City, ready to travel to the newly discovered mines. The North Park stage left Laramie City every Monday morning carrying mail and passengers for Cummins City and the North Park mines.

Early in May, Cummins and his partners traveled to Denver with several sacks of ore for assay. The *Denver Times* May 4, 1880, reported:

> Messrs. John Cummins, J. P. Landt, H. F. Clark, and Dr. Smith . . . exhibited at the *Times* office this morning, some very rich and handsome specimens of free gold quartz. They are from Jelm mountain in the Bramel district some thirty miles south of Laramie City. These ores in practical mill tests run fourteen ounces of gold ($280.00) and two ounces of silver.[34]

By May 8, Cummins was back in camp with his now exaggerated assay results claiming $350.00 per ton in gold and "considerable silver." If this truly was typical ore from the Big Laramie Mines, even the lower assay would be a bonanza. Events following the publishing of the inflated assays led to accusations of fraud and salted mines. Strangely the same events were also a catalyst for the gold rush already in progress. Prospecting quickly gained momentum and spread to a distance of some ten miles around Cummins City. The *Laramie Sentinel* reported the miners were finding precious metals everywhere.[35]

Cummins' dealings for the following few months became rather involved but official records give a fair indication of his transactions. According to sworn testimony, Cummins had purchased several mines for his partners, charging them a total of $5,500 for a group of mines which he had actually purchased for $550. Cummins carried ore samples with him wherever he went, using the samples to convince prospective investors. After convincing his partners of the richness, he then convinced even the assayer of the richness of the newly purchased mines. Word of the assays spread to the East and B. B. Miner of New York sent word that he was traveling west to inspect and purchase the mines. Miner brought with him a mining expert who, after a quick inspection, proclaimed that the ore could not possibly have come from the reported mines. Mr. Miner withdrew his offer to buy and Cummins' partners withdrew their financial support. Cummins proceeded to sue the partners for failure to pay their agreed share of the enterprise. One of the partners, Hamlin Clark, had already sold his share, but the other two, Pierce T. Smith and Jeremiah Landt brought counter charges against Cummins for obtaining money under false pretenses. Smith and Landt gave their testimony by deposition in Denver, failing for whatever reason to return to Wyoming to appear in court. After due consideration, Judge Jacob B.

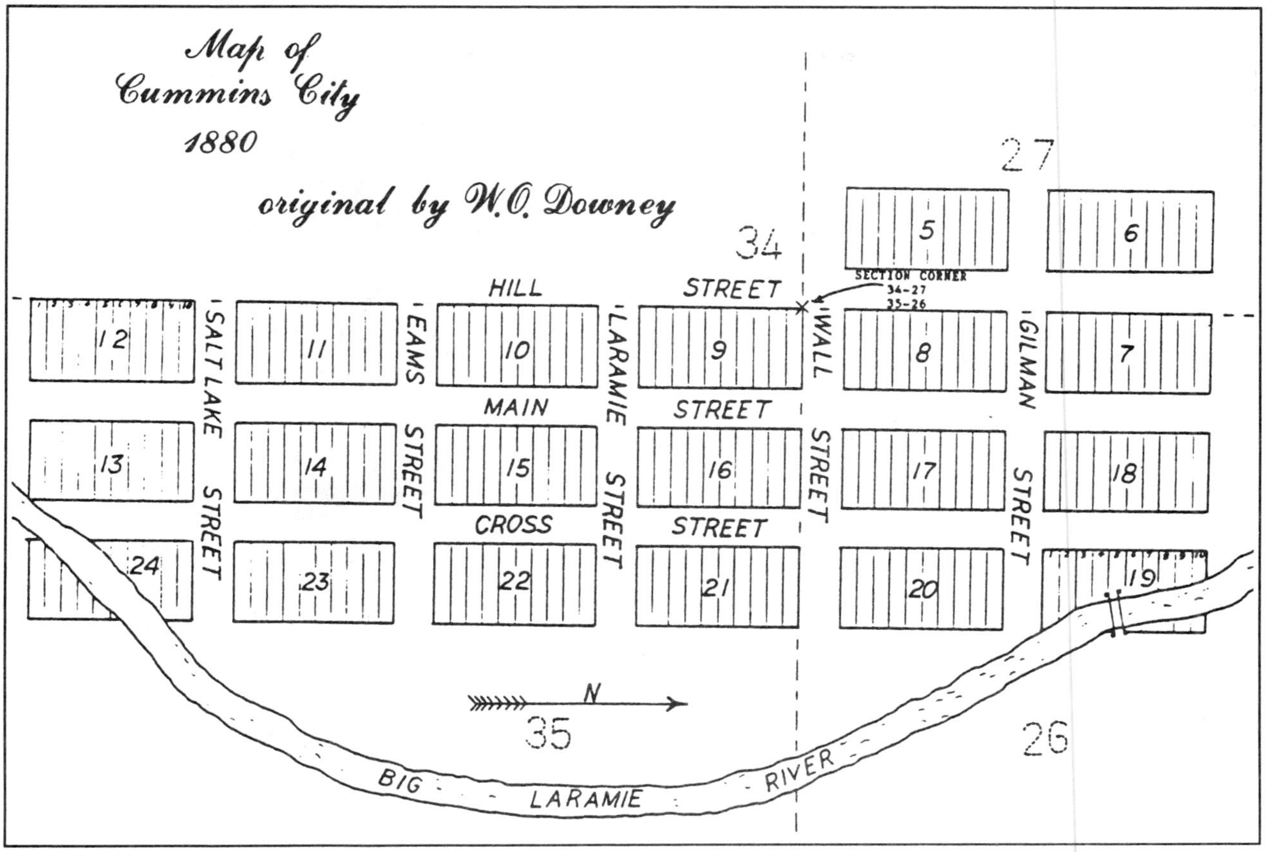

Blair dismissed the case. The partners chose to stay in Denver, apparently never returning to Wyoming.[36] John Cummins, apparently undaunted by his temporary setback, went on the prowl for new partners and continued to promote the mines and his namesake town.

One of his first actions was to hire William Downey to survey and plat the proposed city. Downey, Cummins and W.H. Clark (County Clerk) certified the survey on May 18, 1880. The Cummins City Plat revealed 24 blocks of ten lots each, 25 feet wide and 120 feet long. The Big Laramie River ran directly through several of the lots. The old McGreavey Bridge across the river was shown in Block 19. The east-west streets were surveyed 60 feet wide and named Gilman, Wall, Laramie, Eams and Salt Lake. Running north and south were Hill, Main and Cross. The northwest corner of Lot 9 coincided with the northwest corner of Section 34, T13N, R77W. Lots went on sale; $50.00 for a corner lot, and $25.00 for an inside lot, all payable to John Cummins.[37]

The laying out of the new town and reports of new gold discoveries, all under the direction of entrepreneur Cummins, started another gold rush to the Medicine Bow Mountains. Certainly not a gold rush such as those to California, Colorado or the Black Hills, but by Wyoming standards, a gold rush nevertheless. By mid-May claims were rapidly being staked out, divided, sold and re-sold in a myriad of dealing that must have had even John Cummins reeling.

George Moore, who had been running a small boarding house for an average of twenty-five boarders, immediately made plans to go into partnership with Mrs. Sarah Bieger and erect a new boarding and lodging house 24 by 40 feet and one and one-half stories high. In this building, the size of a small modern home, he planned to feed as many as sixty miners on the lower level and sleep thirty in the half-story loft. The building was to be completed in ten days and would be known as "Moore House."[38]

It was reported by the *Laramie Sentinel*, that Mrs. Beiger owned forty-one lots in the new town.[39]

In May, 1880, the *Laramie Sentinel* reported 250 to 300 people in the camp. On May 26, the Cummins City Post Office was designated and George Moore became the first postmaster. On May 25, John Cummins filed another claim, the "General Miles," located east of Cummins City in Township 13 North, Range 76 West.[40]

By the end of May, the *Laramie Sentinel* began printing the "Cummins City Clatter" as a regular column, sometimes signed by "Antelope," sometimes "Argonaut" and sometimes unsigned. This column provided a good look at the growing mining camp:

> Johnson and Bacon are putting up a residence. Poole, Brown and Ashton have their blacksmith shop going full blast.
>
> Cummins and company have a 30 by 90 foot livery stable in successful operation.
>
> Major Downey has struck a rich lead on Sugar Loaf mountain.
>
> Moore and Smith are running a first class meat and vegetable market.
>
> Lyons and Co. have found a good vein in the foot hills of the Medicine Bow range.
>
> Mr. Eames has his church completed and holds services every Sunday.
>
> Mrs. Beiger is putting up a restaurant, 24 by 40, which will be completed next week.
>
> G. W. Moore is building a restaurant 24 by 40, two stories high and will have it in operation next week.

[Actually Moore and Mrs. Beiger were partners on the same restaurant.]

> Weinspaugh & Son will put in a ten-stamp mill as soon as they can get the machinery from the East.
>
> Cummins & Co. have the rails, cars, etc. for tunneling in the Western and will shortly put in a ten-stamp mill.
>
> Some eight or ten families now reside at Cummins, and a school will soon be opened for the benefit of their children.
>
> The Homestake company's ditch is progressing finely. Water has been turned on, and the enterprise promises to be a success.
>
> Mr. Fred Weinspaugh is a practical and thorough assayer, and does assaying for all who call upon him, making accurate returns.
>
> John Cummins, the father of the camp, is erecting for himself a substantial residence, 18 by 20, in which he will be domiciled soon.
>
> Miller & Blackbourne, report says, have sold their mining interests at Jelm mountain, to Messrs. Weinspaugh & Son of Denver. Consideration, $50,000 - $10,000 down, and balance in monthly installments of $10,000 each.[41]

The *Laramie Sentinel* under Editor Hayford, seldom reported any discouraging mining news but the *Cheyenne Daily Leader* was not quite as benevolent and on occasion had a pessimistic remark. On June 5, 1880, the *Daily Leader* reported, "An expert who has visited Cummins City says the boom there is a fraud, and that the Mill Creek carbonates were brought from Leadville. Laramie City should keep a watchful eye on the prospectors who prowl about her adjacent hills with fraudulent intent."

E. W. (Bill) Nye
Wyoming State Museum

Assistant editor of the *Sentinel*, E. W. "Bill" Nye, was one of the more casual miners. Nye had come to Wyoming in 1876 and gone to work under Hayford on the *Daily Sentinel*. The daily newspaper had ceased publication in favor of a weekly on January 1, 1879. This apparently left Nye with time to do some prospecting. During 1879 he travelled into North Park and staked a claim near what became Teller City. However, the Indian scare convinced him to leave the park without developing his claim. In the spring of 1880 he made another claim on Mill Creek near the Nellis Ranch and one in the valley of the Big

Laramie near the new camp of Cummins City. At the time, Nye was known only locally, for the humor which later made him famous. The mining camps and his mule "Boomerang" were the basis for many of his stories which were later published in *"Bill Nye and Boomerang."*

In May of 1880, he wrote:

> There is a new camp in the valley of the Big Laramie River, near the dividing line between Wyoming and Colorado. A few weeks ago the murmur of the rapid river down the canyon and the cheerful solo of the coyote alone were heard. Now several hundred anxious excited miners are prospecting for gold, and the tent-town grows apace. Up and down the sides of the river and over the side of the mountain every little way a notice greets the eye announcing that 'the undersigned claim 1,500 feet in length by 300 feet in width upon' the lode known as the Pauper's Dream, or the Blue Tail Fly, or the Blind Tom, or the Captain Kidd, or the Pigeon-Toed Pete, with all the dips, spurs, angles, gold and silver bearing rock or earth therein contained.

On June 15, 1880, he wrote:

> The camp has about 300 miners and prospectors now within the city limits. All up and down the picturesque valley of the swift-flowing river the low cabin and white tent dot the green sward, and far above the everlasting hills rear their heads on high, torn by the Titanic power of giant heat in the days of long ago.

On June 18, 1880, he wrote:

> Cummins City still booms. Building lots have gone up to $100 each. This for a place where a few weeks ago the song of the coyote was heard in the land, and where the valley of the river, and the bald sides of the rugged mountains were unscarred, is a good showing. The magical power of a mineral excitement to transform the bleak prairie and the rocky canyon into a thriving village at once, is something to command our admiration and wonder.

His writings then take on some of the humor for which he later became famous:

> Two months ago, I might say, the little village of Cummins City was nothing but a little caucus of prairie dogs, and a ward meeting of woodticks. Now look at it. Opera houses, orphan asylums, hurdy-gurdies, churches, barber shops, ice-cream saloons, dog fights, musical soirees, spruce gum, bowling alleys, salvation, and three card monte. Everything in fact that the heart of man could yearn after.

Nye started his own paper, the *Boomerang*, in March of 1881, which he ran until 1883. In that year he left Laramie for health reasons and after a short time in Greeley, Colorado, moved to the East where his fame as a humorist gained him a worldwide reputation.

John Cummins continued his wheeling and dealing and on June 5, 1880, he sold all his claims, whether specified or not, owned by himself as a member of Cummins and Company. The sale included all ore sacks, car tracks, cabins and improvements and an undivided one-fourth interest in the "Gold Mountain" and the "Western Lode." Buyer of the property was James Adams. The true price is in doubt but the declared price was $10,000.[42]

Cummins was not yet through and continued to sell lots in his namesake city and numerous claims either made by himself or recently purchased. These included his remaining shares of the "Western," "Yellow Hammer," "Southern,"and "Rob Roy," which he also sold to James Adams who later became trustee and director of the Union Mining Company.[43]

On June 17, 1880, Armand Ruschpler conducted the census of the inhabitants in the "Gillem Mountains in the County of Albany." Apparently not all of the people reported to be in or around Cummins City were enumerated. However, the census did list fifty-three people. Twenty were listed as miners, ranging from seventeen year old William Lony to sixty-three year old Mathias Druly. George Hildreth was listed as clerk in the Recorders Office, Sam Woods as storekeeper and Clarence Haines as a store clerk. Sarah Picer was keeping a boarding house and Frank Schenk was cook for one of the mines. Two carpenters were listed and most of the remainder were wives and children, the youngest was two year old Andrew Moore. Ruschpler concluded his census with, "Here ends the Mining Camp of 'Jelmn' Mountain known as Cummins City."

There is a noticeable difference between Ruschpler's count of fifty-three people and the *Laramie Sentinel* estimate of 250 to 300 people only a few weeks earlier. Perhaps Ruschpler counted only the people "at home" in the camp. Perhaps the Sentinel reporter exaggerated the number, more likely a combination of both.

One small difference in the count was the result of a group of miners from the Bramel

Drawing by
By Dan Swift

District (including Cummins, Bramel and Zollman) who had earlier traveled to North Park for several weeks of prospecting.

Another of the Cummins City residents who missed the census was Herman V. S. Groesbeck who later went on to become a judge, served on the Eighth Legislative Assembly and eventually became State Supreme Court Justice. Groesbeck came to Wyoming in July, 1880, and went to Cummins City where he hoped to gain work as an attorney. Finding no need for his legal talents, he went to work for Sam Wood, tending bar, which he said consisted of a plank across two whiskey barrels. Wages were $1.50 a day.[44] Groesbeck remained at Cummins City in various roles, including school teacher, until 1882, when he moved to Laramie City and became associated with the law office of Stephen W. Downey.

In North Park, Colorado, the Teller and Independence mining camps continued to attract prospectors. Laramie City was the primary outfitting and supply point for these camps as well as the Medicine Bow mines. At that time the road from Laramie City to North Park went first to Cummins City, up the Big Laramie River to Beaver Creek, up this creek to the "big beaver dams" and then into North Park. This route was destined to change several times in the next few years, but Cummins City remained as a main stage stop on each of the routes and passengers often remarked about the friendliness of Cummins City. The overnight stop for the stage was at Berry's Ranch on Pinkham Creek in North Park. Two of the early drivers for the stage were Samuel Hanworth and Charlie Mayo. Freighters using six or eight horse teams drawing two and sometimes three wagons also used this route through Cummins City to North Park. Among the early freighters were Jack and Tom Keenan, George Post, Jules Musgrove, Jerome Decker and his father, Sam Decker.

During August, 1880, Dr. John C. Watkins purchased an option on the old Centennial stamp mill which had the reputation of recovering 80 percent of the assayed ore value. Dealing for the mill sight unseen, he was pleasantly surprised to find it was steam driven, not water driven as he supposed, and that it was a ten-stamp mill, not a five-stamp as he had thought. The mill was hauled into Cummins City by a thirty-six-oxen team and the entire camp turned out to watch the arrival. Watkins planned to have the mill in operation by the first of September, and planned to mill ore from the "Gold Mountain" claim on his first operation.[45]

Sometime in September or October, a water operated stamp mill was put into operation at Cummins City but due to a lack of water, only five of the ten stamps were operating. In October, James Adams visited the *Laramie Times* office and displayed a sample of gold processed from the mill. The ore from the Gold Hill mine reportedly showed one and one half ounces of gold or about $30.00 per ton. Adams took the gold sample, some of the ore and some of the mill tailings to Denver to ascertain what percentage of gold the mill was missing. He apparently did not disclose the results of his Denver trip.[46]

The mining camp of Cummins City continued to grow and the Laramie City newspapers continued to report every little event under the dateline, "Cummins City Clatter":

CUMMINS CITY CLATTER.

Interesting Notes from the New Mining Camp.

From Our Special Correspondent

CUMMINS CITY, WYOMING, August 5, 1880.

Mr. Sonder has opened a paint shop, which is something long needed.

A butcher shop would prove a profitable venture, as all are heartily tired of ham and bacon, interspersed with an occasional ancient antelope.

A praiseworthy movement is on foot to establish a free school. One of our merchants offers to build the school house, and several prominent citizens have pledged themselves for the pay of a teacher.

A petition is being circulated praying the honorable board of county commissioners to establish a road from your city to this place, via Burg's ranch and the canyon route. Such highway is undoubtedly the most direct and accessible route from Laramie City, and the establishment would facilitate the speedy delivery of heavy freights as well as the mails and lighter articles.

Whiskey is an indispensable adjunct to a lively mining camp, but the recent riotous proceedings of some of the boys will doubtless result in the appointment of temporary officers of the law, who will probably be elected at the fall elections.

Mr. I. H. Wilson is prominently spoken of as a candidate for justice of the peace, and Mr. Gilman as constable.[47]

On September 18, 1880, Albany County School District Number 8 was organized at Cummins City. S. S. Wood, J. C. Watkins and George Hildreth were elected as trustees. In 1880, there were only thirty-six schools and forty-nine teachers in the entire Territory of Wyoming, and only four schools and nine teachers in Albany County.[48] Establishment of a school and a school district was for that time a notable achievement.

The old McGreavey-Dawson Precinct was renamed the Cummins City precinct and on November 2, 1880, elections were held with polls at the recorder's office. John C. Watkins, Ed Gilman and George Moore were designated as election judges and sixty-eight votes were cast, making it the third largest precinct in Albany County. Besides voting for Delegate to Congress, the precinct also elected George Hildreth as Justice of the Peace and Mike Kendall as Constable.[49]

As the 1880 mining season drew to a close, the Union Mining Company was formed and incorporated under Article of Incorporation, Number 104, dated November 1880. Seven trustees were named to manage the company: Alexander H. Sawn, M. C. Keith, Edgar P. Snow, James Adams, John H. Donnelly, Joseph Frank and Amasa R. Converse. The stated purpose of the company was generally to transact mining business in Laramie and Albany counties. Par value of shares in the $500,000 company was set at $100 per share. Business was to be carried out in Cheyenne, Laramie and Cummins City. H. V. S. Groesbeck was the company secretary and acting superintendent of the companies mines; the

"Southern," the "Western," the "Silver Prince" and the "Gold Hill" along with its water-powered ten stamp mill.

According to "official sources," Wyoming's gold production for 1880 was $20,000. This of course included the entire territory, and the Bramel District could be only a fraction if any of this total. In contrast, Colorado's gold production for 1880 was $3,200,000 and silver production $17,000,000.[50]

• 1881

Although the Cummins City residents often boasted of their good weather, it appears that not much mining was accomplished during the winter months.

On January 26, 1881, Sam Wood replaced George Moore as postmaster. (Possibly the forty-one year old Moore had a premonition, or perhaps he was following the actions of John Cummins, for he resigned his postal duties, sold his boarding house and his interest in some seven mining claims to Andrew Riley and Milton Woods.)

Young English adventurer, Robert Mills, alias Robert Glyndwr who was ". . . living with a couple of fellers in a log cabin close to Cummins" wrote:

> This camp is booming. Last spring the deer, elk, bison, bear and wolves used to frolic here, and now we have a camp of between 2 and 3 hundred inhabitants, with 4 saloons, 5 stores, a public building where political, religious, and other meetings are held, and a circulating library of 100 volumes. There is a J.P., [George Hildreth], sheriff, [N. K. Boswell, Albany County Sheriff, frequented the camp but did not live there], constable, [Mike Kendall] and a jail is talked of. [51]

Justice of the Peace, George H. Hildreth held court in Cummins City from January 1881 to March 1886, presiding over the trials for the petty crimes of the valley. His first recorded trial was a complaint against John McLewan and Charles Shaffer for willfully and maliciously kicking and breaking in the front door of Wood's and Coolbroth's saloon. The case was continued until the next day, ". . . on account of defendant being intoxicated and most of the witnesses the same." McLewan pled guilty and was fined $5.00. Shaffer however, stood trial, was found guilty and fined $15.00. He refused to pay and was committed to the county jail.[52]

The March 29, 1881, correspondence to the editor of the *Sentinel* stated that Cummins City, not quite one year old, now had thirty-one houses occupied, also four saloons and four stores, all of which were doing a "rushing" business. Beard and Thomas, proprietors of the steam saw mill, were constructing a hotel 60 by 100 feet. (According to county records, Martha Beard and James I. Thomas were doing business under the firm name of "Beard & Thomas.") Haynes and Woods were doing a lively business in groceries and mining supplies. E. W. Harwood, a young lawyer from Missouri, was running a small store selling household furnishings and W. E. Hilton was building a large store on the corner of Main and Gilman. Within the Bramel Mining District numerous claims were being made up and down the Big Laramie River and in the surrounding mountains. According to the correspondent, more than 500 claims had been recorded.

The Union Mining Company under the management of Herman Groesbeck was stockpiling ore, awaiting the arrival of a forty-stamp mill and concentrating works (which never arrived).[53]

In April, with the first signs of spring, Louis Miller, Laramie City jeweler turned miner, sent his team and buggy to Laramie City to pick up the *Laramie Sentinel* reporters (which very likely included editor, J.H. Hayford) for a visit to Cummins City. They reported a pleasant ride of four hours brought them to Cummins City where they stayed at the home of Dr. Watkins.

Doctor John C. Watkins was the civic leader of Cummins City. As a young man he had served in the Civil War. After the war he served in the Washington D. C. area and graduated from Georgetown Medical College. In 1868 he was appointed Assistant Surgeon U.S. Army and served at Fort Scott, Nevada. A true medical doctor turned miner, Watkins came to Cummins City in May, 1880 and was followed by his wife, Harriet, two sons and his sister, Mary S. Watkins. Harriet and Mary were accomplished teachers and were instrumental in establishing the Cummins City school system.

On Sunday, the reporters attended the Sabbath School and evening religious services given by Mr. Cole and Miss Mary S. Watkins.

Doing their best to shed good light on the camp, they reported the camp contained "... about one hundred houses, mostly cabins of course, but generally built with much neatness and taste." (Judging from the report of less than a month before of thirty-one cabins, four saloons and four stores, it seems doubtful that there were as many as 100 buildings counting every cabin, shed and outhouse in the camp). They went on to say:

> There are some pretentious buildings and Messrs. Baird [Beard] and Thomas are actually putting up a hotel 60 by 100 feet in size and two stories in height. It will be the largest hotel in Wyoming Territory....
>
> There are in Cummins, we judge, something over one hundred men, from a dozen to twenty woman and as many children. The population consists of many enterprising and intelligent citizens including four doctors, several lawyers and school teachers and other professional and educated men and women. They have clubbed together, contributed money and labor and built quite a comfortable and commodious school house and furnished it with seats, tables, desks etc. It is used for a school on week days, for Sunday school and religious services on the Sabbath day and for a reading room evenings.

When the reporters examined the district recorder's books, they confirmed the earlier report that almost 500 claims had been recorded, and added that, of these, 250 had work done on them, ranging from the minimum required ten foot discovery shaft to the deepest shaft of 160 feet.

They visited the "Betsy Jane, ...one of the first mines discovered," where it was claimed the richest ore in camp was found. Next they were shown a vein some twelve to fifteen feet in width, which contained gold, silver and copper assaying "... $15.00 upward." They were also shown the placer deposits of the valley where they were told "Every time they dig a well or cellar or even a posthole, they can pan out a fair prospect from the gravel. Even the dirt floors of their cabins will prospect fairly." [54]

Whether it was the hospitality, the mines, or both, the impression left upon the reporters was quite favorable and in the weeks to follow the *Laramie Sentinel* continued to report the embellished mining news under the dateline, Cummins City Clatter, May 12, 1881:

> New discoveries are being made almost every day, and prospecting is on the increase since the fine weather set in.
>
> Mr. Adams, superintendent of the Union Mining Company, is now in Cheyenne, and is rumored that upon his return, the work will commence on all the mines owned by that company, some English capitalists having become heavily interested.
>
> The Jelm Mountain Company's committee upon examination and appraisement of claims offered for acceptance are still at work. So far they have accepted six claims and they are among the richest of the district.
>
> John Cummins arrived on May 11th and went into camp near his Gold Blossom mine, about a mile west of Cummins City. (Apparently Cummins had spent most of the winter in Denver).
>
> Several families have lately arrived and taken up their abode permanently.
>
> W. S. Bramel has made a wonderful strike of very rich free gold quartz in his Last Chance claim.

On Saturday evening, May 2, 1881, the citizens of Laramie City met and concluded arrangements with the Patrick Brothers Stage Line for putting on a tri-weekly line of coaches from Laramie City to North Park via Cummins City. A. S. Patrick planned to have the line running within ten days and to continue under the contract for at least a year. The schedule from Laramie to Teller was to take about fifteen hours without stops except for meals and to change horses. Some work was required on the road south from Cummins City and also between the new bridge at Gus Burg's ranch and Cummins City. Under the contract every stage was required to pass through Cummins City.

> May 25, 1881:
> Both lines of stages are making regular trips and come in crowded with passengers and express matter. We understand that a party from Cheyenne will start in the meat business. We have needed a butcher shop here for some time as Messrs. Haynes and Wood use two beeves per week in their hotel alone and will use much more in the future as both the stage lines make this hotel their headquarters.
>
> Mr. George Fales has organized a local company with 100 shares at $10 each. Mr. Fales has found no difficulty in disposing of the stock and has put on a force of miners at work to sink a shaft 200 feet.
>
> Ed Ivinson, Judge Brown and Mr. Blackbourne are here looking after their mining interests. They attended the citizens meeting last night, which was held in the interest of the new road from this place to Shipman's Park, shortening the distance to the park about ten miles.

The work on the new hotel under the direction of Messrs. Baird [Beard] and Thomas, is progressing slowly, owing to the fact of the scarcity of mechanics.

Mr. Harwood has moved into his new store, which has the most attractive interior of any business place in the town.

Mr. McCray's new restaurant is almost completed.

Louis Miller will finish the county work on the river road today.

S. S. Wood is about to enlarge his hotel and stabling facilities to meet the demands of increasing travel.

Also in May, 1881, the *Sentinel* reported that two more companies of the Bramel Mining District had organized and incorporated, the Jelm Mountain Gold and Silver Mining and Milling Company and the Wyoming Consolidated Gold and Silver Mining Company.[55]

The Jelm Mountain Gold and Silver Mining and Milling Company was incorporated for $1,000,000 with 40,000 shares at $25.00 a share par value.[56]

The Wyoming Consolidated Gold and Silver Mining Company incorporated at $2,500,000 divided into 250,000 shares with a par value of ten dollars each.[57] The stock of both companies was being offered to the public at 25 percent of par value.

The Wyoming Consolidated company trustees were some of the more affluent men of the Laramie area: Luther Filmore, Edward Ivinson, C. W. Root, A. S. Peabody, A. S. Blackbourne, Louis Miller and M. C. Brown.

The Jelm Mountain company trustees consisted of mostly miners: James Thomas, Walter Bramel, Edgar Harwood, Francis Leavins, Lewis Zolman, Louis Miller, William Tilton, John Watkins, George Fales, Alonzo Dunn and A. S. Peabody. Upon arrival of the company seal at his store and office, Harwood, the

—THE—

WYOMING CONSOLIDATED

Gold and Silver

Mining Company

MINES LOCATED IN

BRAMEL MINING DISTRICT,

Wyoming Territory.

OFFICERS

L. FILLMORE . . . President
ED. IVINSON . . Vice President
A. S. PEABODY . . . Secretary
C. B. ROOT . . . . Treasurer

TRUSTEES:

A. S. Blackburn, Louis Miller, Edward Ivinson, M. C. Brown, L. Fillmore, A. S. Peabody and C. B. Root.

m8 tf

MINING COMPANIES.

THE

JELM MOUNTAIN

Gold AND Silver

Mining and Milling

COMPANY.

STOCK FULLY PAID UP AND NON ASSESSABLE.

Mines located in

BRAMEL MINING DISTRICT.

OFFICERS:

Dr. J. I. Thomas, President, } Cummins
Wm. E. Tilton, Vice-Pres., }
E. N Harwood, Secretary. } Wyo.
A. G. Dunn, Treasurer—Laramie City.

TRUSTEES:

Dr. J. I. Thomas,	E. N. Harwood,
Louis Miller,	Frances Leavens,
W. S. Bramel,	George H. Fales,
A. G. Dunn.	Lewis Zolman,

Dr. J. C. Watkins.

A. G. DUNN, authorized agent for sale of stock. je18 tf

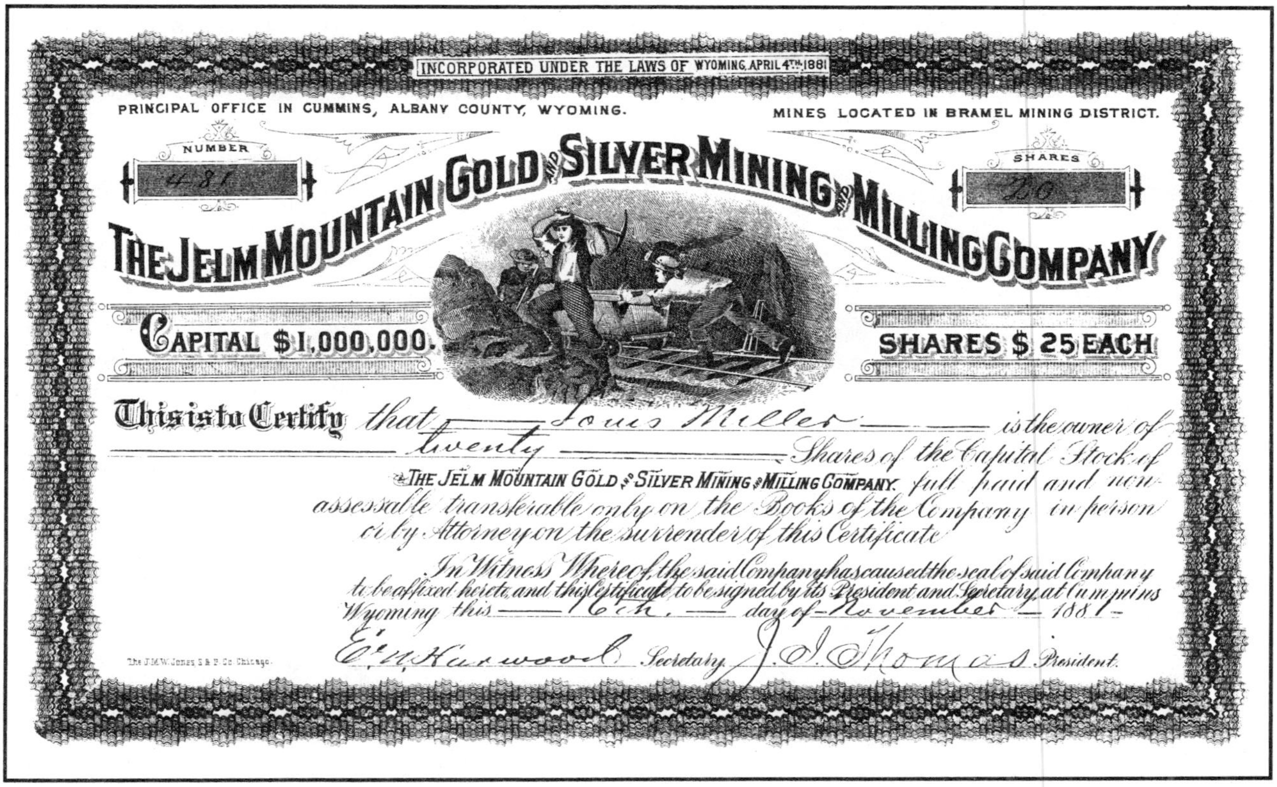

INCORPORATED UNDER THE LAWS OF WYOMING APRIL 4TH 1881

PRINCIPAL OFFICE IN CUMMINS, ALBANY COUNTY, WYOMING. MINES LOCATED IN BRAMEL MINING DISTRICT.

NUMBER 481 SHARES 20

THE JELM MOUNTAIN GOLD AND SILVER MINING AND MILLING COMPANY

CAPITAL $1,000,000. SHARES $ 25 EACH

This is to Certify that Louis Miller is the owner of twenty Shares of the Capital Stock of The Jelm Mountain Gold & Silver Mining & Milling Company full paid and non assessable transferable only on the Books of the Company in person or by Attorney on the surrender of this Certificate

In Witness Whereof the said Company has caused the seal of said Company to be affixed hereto and this Certificate to be signed by its President and Secretary at Cummins Wyoming this 6th day of November 1881

E. N. Harwood Secretary J. I. Thomas President.

company secretary, immediately began making out stock certificates in payment for claims the company had accepted.

The "Cummins City Clatter" continued to report the news of the Big Laramie Mines and on June 16 reported:

> The outlook for Cummins and the Bramel Mining District was never better. Already our merchants and hotel men are beginning to feel the benefit of the North Park travel, which has almost entirely been diverted this way by the completion of the new shortcut from Cummins to Shipmans Park. The stage drivers on the Patrick line pronounce it the best mountain road they ever drove over.

This new road was built to directly compete with the Cameron Pass route into North Park and bring the business north to Laramie City and Cummins City. The stage companies (the Patrick Line and The Haynes and Wood Line) claimed they could make the run from Laramie City to Pinkhamton (in North Park), in six or seven hours and after an overnight stop, on to Teller in another six hours.

After eighteen weeks of labor, the carpentry work on the big Cummins City hotel was completed in late May. Beard and Thomas, however, failed to pay the labor costs and William L. Smith filed a lien on the hotel for $294, the cost of 73½ days of labor at four dollars a day.[58]

In June, 1881, Territorial Governor John Hoyt visited Cummins City and the Big Laramie Mines. Neither the travel nor the interest in mining was unusual for this governor. Throughout his term Hoyt traveled extensively through the territory and displayed a deep interest in mining and agriculture. Even while still in office his name appeared on several mining claims. Judging by his address before the legislature, he believed that unless mining was developed, neither a vigorous agriculture nor a prosperous manufacturing industry could be established. On this trip he was accompanied by Ed Ivinson, Laramie City banker, and Vice President of the Wyoming Consolidated Gold and Silver Mining Company.[59] While Ivinson was conducting the governor around Cummins City, Al Peabody, Secretary of the Consolidated was in Laramie City pushing the sale of stock for the company. J. I. "Doc" Thomas, president of the Jelm Mountain Gold and Silver Mining and Milling Company, was also in Laramie City using this opportunity to promote sales for his company.

In late June and through most of July, 1881, a huge forest fire raged between Cummins City and North Park, threatening the

Territorial Govenor John W. Hoyt
Wyoming State Museum

road and the sawmill of Beard and Thomas. Aside from the threat to personal property, a great many people, among them the miners and prospectors, were not concerned with these fires and even considered them beneficial. For them the fire "... let daylight and sunlight into the dark recesses, burns off the turf and accumulated matter on the surface of the ground clear down to the rocks and gravel thus greatly facilitating prospecting." [60]

At high noon on September 1, 1881, Cummins City received a new saw mill. The huge machinery was pulled through the town by a team of thirty-six oxen.[61]

The "Cummins City Clatter" of September 2, reported that news was not very plentiful. However, rumors had been sent flying by the arrival of the Territorial Geologist, Frederick Stanton, in the company of the Federal Collector of Internal Revenue, Edgar P. Snow of Cheyenne and a Professor Jinks. (Snow was a trustee of the Union Mining Company.) Professor W. W. Cherry, of Colorado also arrived in camp a day later. Stories followed Cherry from Colorado charging him with incompetence and dishonorable practice.

Whatever the rumors or the results of the meeting, the correspondent(s) who had, for more than a year, furnished the information for the "Cummins City Clatter" apparently departed camp. The last "Cummins City Clatter" column was dated September 2, 1881, and signed "M."

However, starting in August, a correspondent calling himself "Saxony," had begun corresponding with the *Cheyenne Daily Sun*, August 8, 1881:

> Knowing that as a business view your city is as deeply interested as the neighboring city of Laramie, it is but just that you receive for the benefit of your readers a reliable report from time to time of the doings in camp.
>
> We have got over the idea of this being a second Leadville and only know that it is an established fact which will bring lots of people and money into the eastern part of the Territory of Wyoming.

Toward the end of summer, a group of disheartened miners left Cummins City to investigate a report of an old Spanish Grant of what was called rich mining land in North Park at the south end of Independence Mountain.[62]

As the good mining weather deteriorated, so did the mining news. The October news from Cummins City dealt with school and church development with only a mention of one mine, that of Theodore Bath. The big hotel was nearing completion "... little remains to be done except the plastering."

A special election was held in October, 1881. The Cummins City precinct had polls located in the recorder's office of George Hildreth. Hildreth, Thomas and Harwood were appointed election judges.[63]

By November and December only the North Park road through Cummins City appeared newsworthy and the only mining news was coming from North Park, the old Centennial District and the Keystone Mine.

The settlers of North Park who did their trading in Laramie City were complaining bitterly about the road to North Park. They regarded the road along Beaver Creek (which they identified as situated in both Wyoming and Colorado) as being the best route but needing about $400 expended on the Wyoming

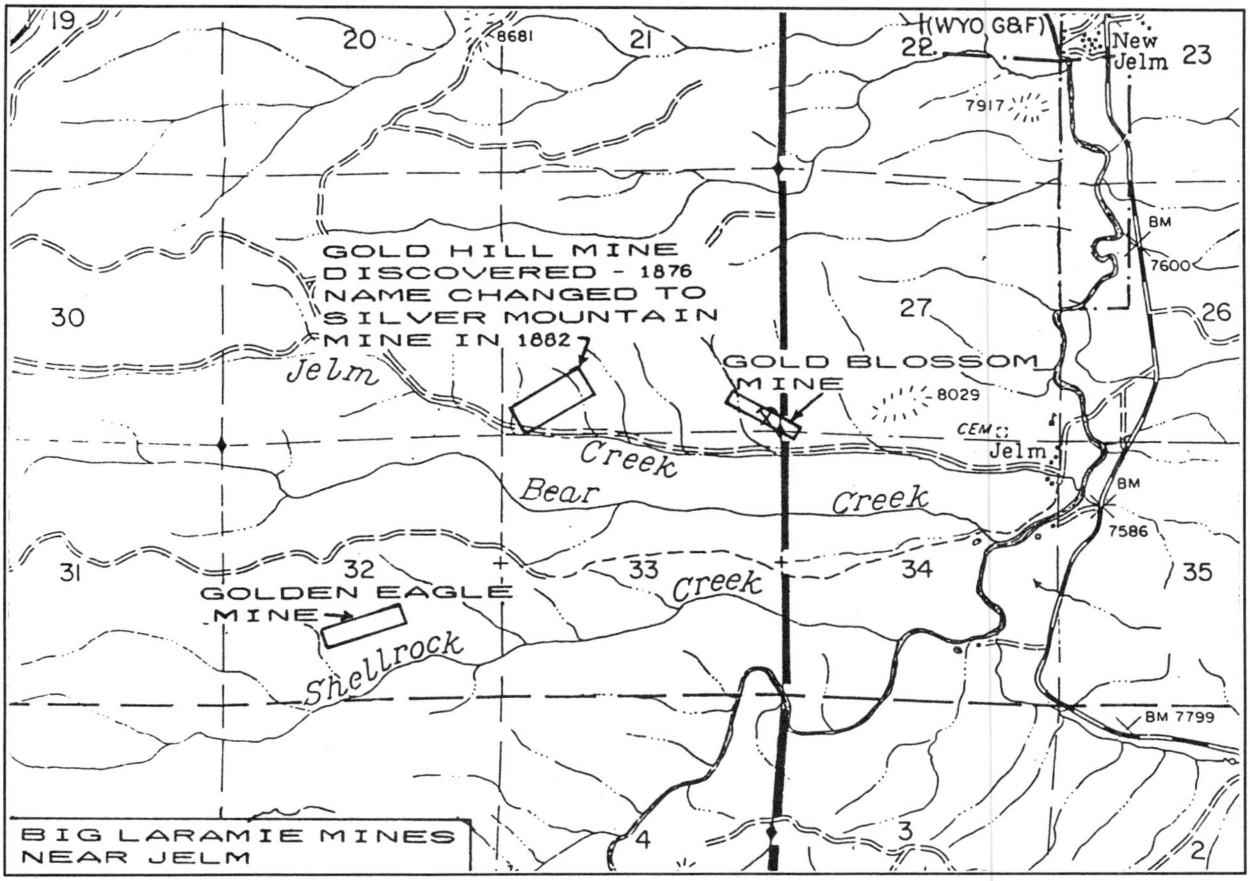

portion of the road to allow a four horse team to pass with ease. Apparently the County Commissioners were slow to act but eventually allowed an expenditure of $300, but only with the provision that the county supervise the work.

• 1882

By 1882, the mining activity along the Big Laramie River had reached a peak and leveled off. Enough ore had been removed to determine whether this district was going to boom or bust, and all that remained was to process the ore for the final analysis. It appears that most of the people in and around Cummins City were hanging on to see the result when the stamp mills began full time operation.

The flow of news from Cummins City slowed to a trickle then almost ceased completely. Editors of the Laramie City newspapers, Hayford of the *Sentinel* and Nye of the *Boomerang,* had been quick to report the good mining news but lethargic to totally absent in reporting the bad. Editor Slack of the *Cheyenne Sun,* however, contrary to his past actions, did print some promotional stories of the Big Laramie Mines, some apparently from the now very rare *Laramie Daily Times* and some on his own. This may have been the result of several of Cheyenne's prominent citizens taking an active financial interest in the mines.

On January 20, 1882, the *Sun* reported the first news that indicated silver was becoming the miners latest or perhaps more correctly their *last* hope:

> Mr. Ferguson of Henry, Illinois, who was here last summer investing largely, will be back in March with machinery for treating silver ores. The Elmira No. 2, owned by S.S. Wood, now down only twenty-six feet, has a vein six feet and a half wide, and assays over 100 ounces per ton in silver and seven dollars in gold.

Copper as a paying ore was seldom mentioned in the miners letters, the newspapers nor the mining claims filed at the recorder's office or the county courthouse.

Gold continued to be the primary objective of the miners and on January 20, the *Sun* reported:

> The Belmont Mine, belonging to L. Miller and D. C. Lyons, continues to improve as the work progresses. Messrs. Moore and Hildreth own the western exten-

sion of the same vein and have commenced sinking a shaft.

The *Sun* also reported twelve claims were recently located by Denver parties and that bids were being received by the Wyoming Consolidated Company to sink a 100 foot shaft on the "Betsy Jane" mine. The Jelm Mountain Company reported they were planning to work a large force of miners as soon as weather would permit. The company claimed to have sufficient funds, realized from eastern sales of stock, to carry out their plans.

Cummins City carried on as a mining camp but a good portion of its longevity was due, not to its mines, but to the stage and mail routes. In January, James Hills was awarded a contract to carry mail from Cummins City to Corlett, a small settlement just north of the Colorado line.[64]

In North Park, the Teller Mines, at the southern end of the Medicine Bow Mountains, continued to attract the miners with reports of promising new discoveries. The majority of traffic to the Teller Mines came from Laramie City and used Cummins City as a midway point on the first leg of their journey to North Park and the mines. The Teller newspaper reported that during the winter, mail on the Fort Collins route had become irregular and had almost ceased altogether.

It appears that John Cummins spent the winter in Denver but returned occasionally to Cummins City. In January, Cummins visited the camp and obtained several hundred pounds of ore from the "Savage" and "Michigan" mines which he took back to Denver.[65]

Several General Land Office (G.L.O.) Mineral Surveys were conducted in 1882. These surveys were required by law for the owners to prove up on their claims and obtain patent. To facilitate these surveys a monument was established at Cummins City by U.S. Deputy Surveyor, M. D. Libby. This monument which he said in his notes; ". . . is centrally located in the camp and its locus made permanent by being the corner stone in the foundation of Wood's Hotel and at the corner of Hill and Wall streets in Cummins City. The hotel is on Lot 10 and N.W. corner of Block 9." [66]

On July 1, 1882, Libby proceeded to complete Survey No. 1 of Mineral District No. 9, the "Gold Eagle" mining claim located 13,340 feet southwest of the Cummins City monument.

On Sunday, July 2, Libby conducted Survey No. 2 on the "Gold Blossom," 6,246 feet west and on July 3, he completed Survey No. 3 on the "General Miles," 11,513.7 feet east and No. 4 on the "Silver Mountain," 9,568 feet west of the monument. These mines were the property of the unincorporated company of Cummins and his new partners, Burnham, Hudson and Still. The "Silver Mountain" was the old "Gold Hill" claim which Cummins had reclaimed and renamed. In August, the Cummins Company bought a steam operated stamp mill from the Hendrie and Bolthoff Foundry in Denver. Bolthoff himself accompanied the mill to Cummins City. George Miller, also of the Hendrie and Bolthoff firm, directed the assembly of the mill and A. A. Burton then took over the final setup and operation of the mill. The Cummins Company teams, under the direction of J. W. Kelly, began hauling ore from their mines to the new mill. George Briggs, an experienced mill operator from Colorado, was also on hand to lend his expertise. Sam Smith gave Briggs a tour of the mines that so impressed him, he immediately proposed putting in a 60 stamp custom mill for the independent miners of the valley.[67]

On the first of August, the Wyoming Consolidated Gold Mining and Milling Company abandoned their property which left only Louis Miller as self appointed caretaker. The property consisted of a frame blacksmith shop with bellows and anvil, valued at $40.00, two log buildings and two mining claims valued at $400.00. Miller went to court a year later to claim wages of $600.00 for watching the property for the year plus $205.00 for assessment and surveying work completed. Luther Filmore answering for the company stated that Miller was acting without company approval and he in turn counter-sued Miller for $1,000.00. Miller won a judgment against the company for $226.10. The company, however, was without capital and a sheriff's sale was later held to dispose of the property. Miller bought the property at the sale for $25.00.[68]

When the November, 1882, election was held, the polling place for the Cummins City Precinct was designated as Beard and

Thomas' Hotel. Election judges were J. I. Thomas, J. C. Watkins and Dr. Baker.[69] Although Cummins City was already in its waning days, the precinct cast 144 votes, second only to Laramie City in Albany County. Several of the Cummins City miners were elected to county as well as local offices. Louis Miller was elected Albany County Sheriff and A. G. Dunn as County Treasurer. George Hildreth was elected Cummins City Justice of the Peace and C. W. Eads was elected Constable.[70]

On December 11, 1882, George Hildreth wrote to his old partner, W. H. Root of Laramie City:

> Big conflagration here this morning at 4 o'clock. Four buildings were burned, the big saloon, the dwelling adjoining, the blacksmith shop and the old recorders office. Wind blowing a terrific gale from the southwest. The big hotel had a narrow escape - all the front scorched. The fire originated among shavings under the floor of the big saloon. Supposed to be the work of an incendiary.

Another letter written the next day, December 12:

> Since writing you yesterday another fire has occurred. A log house at 10 o'clock this morning caught fire and burned, situated on the east side of the river and occupied by Mr. Brocker, blacksmith. The cause of the fire is unknown.[71]

These fires only added to the death knell for Cummins City and apparently no attempt was made to rebuild the lost buildings. The mining camp in the valley of the Big Laramie River had run the boom and bust cycle.

By the end of 1882, the mining boom was over. Cummins City, however, continued on, not as a mining camp, but as a stage stop complete with stores, saloons and post office for the surrounding area.

On December 15, 1882, the Albany County Commissioners directed Justice Hildreth to have a guardhouse built at Cummins City to accommodate prisoners. About a month later Hildreth reported back to the Commissioners; "The building is constructed of two inch plank doubled all around, 12 x 16 feet on the outside, 12 feet high, a 4 foot hall and one room 12 feet square, sealed with 2 inch plank and a hip roof of 1 inch boards doubled, two doors 4 inches thick with grated iron window in each." Total cost for the building was $246.11; $88.00 for carpenter Henry Hunt, $78.48 for lumber and $79.63 for transportation of materials.

• 1883 and LATER

Although the population of the mining camp of Cummins City may remain forever in question, there is no doubt that several hundred people were associated with the camp and the mines.

Many of Laramie City's businessmen were drawn into the rush to Cummins City, some to set up a branch business, some to grubstake the itinerant miner and some to take up pick and shovel themselves.

More than 500 claims were said to be on the camps own mining record and several hundred of these are preserved in the county records.

The Bramel Mining District and the mills at Cummins City apparently produced little or no gold, silver or any other mining product to warrant an entry in the remaining records.

In contrast, the other mining camps of the Medicine Bow Mountains made much ado

Samuel S. Wood
By Joe Barros

about every gold button, every mill cleanup and every sale of placer gold from their district. In retrospect, it appears there was no significant amount of precious metal produced at Cummins City during the 1880s gold rush.

Sam Wood continued as the Cummins City postmaster until March, 1884. In October, Wood gave up his hotel business in Cummins City and moved down river and became postmaster at his namesake post office, commissioned simply "Woods," located at the settlement of Woods Landing. Wood later became interested in the Gold Hill excitement and moved his operation to Brush Creek where, along with his son, he ran his saw mill and freight business.

Although the Woods post office was opened just a few miles down river, the Cummins City post office continued to operate under postmaster George Hildreth for a short time, then under Urbina C. Beard until 1888 at which time all the mail was sent to the Woods post office. After a closure of about ten years, the Jelm post office was relocated and reopened. (The "new" Jelm post office operated until 1965.)

In 1884, J. I. "Doc" Thomas was still at Cummins City and in August of that year, the County Commissioners allowed him a payment of $66.85 for care of the Cummins Bridge.

When the 1884 election was held, Cummins Precinct cast a total of sixty-nine votes. George Hildreth was re-elected as Justice of the Peace. During the 1886 election, the total vote dropped to twenty-nine. Hildreth was again elected as Justice and Frank Smith as Constable, however, there are no apparent records of any court proceedings after March, 1886. After 1886 Cummins was no longer a precinct and from 1888 through 1894, the remaining residents were required to travel down river to Woods Landing to cast their votes.

George Hildreth, Cummins City postmaster and Justice of the Peace, died in December, 1886. Hildreth spent Christmas day with J. I. "Doc" Thomas. After spending the night, he left the following morning for his own "ranche." He was found the next day lying on the floor of his cabin. Hildreth had the reputation of being a heavy drinker and speculation was that he had died from the effects of excess alcohol. Doc Thomas, however, testified that he had been sober when he left his house and had no liquor with him. Further supposition was that he had probably died of natural causes. No inquest was held and the body was shipped east for burial. Hildreth, who had once held a federal office in the U.S. Treasury, was apparently well thought of in Cummins City and had easily won every office for which he ran. His past, however, was tarnished by his close association with James Pattee, the Wyoming Lottery King.

T. D. Pearson, who had been one of the earliest miners of the Big Laramie mines, was by 1886, claiming residence in Denver but was still working a placer mine on Lincoln Gulch in the Medicine Bow Mountains.

Louis Miller gave up his mining when he was elected Albany County Sheriff in November 1882, but kept a watchful eye on his Jelm Mountain interests and was to return periodically to dabble in the mines. He was involved in the 1900 mining boom at which time he was manager of the Wyoming Queen Mining Company.

James Adams, of the Union Mining Company, was elected to the Wyoming House of Representatives and served in the Seventh Legislative Assembly.

Doctor John C. Watkins, his wife and his sister left Cummins City and settled in Buffalo, Wyoming. The Watkins ladies became well known for their efforts in the educational field and Mary S. Watkins became Superintendent of Schools in Johnson County. Doctor Watkins became Johnson County Coroner and in 1892 was called to investigate the deaths of Nate Champion and Nick Ray. He died of apoplexy while on his way to view the remains of the men killed at the Carr Ranch. He was buried in Buffalo.

William Tilton of the Jelm Mountain Gold and Silver Mining and Milling Company, moved to the North Platte River valley where he married and settled down to ranching near the mouth of Brush Creek. Tilton would again be drawn into the prospecting and mining business during the Gold Hill rush.

W. S. Bramel moved to Nevada City, Nevada, still pursuing mining.

George W. Moore continued his prospecting

THE SAINT PAUL PIONEER PRESS: SATURDAY, FEBRUARY 2, 1901.

GREAT CLOSING SALE OF DIVIDEND BEARING **MINING SHARES**

A Money-Making Opportunity of Greater Import Than Has Ever Before Crossed Your Path, and the Best That Will Ever Come to You During Your Lifetime.

Only ten days more for "Queen Bee"; only twenty days more for "Kettle-Curlew"; only thirty days more for "Boundary Camp" shares. After that the unsold shares will be distributed free to shareholders in proportion to their holdings. The time has been greatly shortened. The wind-up and distribution will take place March 1 instead of April 1. This is the last announcement that will be made. Dividends equal to 25 per cent on investments guaranteed by May 1 next. Five dividends already paid. Dividends equal to 80 per cent a year absolutely certain. The grandest mining promotion success of any century. We started alone and now have more than thirty imitators in the Twin Cities. Clear titles to all properties; two great city sites; mills running; complete protection to shareholders; enormous earnings certain; investigate.

We have bought, paid for, erected, and will have in operation in a few days extensive stamp mills for the free milling ores of the "Queen Bee" Company in Jelm, Wyoming—these mills will yield $300 a day for "Queen Bee" shareholders.

We have ordered and are constructing a 75-ton matting plant for the copper ores of the "Queen Bee" mines—and will have it in operation in 60 days—that will produce $1,000 a day for "Queen Bee" shareholders. Additional sections of this plant will be erected until the works will reach a capacity of 1,000 tons a day.

We own 42 rich mining claims at and near Jelm, Wyo.—no richer on earth in copper and gold—these properties worth hundreds of millions, paid for in full, except $15,000, and two years to pay that in. Among these claims we have made exceedingly rich strikes on the "Colorado," the "Rising Sun," the "Jesse," the "Helen," the "Copper Queen," etc. Major Hancock, our treasurer and trustee, is now on the ground personally superintending the erection and installation of the plants.

We own the site of Jelm City, over 3,000 lots, paid for; deeds ready, clear title, worth over a million dollars. Jelm City is already a thriving town, and will have a population of 1,000 this year and 10,000 in less than five years.

and mining activities and was involved in the Hurley Mining District.

William O. Downey whose surveys helped spawn the mining camp died in 1883. His brother Stephen, although heavily committed to public service, continued to be deeply involved in the Medicine Bow mines, although not those along the Big Laramie River. Stephen Downey was apparently well aware of the early attempts of John Cummins to promote the mines. He was, in fact, a close personal friend of Judge Jacob Blair and was his partner in the Wyoming Keystone Gold Mining Company. It was Judge Blair who heard the case of Smith vs. Cummins in the District Court. Downey was also well acquainted with the attorneys in the case. Downey, probably because he was privy to this information, was notably absent from the mining scene along the Big Laramie River.

Cummins City continued it's decline and the Big Hotel was torn down around 1890 and the lumber used to build several cottages in Laramie.[72]

In 1896, Frank Smith claimed the "Copper King" and Louis Miller claimed the "Copper Queen," beginning a new series of claims which would lead to the "Copper Rush" of 1900.

Cummins City was again a designated voting precinct in 1896.

• COPPER RUSH OF 1900

Attention was again drawn to the Big Laramie Valley around the turn of the century. The Jelm Mining District was formed, superseding the old Bramel District. Cummins City became Jelm City. Copper supplanted gold as the prime objective of the miners. Several new companies were formed and their stock placed on the market.

Louis Miller was again drawn to the valley

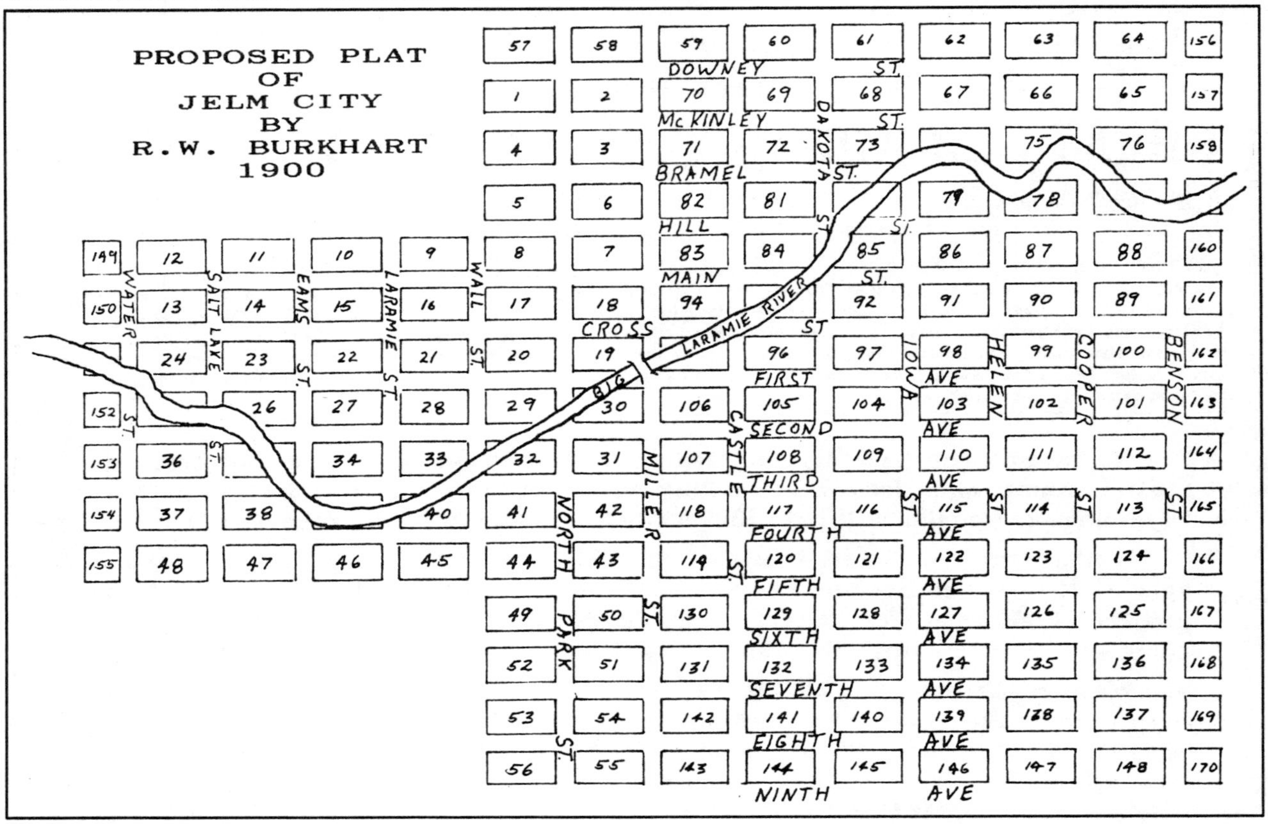
PROPOSED PLAT
OF
JELM CITY
BY
R.W. BURKHART
1900
DOWNEY ST.
McKINLEY ST.
BRAMEL ST.
HILL ST.
MAIN ST.
CROSS ST.
FIRST AVE
SECOND AVE
THIRD AVE
FOURTH AVE
FIFTH AVE
SIXTH AVE
SEVENTH AVE
EIGHTH AVE
NINTH AVE
WATER ST.
SALT LAKE ST.
EAMS ST.
LARAMIE ST.
WALL ST.
NORTH PARK ST.
MILLER ST.
CASTLE ST.
DAKOTA ST.
IOWA ST.
HELEN ST.
COOPER ST.
BENSON ST.
BIG LARAMIE RIVER

Jelm Mountain, circa 1900.

and became President and Manager of the Wyoming Queen Mining Company. Frank Smith along with Doctor C. C. Clark of Salt Lake City organized the Inter-Ocean Water Power and Mining Company.

Joining the miners from the 1880 era was former Surveyor General for Wyoming, "General" John C. Thompson. Charles W. Bramel, (W. S. "Buck's" brother), was part-owner of several mines.

The promise of a railroad apparently inspired some to do considerable new development work on their mines. The railroad, however, did not reach Centennial until 1907 and was then several years building across the Medicine Bow Mountains into North Park. The railroad did not utilize the old Cummins City-North Park route but rather a new route which bypassed not only Jelm City but all of the mining camps except Centennial.

On February 2, 1901, a full page advertisement was placed in *The Saint Paul Pioneer Press* by the American Mining Investment Company advertising:

Only ten days more for 'Queen Bee' . . . shares. . . .

JELM

MOUNTAIN MINING DISTRICT IS SHOWING GOOD RESULTS

The Shaft of the Colorado Mine, Now Down Twenty Feet, Shows Two Feet of Ten Per Cent Copper, Some Gold and Silver—Annie Mine Has Strong Flow of Water—Work In the District.

Colorado Shaft-house and Ore Dump — Wyoming Queen Mining Company, of Jelm, Wyo.

Boston Shaft-house and Ore dump — Wyoming Queen Mining Company, Jelm, Wyo.

> We have bought, paid for, and erected, and will have in operation in a few days, extensive stamp mills . . . . We own the site of Jelm City, over 3,000 lots, paid for; deeds ready, clear title, worth over a million dollars. Jelm City is already a thriving town, and will have a population of 1,000 this year and 10,000 in less than five years.

The 1900 census indicated seventeen people in Jelm City, the 1901 *Directory* listed twenty-one people and the 1906 *Directory*, thirteen people. It is difficult to separate entrepreneurship from outright fraud but in retrospect it appears that it was these claims that led to stories later told of extensive fraud at Cummins City.

Again, no profitable mines were developed and again no records exist showing a production of copper, gold or silver. Even if the reported assays and the greatest hopes of the miners had transpired, deep shaft mining such as carried out in the Medicine Bow Mountains, could not hope to compete with the huge open pit copper mines of Butte, Montana, and the Superior Lake mines of Michigan.

Today, only the scattered and meager remnants of Jelm City remain. Remains of the earlier Cummins City, if they exist, are difficult to find and identify. The surrounding hills still bear the scars of early mining but the stories of early mining are overshadowed by tales of John Cummins and reputed fraud.

"Old" Jelm , 1990

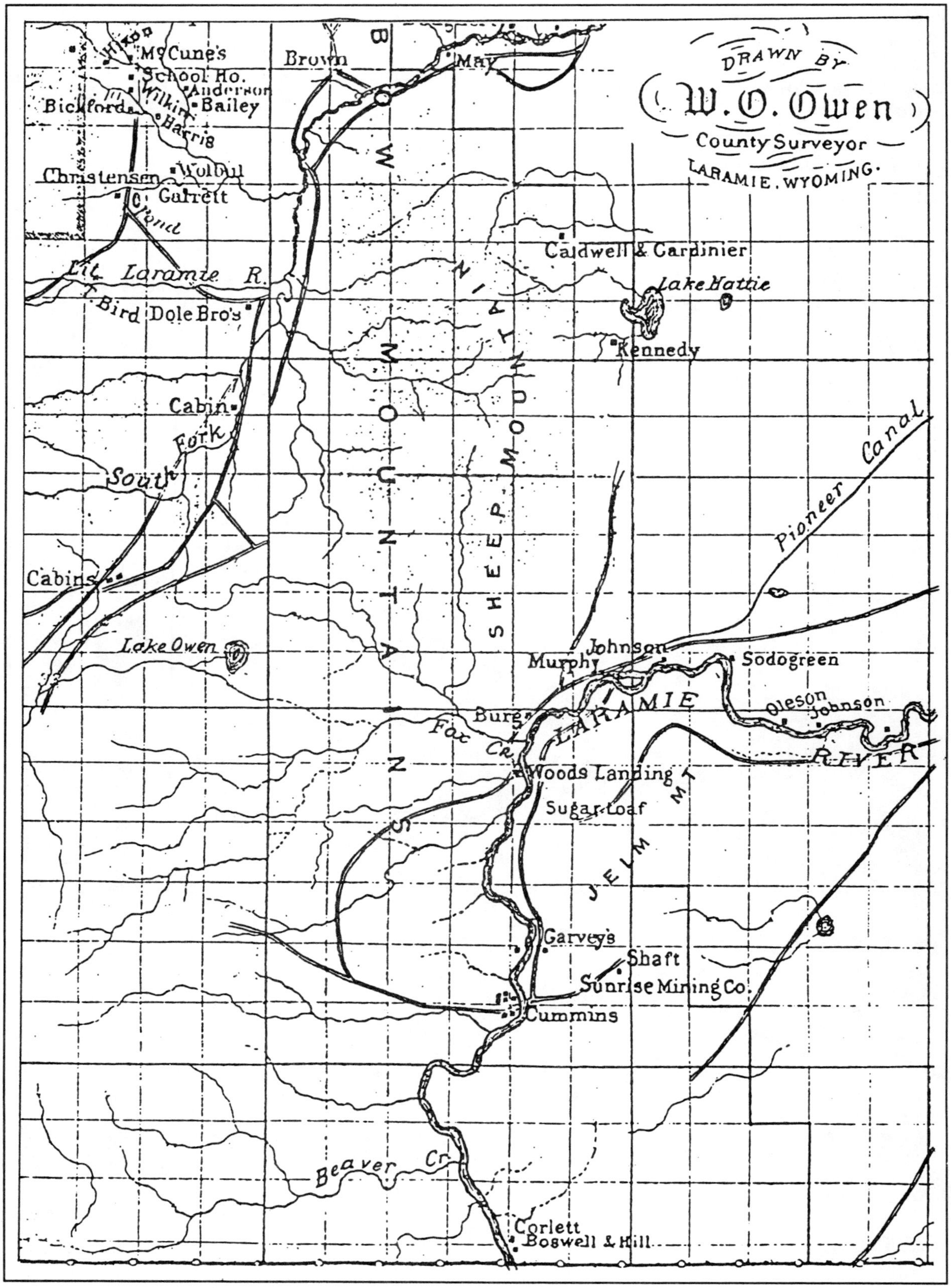

Big Laramie District & Cummins City

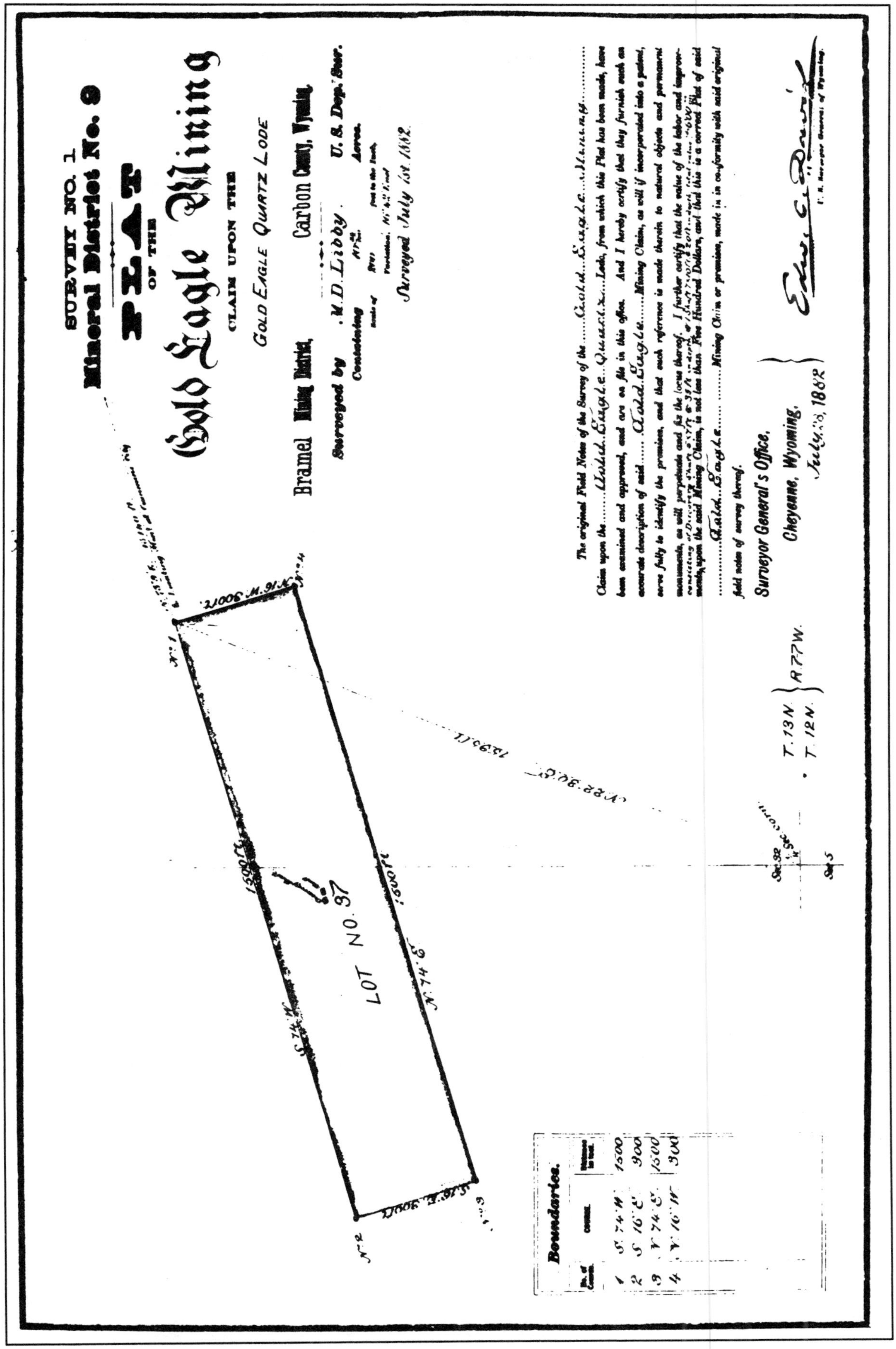
SURVEY NO. 1
Mineral District No. 9
PLAT
OF THE
Gold Eagle Mining
CLAIM UPON THE
GOLD EAGLE QUARTZ LODE
Bramel Mining District, Carbon County, Wyoming.
Surveyed by M.D. Libby U.S. Dep. Sur.
Surveyed July 1st 1882.
LOT NO. 37
T. 13 N.
T. 12 N.
R. 77 W.
Boundaries.
Surveyor General's Office,
Cheyenne, Wyoming,
July 28, 1882.

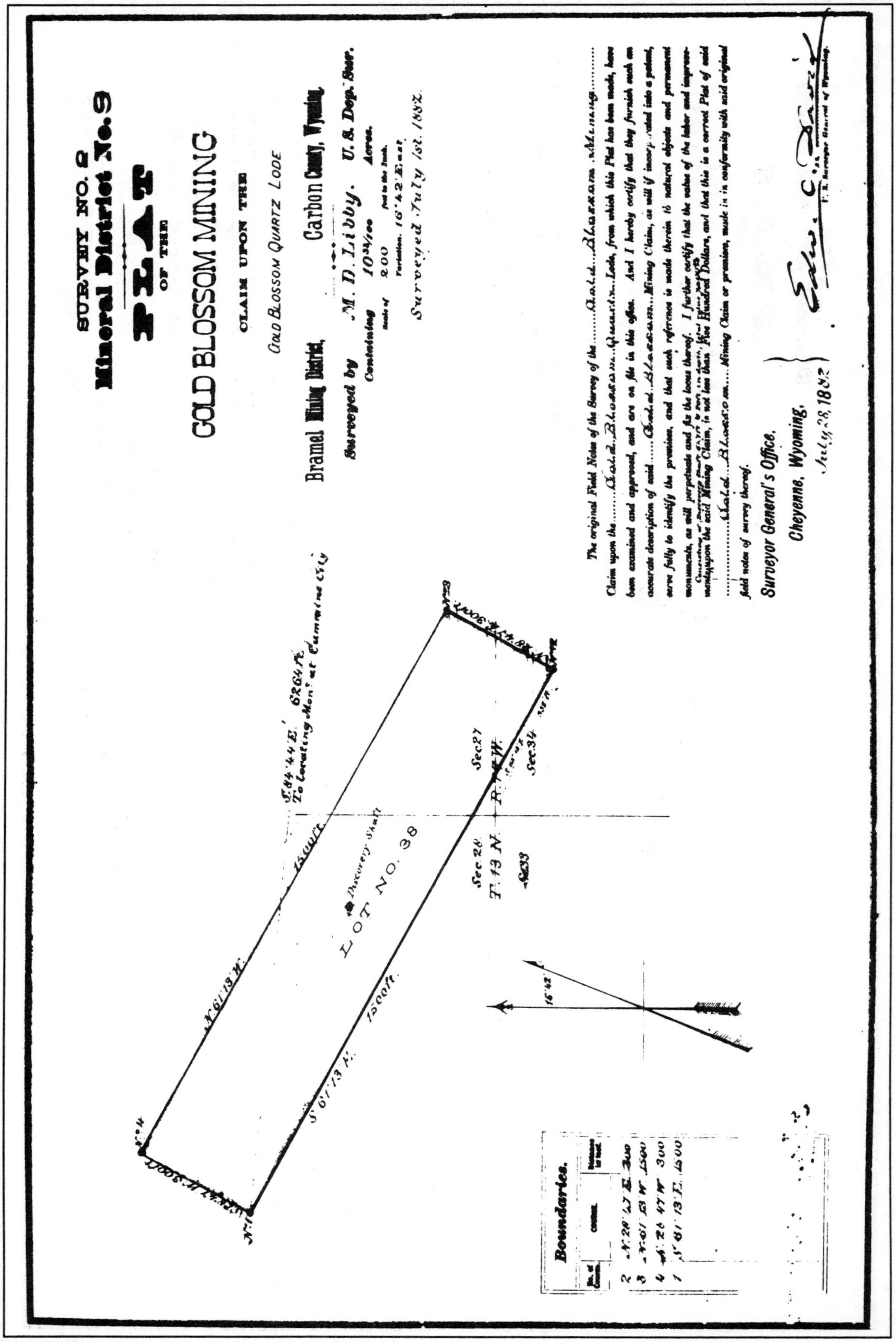
SURVEY NO. 2
Mineral District No. 9
PLAT OF THE
GOLD BLOSSOM MINING
CLAIM UPON THE
GOLD BLOSSOM QUARTZ LODE
Bramel's Mining District, Carbon County, Wyoming,
Surveyed by M. D. Libby. U. S. Dep. Sur.
Containing 10 34/100 Acres.
Variation 16°42' East.
Surveyed July 1st. 1882.
Surveyor General's Office.
Cheyenne, Wyoming, July 28, 1882.
LOT NO. 38
Discovery Shaft
S.84°44'E. 6264 ft. To Locating Mon't at Cummins City
N.61°13'W.
S.61°13'E.
1500 ft.
Sec 28
Sec 27
Sec 33
Sec 34
T.13 N.
16°42'
Boundaries.
2 N.28°47'E. 300
3 N.61°13'W. 1500
4 S.28°47'W. 300
1 S.61°13'E. 1500

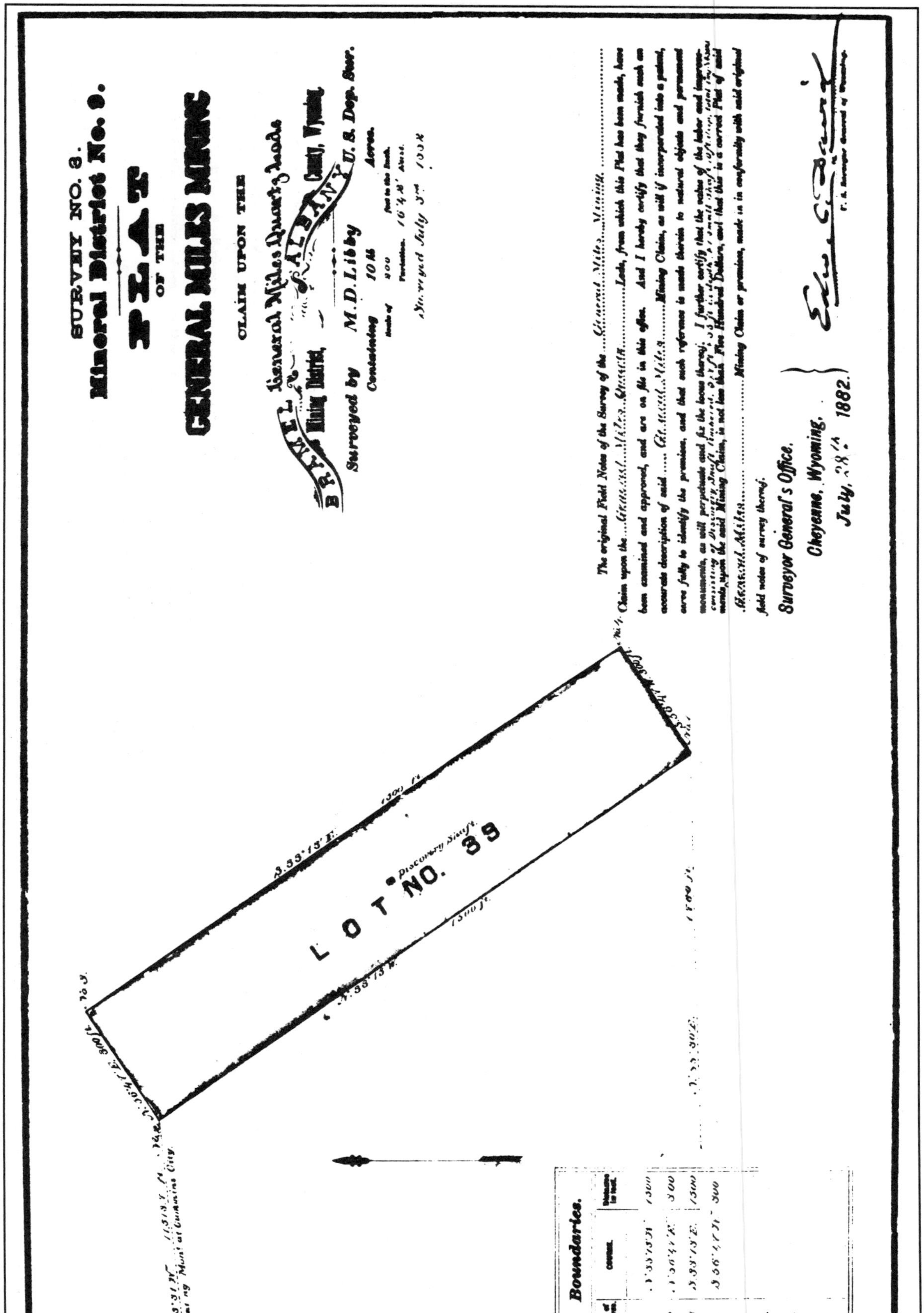
SURVEY NO. 3.
Mineral District No. 9.
PLAT
OF THE
GENERAL MILES MINING
CLAIM UPON THE
General Miles Quartz Lode
BRAMEL Mining District, ALBANY County, Wyoming.
Surveyed by M.D. Libby U.S. Dep. Svr.
Containing 10 Acres.
Scale of 200 feet to the inch.
LOT NO. 39
Discovery Shaft
S.33°13'E.
N.33°13'W.
Boundaries.
No. of Course. Course. Distance in feet.
1 N.33°13'W. 1300
2 N.56°47'E. 300
3 S.33°13'E. 1300
4 S.56°47'W. 300
To locating Mon't at Cummins City.
The original Field Notes of the Survey of the General Miles Mining Claim upon the General Miles Quartz Lode, from which this Plat has been made, have been examined and approved, and are on file in this office. And I hereby certify that they furnish such an accurate description of said General Miles Mining Claim, as will if incorporated into a patent, serve fully to identify the premises, and that such reference is made therein to natural objects and permanent monuments, as will perpetuate and fix the locus thereof. I further certify that the value of the labor and improvements upon the said Mining Claim, is not less than Five Hundred Dollars, and that this is a correct Plat of said General Miles Mining Claim or premises, made in conformity with said original field notes of survey thereof.
Surveyor General's Office,
Cheyenne, Wyoming,
1882.
U.S. Surveyor General of Wyoming.

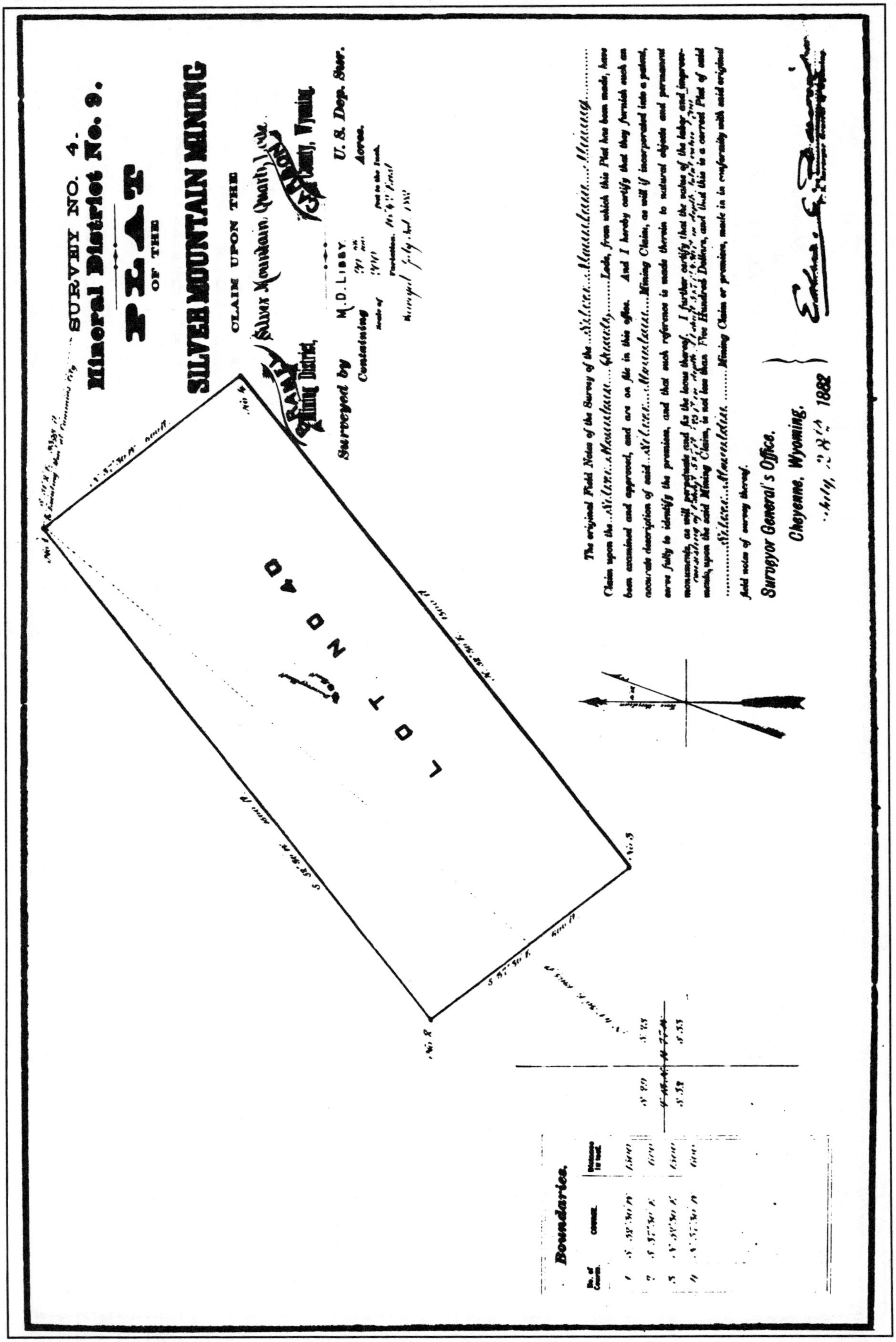
SURVEY NO. 4.
Mineral District No. 9.
PLAT
OF THE
SILVER MOUNTAIN MINING
CLAIM UPON THE
Silver Mountain Quartz Lode
RAMBLER Mining District,
CARBON County, Wyoming.
Surveyed by M.D. LIBBY. U. S. Dep. Sur.
Containing Acres.
Scale of feet to the Inch.
Variation.
LOT NO 40
Boundaries.
The original Field Notes of the Survey of the Silver Mountain Mining Claim upon the Silver Mountain Quartz Lode, from which this Plat has been made, have been examined and approved, and are on file in this office. And I hereby certify that they furnish such an accurate description of said Silver Mountain Mining Claim, as will if incorporated into a patent, serve fully to identify the premises, and that such reference is made therein to natural objects and permanent monuments, as will perpetuate and fix the locus thereof. I further certify that the value of the labor and improvements, upon the said Mining Claim, is not less than Five Hundred Dollars, and that this is a correct Plat of said Silver Mountain Mining Claim or premises, made in in conformity with said original field notes of survey thereof.
Surveyor General's Office.
Cheyenne, Wyoming.
July, 28th 1882

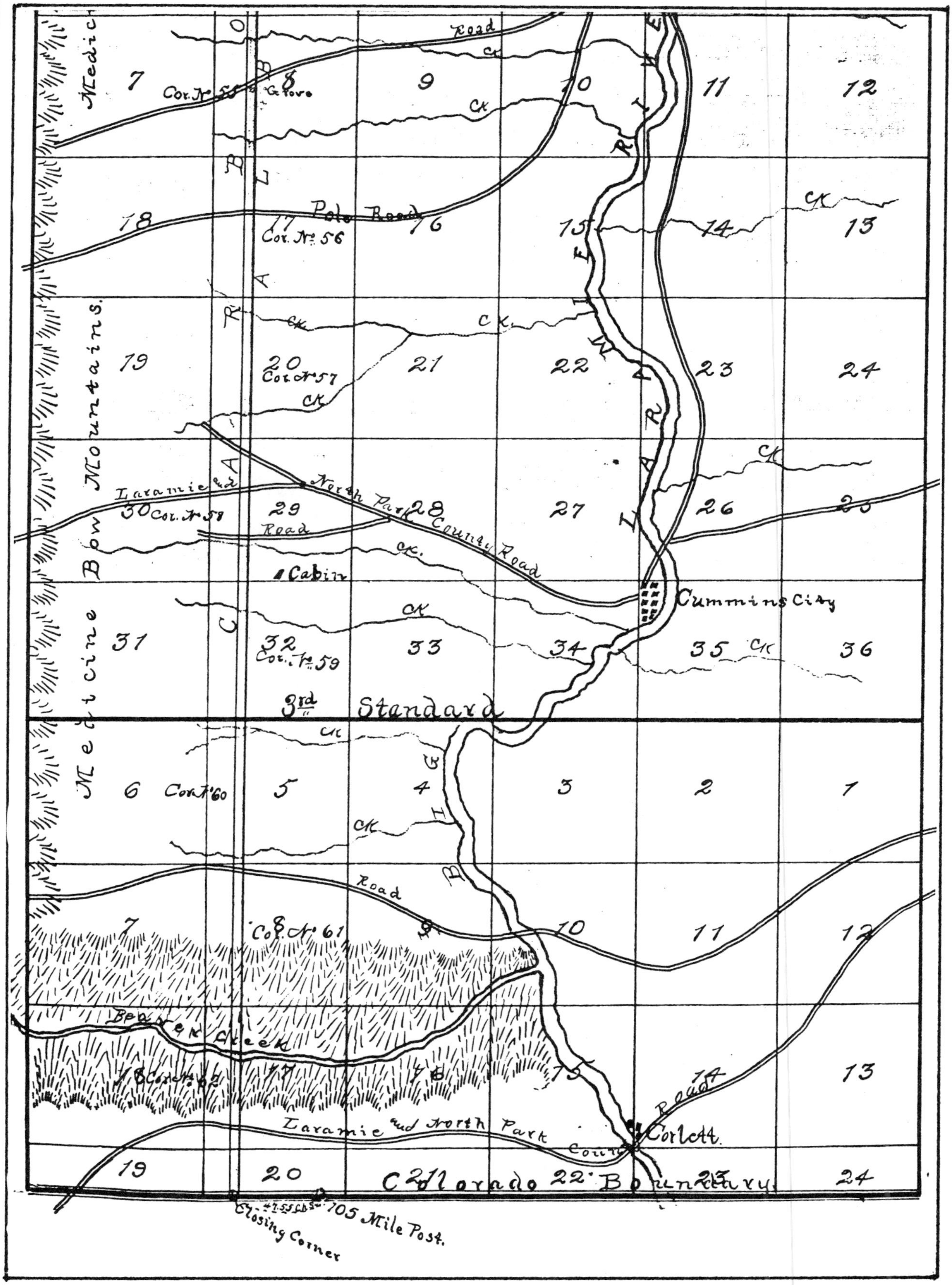

Laramie and North Park roads at Cummins City

"Honest John" Cummins
By Joe Barros

JOHN CUMMINS

V

Some sixty years after the first rush to Cummins City, John C. Thompson, editor of the *Wyoming State Tribune* writing in his August 8, 1941, column, "In Old Wyoming" tells one version of the escapades of John Cummins and the life of Frank Smith.

> . . . Just across the river from the little flat and the flanking bench on which Cummins City was born, briefly flourished, and decayed, abides a living relic, a very virile one, who was present at its accouchement, participated in its lusty growth, saw it falter, wither, fade and disappear. He is 86 year old Frank Smith, who probably sits a horse straighter, can ride it farther in a day and come home comparatively fresh at dusk, than any other of his age in this state; or for that matter, all elsewhere. A remarkable man, Frank Smith; one with whom it is a delight to converse regarding things as they were when Wyoming was much younger and correspondingly less staid.[1] Jelm mountain's cone has been the axis of his operations during 66 years. He has lived every hour of the period and at four-score and six still is not overlooking any bets. Memories crowd upon him as he gazes across the river at where Cummins City was; his reminiscence is engaging.
>
> . . . In 1881, Smith became associated with John Cummins. This white-whiskered confidence man, who looked more like a preacher or son of the soil than a slicker, was inflating the Cummins City boom. He employed the robust young Smith to help him build. Fortified with a large envelope which was presented as being stuffed with "valuable papers," Cummins obtained capital from Laramieites, embarked upon development of the Copper King lode mine. Soon there reared at the Copper King, located up a gulch somewhat less than a mile from the river, perhaps the largest water-wheel ever fabricated in Wyoming, while on the flat beside the river there rose undoubtedly the largest hotel ever built in a Wyoming boom camp. It was a two-story frame structure of 40 or more rooms.
>
> . . . At the height of Cummmins City's career a population of 200 was claimed for the camp; a town was laid out, with lots and streets; a mining district was organized. . . . The 200 population claim, it is Frank Smith's recollection, was inflated about 150 percent; at no time did the actual residents of the "city" exceed 50. Around 30 structures were erected, these including three stamp mills for treatment of ore, a church, a very substantial jail. . . . The whole boom pivoted on the Copper King - and the Copper King was a false alarm. . . . When it became apparent that the metalliferous resources of the district were nil . . . , Cummins City wilted almost as rapidly as it had bloomed. Inhabitants left without taking the trouble to move their household possessions and for years crockery might be observed on cabin shelves, furniture just as it had been left when its owners departed, even blankets and quilts still on beds. . . . The Cumminses and Doctor Thomas lingered on for several years after the camp was deserted by others, living in the otherwise empty big hotel.

Frank Smith.
Wyoming State Museum

Most recent accounts of Cummins City seem to dwell primarily on this article.

Rather than rely on the memory of Smith and the writing of Thompson, a thorough search of existing records probably gives a more accurate account of John Cummins and the boom and bust of Cummins City and Jelm City.

Even before the demise of Cummins City, it appears that John Cummins returned to Colorado. His last transaction at Cummins City appears to be the sale of the "Mayflower" and the "Golden Rule" mines, one-half to Emily M. Neil and one-half to Watson Rockwell on July 12, 1882. However, both Neil and

Rockwell lived in Denver, Colorado and it is quite possible that Cummins conducted these transactions there. [2]

The first word of John Cummins after the Cummins City rush appeared in *The Boomerang*, June 4, 1886:

> Honest Old John Cummins the founder of Cummins City and a miner from away back is in the city after an absence of about two years, visiting his old friends and especially W.H. Root whom he regards almost as a brother and both of whom beheld the glories of Cummins City when her boom was at its zenith.
>
> Mr. Cummins is now on his way to Montana to settle up some mining matters there and after that will return to Laramie and stop for a while to look around; his friends hope to such good effect that he will again pitch his tent in our midst. He is now living in Denver and the cares of life seem to have rested lightly on him for he is the same old man in every respect and does not look a day older than when he went away. He met with a warm welcome here.

At that time there seems to be no suggestion of mis-dealings but later the same year another brief article appeared, imparting a look at one of Cummins' apparent weaknesses. *The Laramie Sentinel*, October 9, 1886; "We met old John Cummins at Ogden. He was on a bridal tour with a young wife."

Several years pass until we again hear of Cummins, this time casting some light on his apparent true character.

The Laramie *Boomerang*, April 18, 1890:

> John Cummins, after whom Cummins City was named, is now in the penitentiary at San Quentin, California. The cause of his incarceration is not known here.

The source of the *Boomerang* report was not revealed, but if indeed he served time in San Quentin prison, the California prison has no record of his imprisonment.[3]

A little more than a year later we again hear of Cummins back in Colorado.

The *Laramie Weekly Sentinel*, June 6, 1891:

> John Cummins has been putting his foot in it again. He persuaded some people to entrust him with money to invest in coal lands and then attempted to abscond with it but was arrested and jailed in Pueblo. John has been cutting a pretty wide swath for some years past and has several criminal charges and indictments standing to his credit. His wife recently caused his arrest at Trinidad for fraudulently procuring a divorce, charging him with perjury before the grand jury. No indictment was found, and then the woman endeavored to take the law into her own hands and shot twice at him. He has had an eventful history since he left here and is a worse wreck than his old mining town up the river.

Later in the fall another article followed along the same line; *The Laramie Weekly Sentinel*, October 10, 1891:

> Old John Cummins who figured so prominently here years ago, seems to have put his foot in it at last in Colorado. He entered into a conspiracy with certain parties in Trinidad to prove up, fraudulently, on certain coal lands at Trinidad. His victims, who claimed to have been swindled out of their money by the oily statements of Cummins, became his prosecutors.
>
> In June last Cummins was examined before a Justice of the Peace of Las Animas County on four separate and distinct charges for obtaining money under false pretenses, and has since been indicted by the grand jury of that county for the offense in some ten cases. After the judgment of the justices binding him over to await action of the grand jury, a writ of habeas corpus was sworn out on the grounds that the whole arrangement was a fraud in the first place, but the Supreme Court refused to see it that way and he was remanded for trial. He is pretty likely to serve a term in the penitentiary.

The *Saratoga Sun*, December 3, 1891:

> John Cummins, founder of a mining camp in Albany County that was called for him, stands a fair chance of getting out of a bad scrape on account of a flaw in the indictments against him for swindling people on Colorado coal lands. The case will be argued in Trinidad Saturday.

The results of this escapade were reported in the *Laramie Sentinel*, April 23, 1892:

> John Cummins was sentenced to four years in the penitentiary, last Saturday at Trinidad, Colorado. The crime for which he was sentenced was for defrauding the people out of money on a coal land scheme and obtaining money under false pretenses. He has been in jail for over a year and at his age a four year sentence is a good long while.

It appears that Cummins did not serve his full sentence, for in June, 1893, the *Boomerang* reported that J. W. Cummins, the founder of Cummins City, was accused of stealing a sack of ore from the mill of W. J. Chamberlin in Georgetown, Colorado. A jury found him guilty but he appealed the case and apparently was able to avoid a jail term.[4]

A mining transaction places both Cummins and J. I. "Doc" Thomas in the Denver, Colorado, area in 1894. On May 31 of that year, Thomas sold to Cummins the old "Gold Nuget" [sic] and the "Green Monster" mining claims located about half way between Woods Landing and Cummins City. The transaction was notarized in Colorado but recorded in Wyoming.[5] It appears that John Cummins was again planning some venture in the old mining district, but there is no apparent evidence that he ever returned to his namesake mining camp.

A few years later his death was reported in *The Laramie Boomerang*, February 25, 1897:

> Trinidad, Colo. Feb. 24, 1897:
>
> John Cummins whose recent death is reported was at one time a resident of Trinidad, leaving a record behind him that will not soon be forgotten by the residents of this city. He came to Trinidad in 1888 with about $30,000 in cash and when he left his earning was reported to be anywhere from $30,000 to $50,000. By a liberal exchange of money he had, and by changing it from one bank to another, he created the impression that he was a heavy capitalist, and then started out after his victims. His scheme was to get the confidence of men who had a little money laid by and then show them how, by the investment of $50 to $75 they could get net returns of $1,500. He pretended to be representing a syndicate that wanted a large body of coal land. For the small sum above mentioned he would pretend to locate his victim on a quarter section of coal land and guarantee each of them that the land would be taken by the syndicate at a figure that would net each claimant $1,500.
>
> He caught victims by the score, some men taking a claim for every member of his family. When his fraud was exposed his victims had Cummins arrested for obtaining money under false pretenses. A bondsman was secured from Denver and for a time Cummins enjoyed his freedom, but finally there was a falling out, the friend withdrew from his bond, and the capitalist and coal land speculator went to the county jail. After several continuances in the district court, a conviction was secured and Cummins was sentenced to the penitentiary, but the decision was reversed by the higher court on the ground that the whole scheme was a fraud against the government and that men who were prosecuting Cummins were parties to the fraud and he was given his liberty.
>
> When Cummins first came to Trinidad he pretended to be a single man and was untiring in his devotion to the fair sex, but later in the game a legal wife appeared on the scene and there was a suit brought by her for divorce. One night there was several shots fired at him through a window and he made the assertion that his wife had tried to assassinate him, but the general belief was that he had hired a man to make the gun play to create sympathy. Since leaving Trinidad all traces of Cummins have been lost.

About a month later the *Boomerang*, March 22, 1897, printed the final chapter on John Cummins.

> **Mr. Cummins Fortune**
>
> The people of this section all remember John Cummins, the old mine prospector, speculator, and the founder of Cummins City, who recently died in California. The following article from Oakland, California, published in the San Francisco Examiner of the 18th will be read with interest.
>
> John Cummins who died suddenly in Alameda on January 19th, reputed millionaire, it now transpires was not worth a dollar.
>
> Cummins was an old acquaintance of E.H. Porter of 448 Central Avenue and several weeks before his death he came here for his health. Mr. Porter had not seen Cummins for many years, but he heard he was a rich mining speculator.
>
> Cummins expired suddenly from heart disease at the Porter residence on the evening of January 19th. An examination of the effects of the deceased confirmed beyond all doubt the belief of the elder Porter that Cummins was wealthy. In a valise, which Cummins always carried about were found four packages supposed to contain $20,000 in currency and sureties representing several hundred thousand dollars. A. P. Smiley, then deputy coroner for the city was notified of the death and also of the apparent wealth of the deceased. Mr. Smiley and attorney R. B. Tappan went to the Porter residence and the valise was examined in the presence of the Porter family. A complete list of the contents was made and everybody was pledged to secrecy. The valise and its contents were carried away by the deputy coroner and the attorney for safe keeping. The valise was hermetically sealed so that none of the wealth might escape.
>
> The next morning the valise was taken to the Bank of Alameda. It was labeled with great precision and Cashier J. E. Baker was particularly requested to take good care of the old leather case as it contained a fortune. J. W. Cummins, the only son of deceased, residing in San Francisco, was notified of his fathers death. He came to this city and was told by the attorney that his father had left a small fortune. A visit was made to the bank. Accompanying the son were Mr. Tappan, the attorney, Mr. Smiley, the coroners deputy, and Archie Porter a friend of the deceased. The valise was called for and taken into the office of the president of the bank for examination. There were four large Wells Fargo & Co. envelopes sealed, each endorsed "$5000 in currency." The envelopes were cut open in trembling expectant haste and found to contain — not currency, not gilt edge notes nor

United States bonds, but carefully cut and prepared brown tissue paper.

The silence of the little crowd gathered around the valise was profound. Everybody, everything, including the "valuable" effects of the deceased millionaire might have been bought upon the spot for a nickel.

John Cummins was a native of Virginia and seventy-two years of age. He had been a successful schemer and it is believed that at one time he had some money. His schemes were of a gigantic nature, and no enterprise was too great for him to undertake. He represented that he owned rich mines in Colorado, Utah and California. Among his papers was found a letter written from Los Angeles offering $30,000 for a certain business block on Broadway, Oakland. When search was made for the property it could not be found and from the description given it is buried in the depths of San Antonio estuary. The letter had evidently been written by John Cummins himself.

So it appears by the record that John Cummins was indeed a scoundrel and even his earlier dealings in Cummins City were probably a little shady, to say the least. However, Thompson's 1941 version contains a few references which are not borne out by existing records. "The whole boom pivoted on the Copper King . . ." is not supported by the many newspaper articles appearing at the time. Of the more than one hundred articles appearing in the region's newspapers at the time of the Cummins City boom, only one makes even a passing mention of copper ore as being of interest. The Copper King Mine was of a later mining boom, the one of 1896 through 1901. In fact, the "Copper King" was claimed by none other than Frank Smith, the source of Thompson's information on Cummins City.[6]

The water wheel, alluded to in Thompson's story, was also part of the later boom.[7]

Cummins' association with James I. "Doc" Thomas seems, in retrospect, to be less than intimate. While Cummins was coming and going from the camp almost constantly, Thomas was a model citizen, serving as precinct election judge, trustee for the Jelm Mountain Gold and Silver Mining and Milling Company and serving on the camp's school board. Thomas stayed on in Cummins City long after John Cummins left the camp and became involved in the Colorado mining schemes. While it is unlikely that Cummins stayed on in the big hotel, it is entirely possible that Thomas did, considering that he was an earlier partner in the hotel. (The hotel was dismantled around 1890 and the lumber used to build several cottages in Laramie).

However, Thompson was privy to information from his father, John Charles Thompson, Sr., Frank Smith and possibly other pioneers of the area and, there is probably more to the story than the printed record reveals. J. C. Thompson, Sr., himself, was involved with mining in the Medicine Bow Mountains and around the turn of the century, had dealings in the Jelm Mountain area.

Although there is no recorded evidence linking Frank Smith to John Cummins or the 1880 rush, records do show Smith was deeply involved in the later 1896-1901 activity. Frank Smith's "Copper King" claim was the beginning of the Jelm City copper commotion around the turn of the century. Smith's memory of events of 50 or 60 years earlier may have been correct but his placement of time seems in error. The use of John Cummins as a scapegoat for the failure for the five incorporated companies, the almost 100 mines and the more than 500 claims seems at this time a little ludicrous. Nevertheless, it is true that it was John Cummins who sold lots in the fledgling city to the miners, investors and hopeful business people. It is also true that "Honest Old John" bought several mining properties, fraudulently presented rich assays and then re-sold them usually for a large profit. Evidently, John Cummins was one of few to realize a profit on the many transactions. Lacking is any evidence that he sold stock in any of the companies, and although he did organize his own little companies, Cummins' name fails to appear as an officer or trustee of any of the incorporated companies.

Several of the Big Laramie miners went on to become prominent or at least well known citizens of the area. Herman V. S. Groesbeck, who had come to Cummins City as a young attorney looking for work, and later became State Supreme Court Justice, later said that the reason behind the "bust" was that the gold was found only in pockets.[8] No mention of fraudulent dealings. His viewpoint may have been a little biased, as Groesbeck himself was involved in some of the transactions and sold John Cummins several claims and land parcels, including the Excelsior Mill Site, which adjoined Cummins City on the south.

H. V. S. Groesbeck
Wyoming State Museum

Groesbeck's mining and business partner, George Hildreth, who at the time was Justice of the Peace and well known for his honesty and integrity was also involved in many of the transactions. He also sold John Cummins several claims and parcels.

Thompson's version contains many verifiable statements, however, the 1880 gold rush and the 1900 copper boom are interwoven into a story resulting, probably in an incorrect conclusion as to John Cummins' culpability.

The mining district was established long before John Cummins happened upon the scene, not as Thompson states, after Cummins arrival.

The population claim of 200 made by the various correspondents appears to be a reasonable estimate although Smith's estimate of 50 could have been a more accurate figure of "permanent" residents.

Cummins' association with "Doc" Thomas is not evident in any of the remaining records.

The old town of Cummins City was not abandoned overnight. Records exist showing a continued although somewhat diminished population for years after the boom and bust of the camp.

True, the potential of the camp was continually exaggerated, probably more by John Cummins than anyone. True, John Cummins was one of few to turn a profit on any of the property. Also true, Cummins was in later years more or less proven to be an unscrupulous entrepreneur and outright crook. His dealings in the North Park mines and then the Big Laramie Mines indicate he took every opportunity to turn a dollar, but to place the blame of failure of the mining camp on Cummins and Thomas seems at this late date to be at the least, inequitable.

In his speech before the Wyoming Industrial Convention of 1901, George W. Fox, himself a miner, made no mention of fraud or salted mines and instead, said that "a marvelous strike was made less than a month ago" thus perpetuating the belief of valuable mines.[9]

Apparently many years then passed before the concept that Cummins City, Jelm City and the surrounding mines were a fraud, was widely circulated.

C. C. Clark
Wyoming State Museum

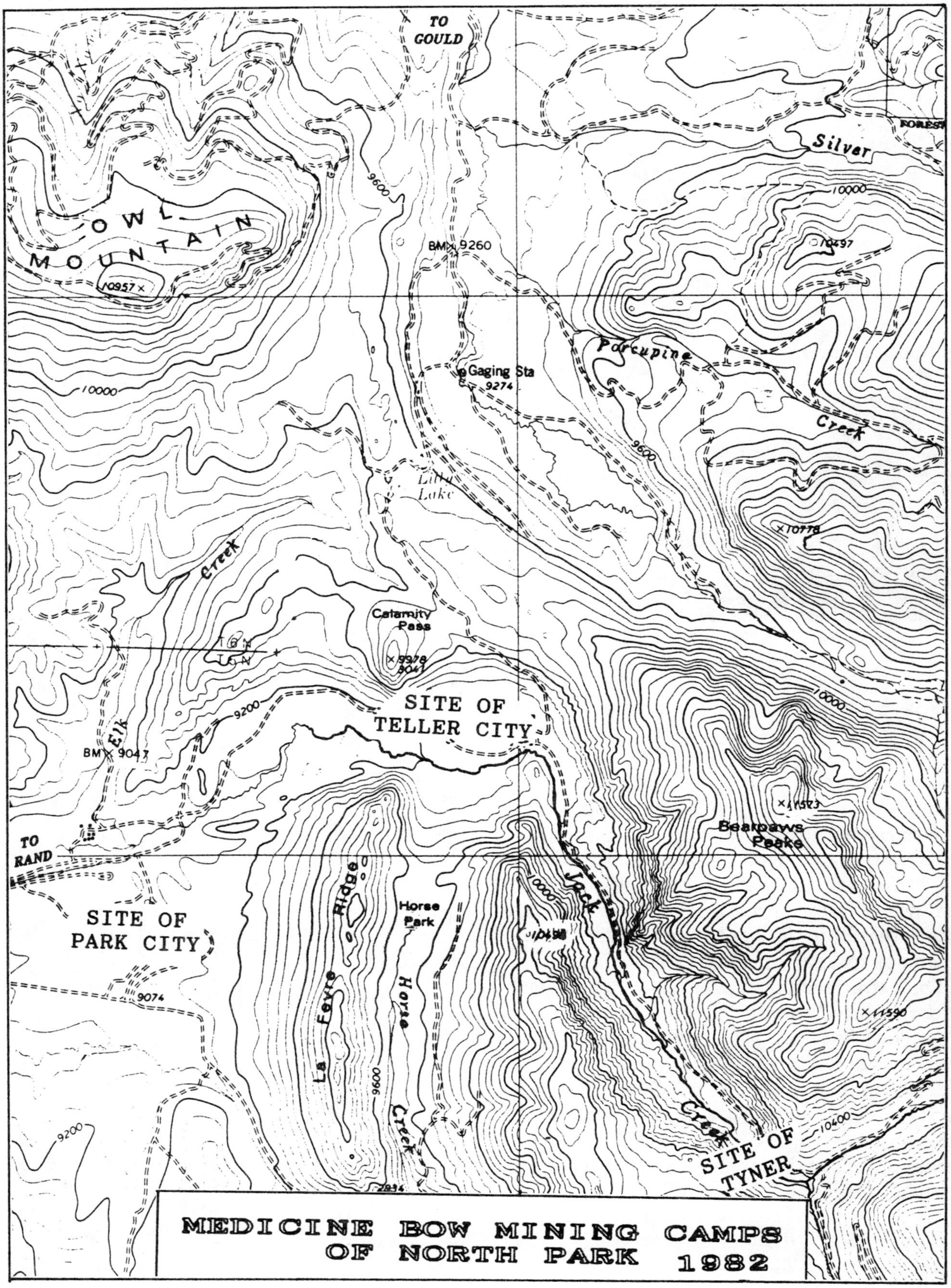

MEDICINE BOW MINING CAMPS OF NORTH PARK 1982

VI THE NORTH PARK AND TELLER MINES VI

• LOCALE

The Medicine Bow Mountains stretch from Elk Mountain on the North, across the present state line to the Continental divide on the south, forming the eastern boundary of North Park. Although located in what is now Colorado, the southern portion of these mountains was subject to prospecting in the early 1870s, mostly by prospectors from the Territory of Wyoming. Several mining camps were established by the end of the decade, the largest being Teller City. Colorado history has practically ignored Teller City and it's unsuccessful mines in favor of the more famous and successful, of which Colorado has an abundance. A visit to the Teller City townsite today, will disclose numerous cabins which have been reduced to barely discernible piles of logs, victims of more than a hundred high country winters. Only the remains of these crude log cabins are now evident, the sawn lumber having long since disappeared. No stone fireplaces or chimneys are visible, it was said at the time that most of the cabins had metal stoves. The forest has reclaimed the streets of Teller. Pine trees eight to ten inches in diameter now grow in the old streets and even from the center of some of the cabin ruins.

To reach the site of the North Park Mines from Wyoming, one travels along much of the same route as used years ago by the miners. From Laramie, Highway 230 leads southwest, across the Colorado border to Walden. From Walden, Highway 125 will take one to Rand and slightly beyond to Old Homestead. A turn to the north leads to a well graded road which branches to the Teller City road within a mile or so. Although not well known in Colorado, the road to Teller City is well marked and can be negotiated by a vehicle with good ground clearance. The site of old Teller City is marked and becomes quickly evident by the remains of cabins on both sides of the road. A walk through the area will quickly convince one that this was most probably the largest camp in the Medicine Bow Mountains.

• EARLY NORTH PARK

Throughout the years, what we now call North Park has variously been called Park Kyak, the Bull Pen, New Park and ultimately North Park. The first White men into the North Park area were trappers, probably around 1816. John C. Fremont passed through the park in 1844 describing it's beauty in glowing terms. The Cherokees traveled through a portion of the park in 1848 and Sir George Gore hunted there in 1855. F. V. Hayden visited the park in 1868 on a side trip from his survey of the West. It was said that gold was discovered by members of the Fremont Expedition of 1844 and the Gore Expedition of 1855.

Some of the earliest prospecting in this part of the Medicine Bow Mountains was carried on even before the Last Chance explorations of 1868. North Park and it's surrounding mountains were said to have been explored and prospected even before the Pikes Peak gold rush of 1859.

Early access to North Park and the western slopes of the southern reaches of the Medicine Bow Mountains was, at the time, considered easiest from the Oregon and Overland Trail system, up the North Platte River to North Park and then into the surrounding mountains. Most of the early mining activities were centered around Independence Mountain, primarily for placer gold. Another area, the Hahns Peak district, also drew prospectors through North Park, to its discoveries.

• LARAMIE TO NORTH PARK TRAIL

As prospecting in North Park gained ground, a new route from Laramie City came into use. This route became increasingly important as explorations for North Park's precious metals increased. Trails leading north from Middle Park were also in use at an early date but weather in the high mountain passes limited their use.

The Laramie City to North Park Trail generally followed the Big Laramie River into the valley south of Jelm Mountain and then

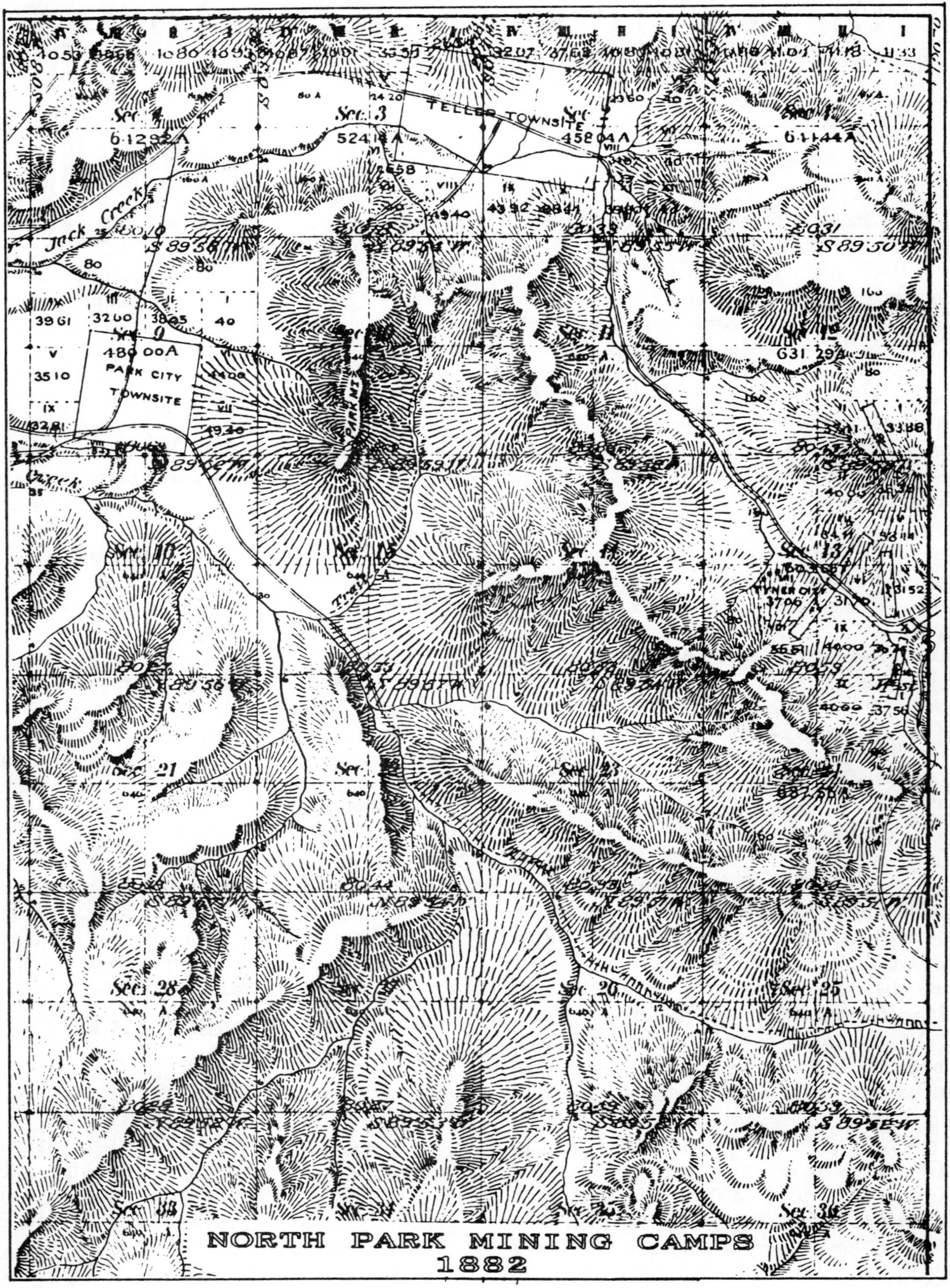
TELLER TOWNSITE
PARK CITY
TOWNSITE
Jack Creek
TYNER CITY
Sec. 3
Sec. 9
Sec. 10
Sec. 21
Sec. 23
Sec. 25
Sec. 28
NORTH PARK MINING CAMPS
1882

crossed the well timbered mountains into North Park, apparently following several different routes. Earliest reference to the trail are rather vague and the exact route is difficult if not impossible to trace. John Henry Mullison, who came to Wyoming in 1867, in later years wrote that in that year he followed John C. Fremont's tracks of 1844. He claimed he found Fremont's cart tracks, camp sites and even found names of his party carved on trees on the trail between the Big Laramie River and North Park.[1] Mullison who later lived in the North Platte River Valley, had explored much of North Park and the Medicine Bow Mountains and was considered one of the most reliable sources of information about the area.

In 1868, an English adventurer, F. Trench Townshend, traveled across the Laramie Plains up the Big Laramie River and crossed over into North Park. He reported a trail existed but said they were required to cut down trees and build corduroy roads across the bogs to get their wagons through. Even with these hardships, they made it across in fourteen hours.[2]

When F. V. Hayden visited North Park in 1868, he followed an existing trail which he said was made by the Cherokees, twenty years earlier.[3]

Acknowledgment that this was at one time a portion of the Cherokee Trail was wide spread. Young English adventurer Robert Mills wrote from Cummins City in November, 1881, "The old Cherokee Trail to California passes over the Big Laramie River here at Cummins and goes through one of these primitive forests to the North Park of Colorado." [4]

The survey team of 1873 traveling along the southern border of Wyoming Territory reported following a well marked "road" into North Park and beyond.[5]

The first settler in the park was probably James O. Pinkham. Pinkham was said to have come into the park with one of the government survey teams and stayed on, building a small ranch with a blockhouse and eventually establishing a stage stop and post office. Pinkham

Ox Team

Courtesy of Lynn Benson

was born in 1835 in Canada of English parents and claimed to be Irish-Canadian. From the time he first came into the park, Pinkham and others were panning for gold in the streams of the area. As early as 1875, there were reported to be as many as 100 prospectors in the North Park area. The Pinkhampton post office was established in 1879 and Pinkham became the first postmaster.[6]

As mining traffic increased, the Wyoming business men became increasingly concerned about the road and frequently supported action to improve the route. Laramie City became involved with the promotion of the route and on July 13, 1876, the Albany County Commissioners directed a survey and work party to locate and construct a road from Laramie City through North Park to Hahns Peak. George W. Fox, Mortimer N. Grant and Owen D. Downey were chosen to head up the small party who took with them two teams of mules and six laborers. In early October the group reported the road was complete, 112 1/2 miles, to Bug Town, the leading settlement at Hahns Peak.[7]

George Bird Grinnell traveled a portion of this route in 1879 and reported a good wagon road led them over the spur of mountains into the "Neck of the park" and Pinkhampton. Grinnell reported only two "permanent" settlements in the park, Pinkhams and a "ranche" close by. The only other "resident" they found in the park was a trapper named Frank Kosier. However, Grinnell noted that there was considerable placer mining on Owl Creek, ". . . dignified by the pretentious title of the Golden Gulch" and being mined by the Golden Gulch Mining Company. After about two weeks in the park, Grinnell returned to Laramie City by the same road, stopping again at Pinkhampton.[8]

Several deviations of the route were used between 1876 and 1880. The preferred route, however, crossed the Big Laramie near the Colorado Line (known then as Corlett and later as the Boswell Ranch) and traveled up the ridge parallel to Beaver Creek to what was then known as "Big Beaver Dam Creek" (near present Boswell Creek Campground). From there the road turned south-southwest to "Shipmans cabin" before swinging westward another four and one-half miles to "Pinkhams Ranche."[9]

At one time, in an attempt to shorten the distance to the mines in the southeast corner of the park, a stage stop was built in Shipman Park and Ute Pass was utilized for a short time, bypassing Pinkhampton. However, the route was impassable during the winter, steep and rough in the summer and was abandoned in favor of the Beaver Creek Road, a portion of which is still in use today.[10]

During the 1880s, the Keystone Toll Road Company laid out another road, which roughly paralleled Wood's Creek to the summit, a portion of which was used for several years.[11]

It was some years later that the route utilized the Wood's Creek Road, (the present route of Wyoming Highway 230), traveling up the canyon from Woods Landing, passing the Mountain Home Ranch and then into North Park, still by way of Pinkhampton.

In May, 1879, the Cache la Poudre and North Park Toll Road Company of Colorado was incorporated to build a road from Fort Collins to Teller. The road was completed as far as Chambers Lake in 1880 and into Teller in 1881.[12]

• THE MINES

Some mining and prospecting was carried on in the park during the early 1870s. Among the early prospectors were George Marquette, Cuthbert T. Waldron, W. H. Sebastion and J. H. Mullison. Mullison came close to being involved in the Shipman Cabin Massacre but had left the area after being warned by the Utes. Later he spent time at Meeker, again leaving just in time to miss the massacre. His own account states that he was responsible for the naming of Independence Mountain.[13]

The discovery of what was thought to be good indications of gold and silver in the southeast end of the park during 1879 was the precipitating factor in starting the 1880 gold and silver rush. The first recorded claim in the area, that was soon to become known as the Lead Mountain Mining District, was made on June 9, 1879, by Alonzo Rogers near the headwaters of Illinois Creek.[14] Shortly thereafter, on June 17, Henry Frisins staked out his "Fairview Lode" on the headwaters of Jack Creek.[15] In the same area three Cheyenne men (J.S. Taylor, J.E. Tuttle and Russell

Thorp, the well known Wyoming stage line proprietor) joined with John Cummins and claimed the "Cheyenne Lode" on June 26.[16] This same group also claimed five acres for a mill site.

By the first of July, numerous claims were being made up and down the slopes of what the prospectors called Lucky Mountain. Several Cheyenne businessmen, who had been involved with the Black Hills gold rush of a few years before, were actively staking out claims as fast as they could with little regard to the mining law which required the minimum of a ten foot discovery hole. A good portion of these prospectors were from Laramie and had recently been prospecting the northern reaches of the Medicine Bow Mountains. Joining them were miners and prospectors, most recently from Fort Collins, and also a large group from Illinois. Another group from the Salt Lake area was also among the prospectors.

John LaFevre, who had been prospecting in the Big Laramie River valley, made claim to the "Gaslight" lode on July 14.[17] LaFevre was using a cabin built some years earlier by gold prospectors and discovered a ledge of "Ruby Silver" ore, apparently ignored by the previous occupants, directly in front of the cabin.[18] The "Gaslight" soon became the outstanding mine in area and was at least partially responsible for the following season's rush.

Mons C. Jahren of Laramie claimed the "Alpha Lode" August 5, 1879.[19] Jahren was a candidate for the Wyoming Territorial House of Representatives in the September 1, 1879 election. He was successful and served in the Sixth Legislative Assembly which met in Cheyenne November 4, 1879. Jahren was to return the following season and became quite well known in the Colorado camps.

W. S. "Buck" Bramel, who had also been mining on the Big Laramie River, along with John Gibbons, made claim to the "Mountain Boy" on Jack Creek August 26, 1879.[20]

John Cummins, who would later become known for his dealings at his namesake Cummins City, was actively claiming and selling mines in North Park the year before his escapades in Cummins City. In September, 1879, Cummins sold one of his first claims to Stephen Farwell and A. Underwood of Cheyenne. On October 28, he sold Louis Zimmerman one half of the "Daisey" and in November, he sold the other one half of the "Daisey" and one half of the "Mary" and "Wallace" claims to Doctor E. B. Graham of Cheyenne.[21] During the following few months, Cummins divested himself of most of his claims and would in the future direct his efforts toward the mines across the border in Wyoming Territory.

The first mining camp, called Tyner, was established on the upper limits of Jack Creek and a post office was established with John LaFevre as postmaster. Although many claims were made during the '79 mining season, due at least partially to the Meeker Massacre, (September, 1879), few of the miners chose to remain for the winter.

• 1880

Even before the beginning of the 1880 mining season, the Laramie City, Cheyenne and Fort Collins newspapers began boosting the North Park mines. Each city was promoting their own route to the mines and each lauded the rich discoveries and expounded the theory that the North Park mines were lying on the same great mineral belt as Leadville.

With the memory of the Meeker Massacre less than a year old, fresh Indian scares during the spring of 1880 were repeated in the area newspapers and had, perhaps, the desired effect of directing some miners to the Big Laramie mines near the future site of Cummins City and a little closer to civilization. John Cummins was apparently the source of at least a portion of these scares. Two of the North Park miners, George Carr and Prentice Dow, were reported by William Burlingame as being killed by the Indians and this too was dutifully reported by several newspapers.[22] Troops, under the command of Colonel Carlson, were sent into the park to protect the miners but found no evidence of the deaths. Burlingame was subsequently accused of the murders. Two weeks later, however, Carr and Dow were reported alive and well and mining in North Park.[23] No actual Indian depredations were verified and the Indian scare faded away.

With the promise of rich mines, the repudiation of the Indian scare and the seeming arrival of spring, many miners started for the North Park mines as early as March. Spring weather on the plains comes months before good weather in the high country and several groups were soon bogged down in the soft snows of spring. One Cheyenne party, led by Solon Burgess, was twelve days struggling to make it to the camp of Madore Cushman on the Michigan River, which was still several miles from even the lower discoveries. The group ran into deep, soft snow along the east side of Owl Creek and were forced to abandon their wagons, build sleds and make "snowshoes" to get into the lower camp of Cushman. By the time they arrived on the Michigan, most of the group suffered from snow blindness.[24]

Mons C. Jahren, who had been in North Park the previous season and staked out his own claim, the "Alpha Lode" in August, was planning a return trip to the North Park mines before the snow was gone. He had received a pair of "Norwegian snowshoes" (now known as Nordic skis) from the old country which much impressed his associates in Laramie City. The Laramie paper said of them, "Hon. M.C. Jahren has received a handsome pair of snowshoes, similar to those so much used on the Sierra Nevada, on which he will sail into North Park shortly. They are stunners." [25]

The road to North Park, although much traveled, proved to be treacherous especially in the spring time. During April, 1880, three prospectors lost their way in a snowstorm and were frozen to death.[26]

Several new camps were established during the spring of 1880, with no indication as to which one would become more predominant. On April 18, the Postmaster General ordered the establishment of a post office at Crescent City with W. J. Sturgis of Cheyenne designated as postmaster.[27]

During June, 1880, the Colorado toll road of S. B. Stewart, was pushed beyond Cameron Pass to the forks of the Michigan River. Here, Stewart and Joseph Mason claimed 640 acres for a townsite and H. P. Handy and A. M. Crafts began laying out the town which they christened "Mason City."[28] By July, there were said to be 75 people but only three buildings in the camp, one of which was a hotel and restaurant owned by a Mr. Grant. For a short time it appeared that Mason City would become the preeminent town among the North Park mining camps and the first stage road from Pinkhampton ran up the Michigan River to the aspiring camp. Scarcely a month later, however, on August 12, 1880, the *Larimer County Express* reported that Mason City consisted of three acres of surveying stakes and four unfinished cabins, but not a sign of human life. (Joseph Mason, for whom the camp was named, died in Fort Collins February 17, 1881. Mason was a pioneer of the west and had been a member of F. V. Hayden's Yellowstone expedition.)

In June, 1880, the *Cheyenne Daily Leader* reported the new mining region located in the southeastern corner of the park along Illinois Creek and Jack Creek was now known as the Lead Mountain Mining District. They also reported:

> The district was opened during the past year, and many claims were made; but owing to the Cumming's [sic] Indian scare, and the severity of the winter, but little work was done. . . . Two towns have been built up, one on Illinois Creek called Park City and the other Jack City or Teller City on Jack Creek. The former containing over a hundred houses nearly finished, the latter nearly fifty buildings completed.[29]

I. S. Bartlett, in his *History of Wyoming*, told of a trip into North Park with his wife and children during Teller's building boom:

> Incidental to this trip we visited Teller City, a new mining camp where great gold discoveries had just been reported. There was a great rush there. A town had been laid out in the timber and many houses, shacks and cabins were being built. A new hotel, roughly constructed of pine boards, was being built and I applied for a room. The proprietor said all the completed rooms were taken, but explained that the carpenters would have another room enclosed by night if we wanted to take it. We took it and the room was built round us during the day. The windows and door were put in and the boarding of the walls completed while we occupied the room. It was the first time I had ever seen a room 'built around you while you wait.' That night we had a grand reception. The mountains to the east were lit up by a great forest fire making a scene almost terrific in grandeur.
>
> Another thing - in laying out the streets a great many pine trees had to be cut down. Therefore the streets and roadways were full of stumps and it required the utmost skill in a driver to get through without smashing a wagon or breaking the horse's legs.[30]

On June 26, 1880, the *Laramie Sentinel*

printed a letter from Jack City signed by " 'F' Sharp":

> Jack City has a population of about two hundred; some three or four having brought their wives and children, have established themselves in good substantial cabins, with the evident intention of staying right here. . . . Of affairs at Park City, which is some three miles from this place, your correspondent can say but little, as he has had no opportunity of visiting that place; but I learn there are about 125 men there at present, also, that the name of the town is to be changed from Park City to Aurora, as there is already a Park City in the South Park.

Apparently the town was abandoned before the name change could occur.

> At an election held here [Teller City] Saturday evening, J. G. Mills, of Cheyenne was elected city recorder. The number of trustees was increased from three to five, and two gentlemen whose names I could not learn were elected to fill the vacancy.

When the 1880 census was taken on June 28 & 29, enumerator Jules Harmon counted 177 people in North Park, which at the time he considered to be in Grand County.

On July 31, 1880, the *Laramie Sentinel* printed a letter from Jack City signed by M. C. Jahren. He remarked about the large number of cabins not yet completed and of the sawmill of Picket, Riffle and Towner which he said was already doing a good business. He had traveled up Jack Creek gulch and found:

> . . . there are cabins and tents scattered all along from the town to the head of Jack Creek. . . . As near as I can count, there are about 100 miners in Jack Creek gulch alone. One thing lacking and that is a stamp mill, or works to get the precious metals out of the ore, which is now either piled up on the dump or sacked away and stored in cabins.

By late summer and early fall of 1880, Jack City or Teller City as it was becoming known, was rapidly becoming the leading camp of the region. A post office was established July 19, 1880, with Frank Yates as postmaster. Aside from Teller City, Park City and Mason City, several other camps were established about the same time. Tyner was located above Teller and was located closest to the mines on upper Jack Creek. Crescent City was said to be located nearby on Illinois Creek but the exact location is not apparent. Also nearby, located across the divide and not in the Medicine Bow Mountains, were the mining camps of Lulu, Dutchtown and Gaskill.

By the end of the summer, Park City, although well laid out, was almost deserted, many of the residents having moved to Teller City and closer to the mines. Teller City, named after Colorado Senator Henry Teller, quickly became the major mining camp of the district and the residents were soon electing their own officials.

The September 1, 1880, issue of the *Cheyenne Daily Leader* reported the sale of the Gaslight Lode by John LaFevre to the Charles E. Adams Company of Chicago for the consideration of $20,000. The same issue reported an assay on the Sulpheret mine ore which ran 225 ounces to the ton. The "Sulpheret," they went on to say, was the mine that John Cummins sold to Thorp, Tuttle, Pease and Taylor. Also, the paper printed the first note of controversy on whether the North Park mining camps were located in Grand County or in Larimer County.

On September 23, 1880, the *Laramie Sentinel* reported the forming of the North Park and Vandalia Mining and Smelting Company. Dr. G. W. Bassett of Vandalia, Illinois, was named president; Sam Moffett, vice president; Thomas Hunter, secretary; Frank Yates, treasurer; B. W. Towner, general superintendent; John LaFevre, superintendent of mines; and also a board of seven directors. The company reported that they owned twenty three mines and had a capital stock of $802,000.

The first marriage in the mining camp of Teller occurred September 25, when Thomas Downing married Mary Riffle. Ben Towner and his wife served as best man and lady.[31]

During the peak of the mining season, 300 to 400 men were reported in the area but by October, the end of good weather, there were only 150 to 200 men who planned to spend the winter. When the election of 1880 was held on November 11, there were only two precincts in North Park; Pinkhampton which reported 10 votes and Teller which reported 17 votes.[32] The debate on whether Teller was located in Grand County or Larimer County continued unresolved but the votes were recorded in Larimer County.

As winter settled into the high mountains, John Lytle of Teller reported the accidental death of two of the miners. Five miners had left Georgetown to work on the Gaslight Lode near Teller. While crossing the Continental

Divide, the group started a snow slide that carried the entire party down the mountain. C. H. Eaton and Thomas Grey were killed. Thomas Frazer had his thighbone fractured in two places. Nelson and Sandler escaped with minor injuries. At the time, the body of Grey could not be found.[33]

By the end of 1880, Teller, the principal mining camp, was reporting a population of 150. In addition, the smaller camps of the area were still holding a few miners. Most of the mines remained unworked during the winter. The deep snow had come early and unexpectedly, consequently the shaft houses were not prepared and work was abandoned for the winter.

• 1881

By March, 1881, the miners began returning to Teller and 300 people were reported in the district. Teller, it was said contained three hundred cabins. Reports were of a "lavish" hotel and many businesses. Twenty-seven saloons and numerous parlor houses were also said to be located in the mining town. Some of these reports may have been exaggerated, but the fact remains this was a thriving mining camp by 1881.

On April 13, 1881, Yates turned over his postmaster duties to Eugene Tompkins and devoted full time to his hotel business.

The *Laramie Times,* May 2, 1881, said:

> On Saturday evening the citizens of Laramie concluded their final arrangements with Messrs. Patrick Bros. for putting on a tri-weekly line of coaches from Laramie to the North Park via Cummins City. Under the arrangements made, the Patrick Bros. within the next twenty days, are to put on a tri-weekly line of Concord coaches and run the same for at least a year. Mr. A. S. Patrick informed the writer that he intended to have the line in running order within ten days, as he is desirous of having the route well established and advertised as quick as possible.

The camp had a crude road into Middle Park connecting them to Georgetown and later a toll road from Fort Collins, but throughout its short life the primary route to the outside world was the stage road to Laramie City. Through the boom of Teller, the stage line continued to improve both their route and their equipment. The scheduled time from Laramie to Teller was about fifteen hours, without any stops except for meals and to change horses. Fare from Laramie City was eleven dollars.

On May 17, 1881, an election was held in Teller to elect the town officials. Polls were at J. G. Mills assay office and commissioners of election were John LaFevre, E. Tompkins, E. M. Wood, Edward Gleason and Samuel Moffitt.

Shortly after the election, the new officials of Teller City assessed each male resident between the ages of 21 and 45, three dollars in cash or two days of work on the roads.[34] (Miners wages at the time were $3.00 a day).

By 1881, Teller had advanced toward civilization enough to warrant a school taught by Miss Clark.[35]

There was, no doubt, work for a good lawyer and John Mills was considered the most prominent in the camp.

According to the *Grand County Times* estimate, there were about 650 persons in the Teller district in May, 1881. This may well have been, although exaggeration of populations was common among the mining camps. Study of the existing records indicate that there were indeed quite a large number of miners in the area. In later years, various accounts of the camp have indicated a peak population of anywhere between 300 to 2,000 residents. Although not verifiable, the lower numbers appear to be most correct, however, still leaving Teller as the largest mining camp of the Medicine Bow Mountains.

The establishment of the newspapers is somewhat of an enigma, due at least partially to the universal habit of 19th century reporters referring to any and all newspapers as the "Times." The *Grand County Times* was indeed printed in Teller, it's first edition appearing May 12, 1881, with W. H. Clark as editor. Another paper, *The North Park Miner* was also printed in Teller and several editions exist. However, the "Teller Times" occasionally referenced, does not appear to have been printed in the camp.

By summer, 1881, Frank Yates had his hotel in full operation. The 40 by 70 foot two story building was home to about forty miners. Boasting a dining room with a Grand piano, it was the predominant hotel of the camp. Just opposite the hotel, Ben Towner ran a livery stable and nearby was J. G. Springer's furniture store. The *Colorado Business Direc-*

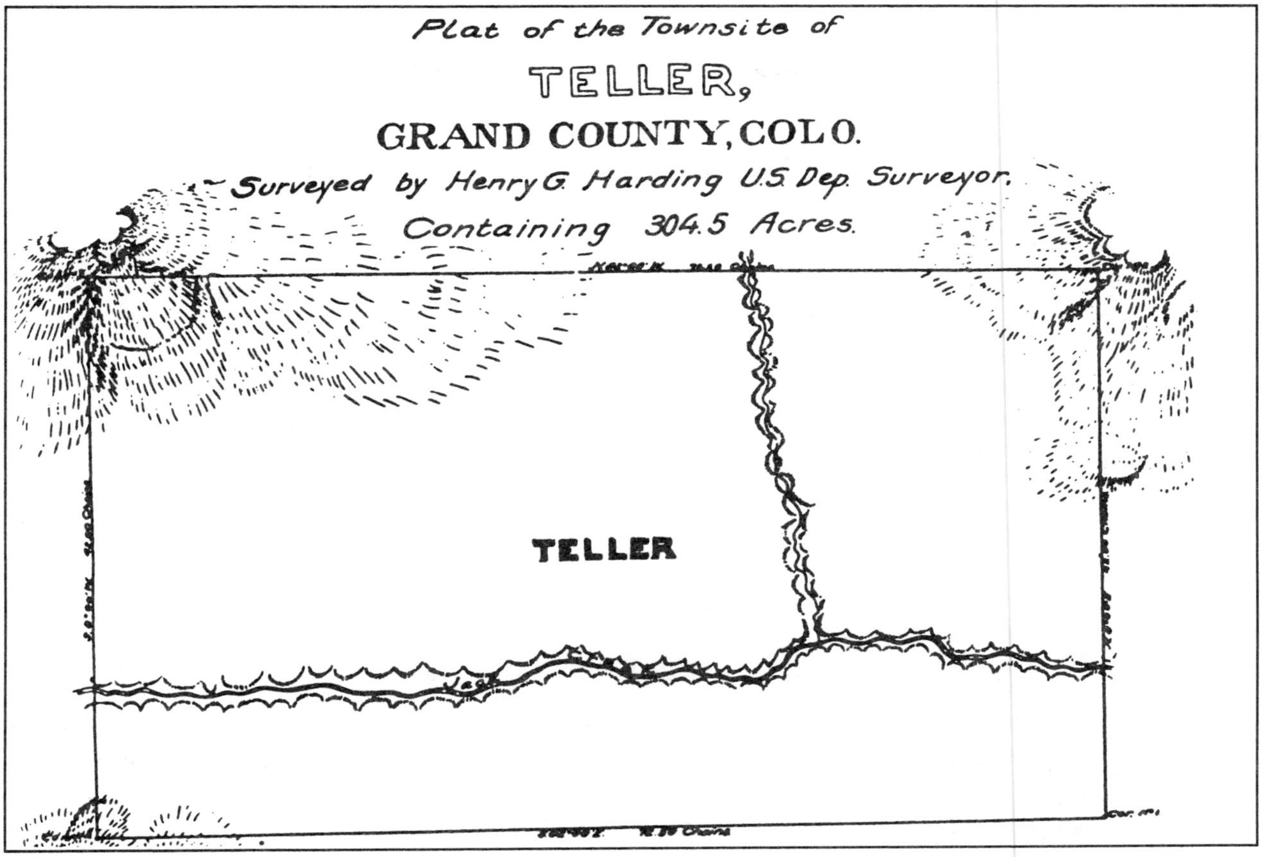

tory, 1881, listed Teller's more prominent establishments as: L. Browner, grocery and drug store; Cohen and Company, General Merchandise (Cohen's general store also served as post office from 1881 - 1883); W. H. Redfield, grocery and liquor store and A. Ringolsky, a saloon.

Up Jack Creek canyon, the "Gaslight," owned by the Chicago and North Park Mining Company, continued to be the most developed mine and had a dozen men working under the supervision of Mr. Adams. Adams operated a store for the local miners and was also the Tyner postmaster. Other notable mines on Jack Creek were the "Yellow Jacket" owned by J. W. Lytle, the "Pennsylvania," and the "Endomile." The Endomile was discovered by Alex LaFevre and John Moore in 1879. By 1880, the Endomile had become the most developed mine of the district under superintendent Charles Smith. The mine had a large cabin warmed by two stoves, outfitted with bunks and stocked with groceries. In 1881, the Endomile was jointly owned by Otto Gramm, Charles Smith, G. P. Atchison, M. Carroll, Louis Miller and Phillip Mandell, all of Laramie.[36]

• 1882

When surveyed in 1882, the surveyor, Henry G. Harding, wrote in his field notes a general description of Teller:

> Teller is situated about 2 1/2 miles from Park City in a direct line. An almost inaccessible mountain intervenes. Teller bears about northeast from Park City. The site is comparatively level. There are about 100 log cabins and about 200 inhabitants. Jack Creek, a small stream, runs through the south portion.[37]

As with most burgeoning camps, the population of Teller was the subject of a wide variety of estimates, however, the surveyors field notes are probably the most valid.

The winter of 1882-1883 proved not as severe as the two previous years and Sam Hanworth carried mail to Teller all winter long, although sometimes on "snowshoes."

• 1883

On June 9, 1883, articles of incorporation were drawn up for the Rarus and Maude S. Mining Company. Incorporators were Isaac Burton, B. W. Towner, Monroe Wyeth, Wiley Morrill and M. C. Jahren.[38]

The weekly newspaper, *The North Park*

Benjamin W. Towner
Photo courtesy of Joe Barros

Miner, September 6, 1883, listed the Teller City Officers: Mayor, M. C. Jahren; Recorder, T. L. McKee; Trustees, W. D. Beckwith, B. W. Towner, J. E. Shepard and J. G. Springer; and Marshal, Max James. Teller justices were J. E. Shepard and F. D. Gove. The newspaper was still running advertisements for the local businessmen: C. F. Shelton had an assay office on Fifth Street; S. Bon, "the pioneer boot and shoe man of Wyoming" was apparently doing business in Teller; W. H. Germain ran a meat shop and Yates and Germain were the hotel proprietors.

Even though many of the mines continued to work into the fall of 1883, there was still no heavy machinery in the camp, not only because of the obvious transportation problem, (miners had many times over, proven they could transport huge machinery into almost impossible areas) but also because the ore was not rich enough to warrant the expense.[39]

• 1884

By January, 1884, the *North Park Miner* had ceased publication, although it promised to resume in the spring under new management. The mines were being declared as "first class" but the majority of printed material appearing in the Fort Collins and Laramie newspapers about the park, were of ranching and the great summer range. Teller City was beginning a downward slide. The post office, however, continued to operate. John G. Springer became postmaster in January, 1883, Robert Schwartz in May, 1884, and Madore Cushman in November, 1884. The post office was discontinued on December 16, 1885, signaling the last of Teller City.

Teller joins the list of unsuccessful mining camps of the Medicine Bow Mountains. Although the prospectors found encouraging evidence of good silver ore, just as in the northern Medicine Bow Mountains, no paying mines were developed. Problems were compounded by a lack of machinery and economical transportation to the "outside."

• LATER

By 1885, Teller was completely abandoned. After the boom and bust of Teller, ex-mayor Mons Jahren returned to Laramie where he served as Laramie City Attorney in 1888 and 1889. He also ran for Probate Judge of Albany County and served as Clerk of the District Court in 1892. He died in Laramie on January 3, 1902.

Madore Cushman, Teller's last postmaster and one of the last residents, stayed on, claiming most of the townsite as his own.

Towner moved back to Wyoming and settled briefly near Keystone before carrying his search for gold to the remainder of the Medicine Bow Mining Camps.

Several of the Jack Creek Miners gave up their mining ventures but stayed in North Park to establish ranches and settlements in the area. By 1931, only four of the old miners remained in North Park; Pete Munro of Rand, William Kerr and Sam McIntosh of Walden and Dave Hendrickson of Cowdry.[40]

In April, 1886, the property of the North Park and Vandalia Mining and Smelting Company was sold at public auction in Denver to satisfy a mortgage debt of $62,978. The sale included not only the Endomile mine and millsite but also all 25 of the company's mines near Teller.[41]

OWL MOUNTAIN PLACERS.

The Docker Owl Mountain Placer Mining company is the owner of 3,680 acres of placer ground on the famous Owl mountain, in North park, Lorimer county, Colorado, together with sufficient water right to work the same. The ground is located on the slope of the mountain, giving an excellent dump. Water can be conveyed on and above the ground in an open ditch in five miles. This is the most promising placer mining field in the State. The above tract is for sale. Full particulars furnished on application to—

DWIGHT P. SMITH,
Secretary.

DIRECTORS:

GEORGE W. FOX, President. R. J. GARDINIER, Vice President.
D. P. SMITH, Secretary and Treasurer.
A. C. JONES. W. E. HISKEY. U. S. HARRIS. H. O. DOCKER.

Steam powered windlass on Endomile claim, 1990.

Teller cabin sites, 1990.

Blacksmith's cabin on Endomile claim, 1990.

Teller cabin, 1990.

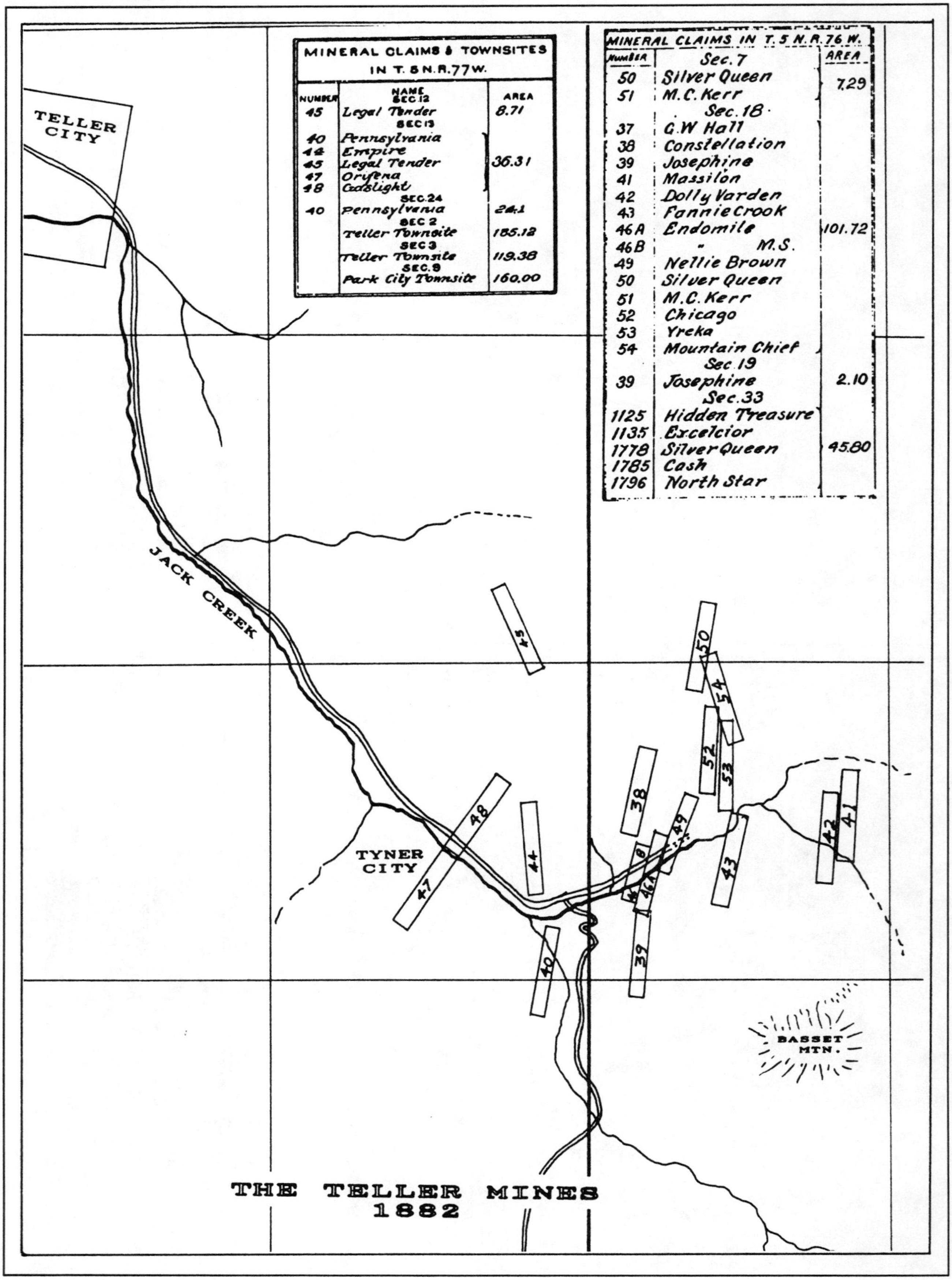
MINERAL CLAIMS & TOWNSITES IN T. 5 N. R. 77 W.
NUMBER NAME AREA
SEC. 12
45 Legal Tender 8.71
SEC. 13
40 Pennsylvania
44 Empire
45 Legal Tender
47 Orifena
48 Gaslight
36.31
SEC. 24
40 Pennsylvania 24.1
SEC. 2
Teller Townsite 185.12
SEC. 3
Teller Townsite 119.38
SEC. 9
Park City Townsite 160.00
MINERAL CLAIMS IN T. 5 N. R. 76 W.
NUMBER AREA
Sec. 7
50 Silver Queen
51 M.C. Kerr
7.29
Sec. 18
37 G.W Hall
38 Constellation
39 Josephine
41 Massilon
42 Dolly Varden
43 Fannie Crook
46A Endomile
46B " M.S.
49 Nellie Brown
50 Silver Queen
51 M.C. Kerr
52 Chicago
53 Yreka
54 Mountain Chief
101.72
Sec. 19
39 Josephine 2.10
Sec. 33
1125 Hidden Treasure
1135 Excelcior
1778 Silver Queen
1785 Cash
1796 North Star
45.80
TELLER CITY
JACK CREEK
TYNER CITY
BASSET MTN.
THE TELLER MINES 1882

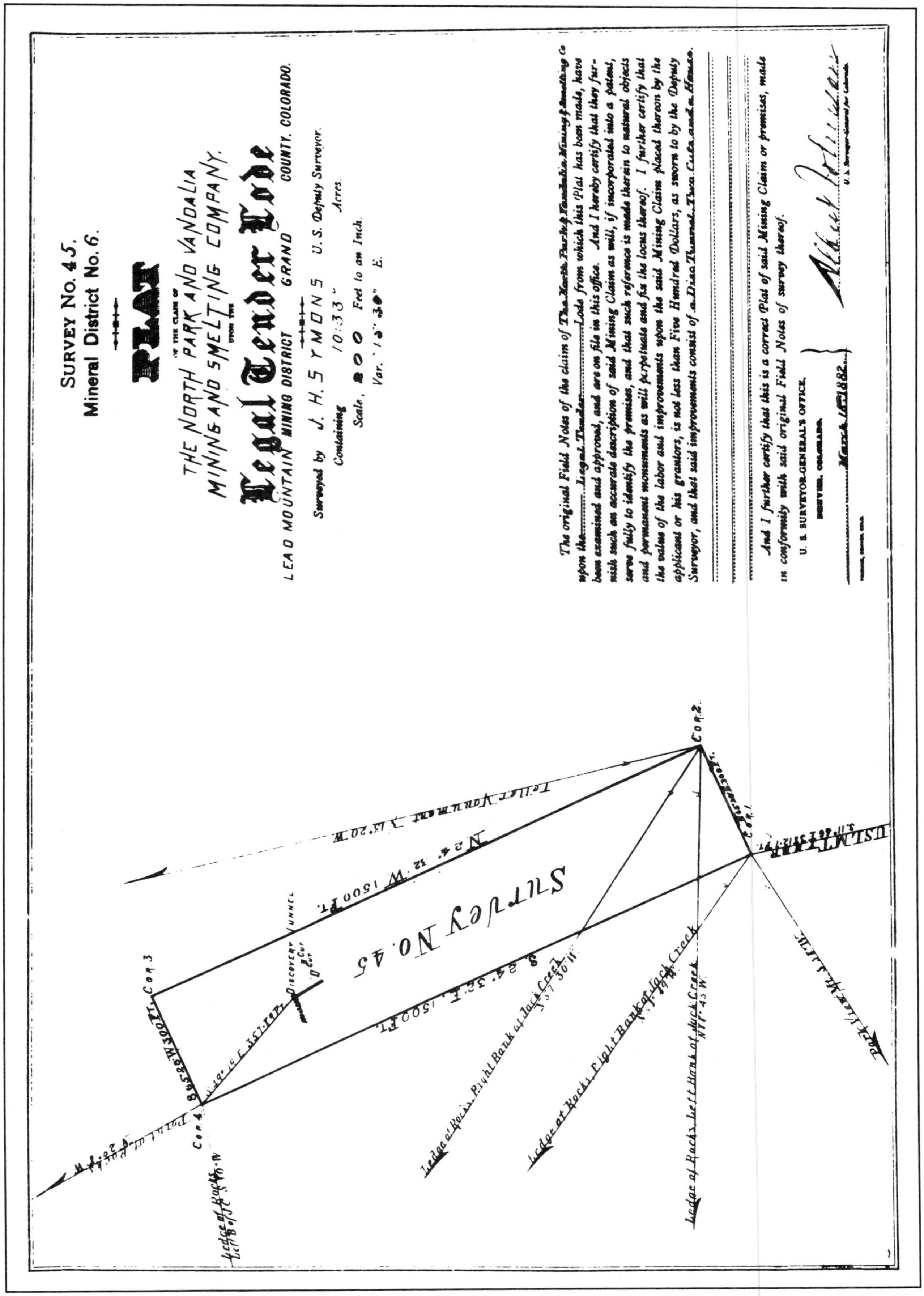
SURVEY No. 45.
Mineral District No. 6.
PLAT
OF THE CLAIM OF
THE NORTH PARK AND VANDALIA
MINING AND SMELTING COMPANY.
UPON THE
Legal Tender Lode
LEAD MOUNTAIN MINING DISTRICT GRAND COUNTY. COLORADO.
Surveyed by J. H. SYMONS U.S. Deputy Surveyor.
Containing 10.33 Acres.
Scale, 200 Feet to an Inch.
Var. 15° 30" E.
The original Field Notes of the claim of The North Park & Vandalia Mining & Smelting Co upon the Legal Tender Lode from which this Plat has been made, have been examined and approved, and are on file in this office. And I hereby certify that they furnish such an accurate description of said Mining Claim as will, if incorporated into a patent, serve fully to identify the premises, and that such reference is made therein to natural objects and permanent monuments as will perpetuate and fix the locus thereof. I further certify that the value of the labor and improvements upon the said Mining Claim placed thereon by the applicant or his grantors, is not less than Five Hundred Dollars, as sworn to by the Deputy Surveyor, and that said improvements consist of a Disc. Tunnel, Two Cuts and a House.
And I further certify that this is a correct Plat of said Mining Claim or premises, made in conformity with said original Field Notes of survey thereof.
U. S. SURVEYOR-GENERAL'S OFFICE,
DENVER, COLORADO,
March 1882.
U. S. Surveyor-General for Colorado.
Survey No. 45
Cor. 1
Cor. 2
Cor. 3
Cor. 4
N 24° 32' W 1500 Ft.
S 24° 32' E 1500 Ft.
Teller Monument
Tunnel
Discovery
Cut
Ledge of Rocks Right Bank of Jack Creek
Ledge of Rocks Left Bank of Jack Creek
Ledge of Rocks

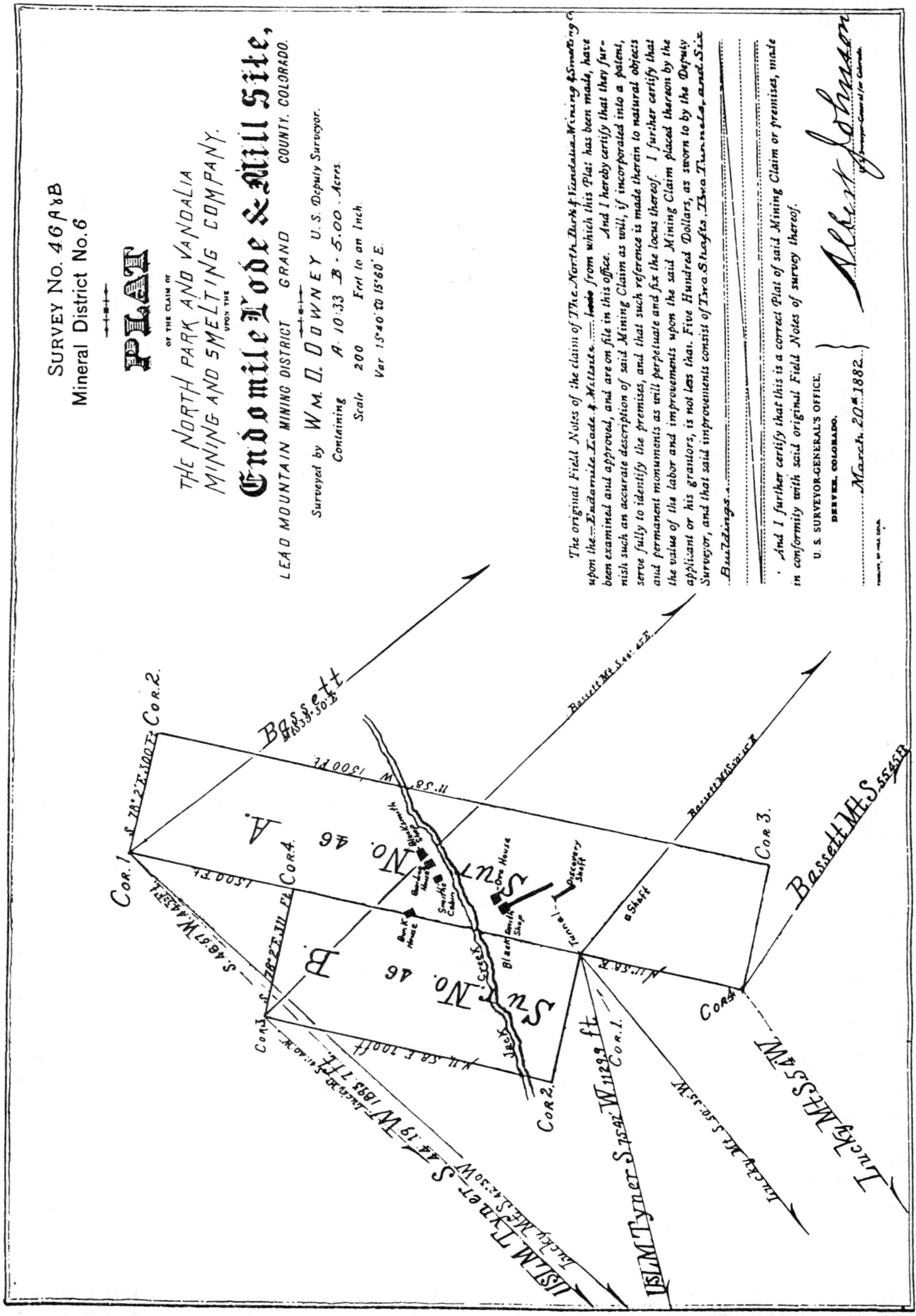
SURVEY No. 46 A & B
Mineral District No. 6
PLAT
OF THE CLAIM OF
THE NORTH PARK AND VANDALIA
MINING AND SMELTING COMPANY.
UPON THE
Endomile Lode & Mill Site,
LEAD MOUNTAIN MINING DISTRICT GRAND COUNTY. COLORADO.
Surveyed by Wm. D. Downey U.S. Deputy Surveyor.
Containing A. 10.33 B. 5.00 Acres.
Scale 200 Feet to an Inch.
Var 15°40' to 15°60' E.
The original Field Notes of the claim of The North Park & Vandalia Mining & Smelting Co upon the Endomile Lode & Millsite from which this Plat has been made, have been examined and approved, and are on file in this office. And I hereby certify that they furnish such an accurate description of said Mining Claim as will, if incorporated into a patent, serve fully to identify the premises, and that such reference is made therein to natural objects and permanent monuments as will perpetuate and fix the locus thereof. I further certify that the value of the labor and improvements upon the said Mining Claim placed thereon by the applicant or his grantors, is not less than Five Hundred Dollars, as sworn to by the Deputy Surveyor, and that said improvements consist of Two Shafts, Two Tunnels, and Six Buildings.
And I further certify that this is a correct Plat of said Mining Claim or premises, made in conformity with said original Field Notes of survey thereof.
U. S. SURVEYOR-GENERAL'S OFFICE,
DENVER, COLORADO.
March 20th 1882.
Albert Johnson
U.S. Surveyor-General for Colorado.
Cor. 1
Cor. 2.
Cor. 3.
Cor. 4
S. 78° 2' E. 300 Ft.
S. 78° 2' E. 311 Ft.
1500 Ft.
N. 58° W. 1500 Ft.
N. 11. 58 E. 1001 ft
S. 46° 67' W. 443 Ft.
A.
B.
Sur. No. 46
Bassett
Bassett Mt S 48° 45' E.
Bassett Mt S 55° 45' E
Lucky Mt S 54 W.
Lucky Mt S 50° 35' W
Lucky Mt S 42° 50' W
U.S.L.M. Tyner S 44° 19' W 1893.7 ft.
U.S.L.M. Tyner S 75° 42' W 1129.9 ft.
Jack Creek
Bunk House
Boarding House
Blacksmith Shop
Smiths Cabin
Ore House
Tunnel
Discovery Shaft
Shaft

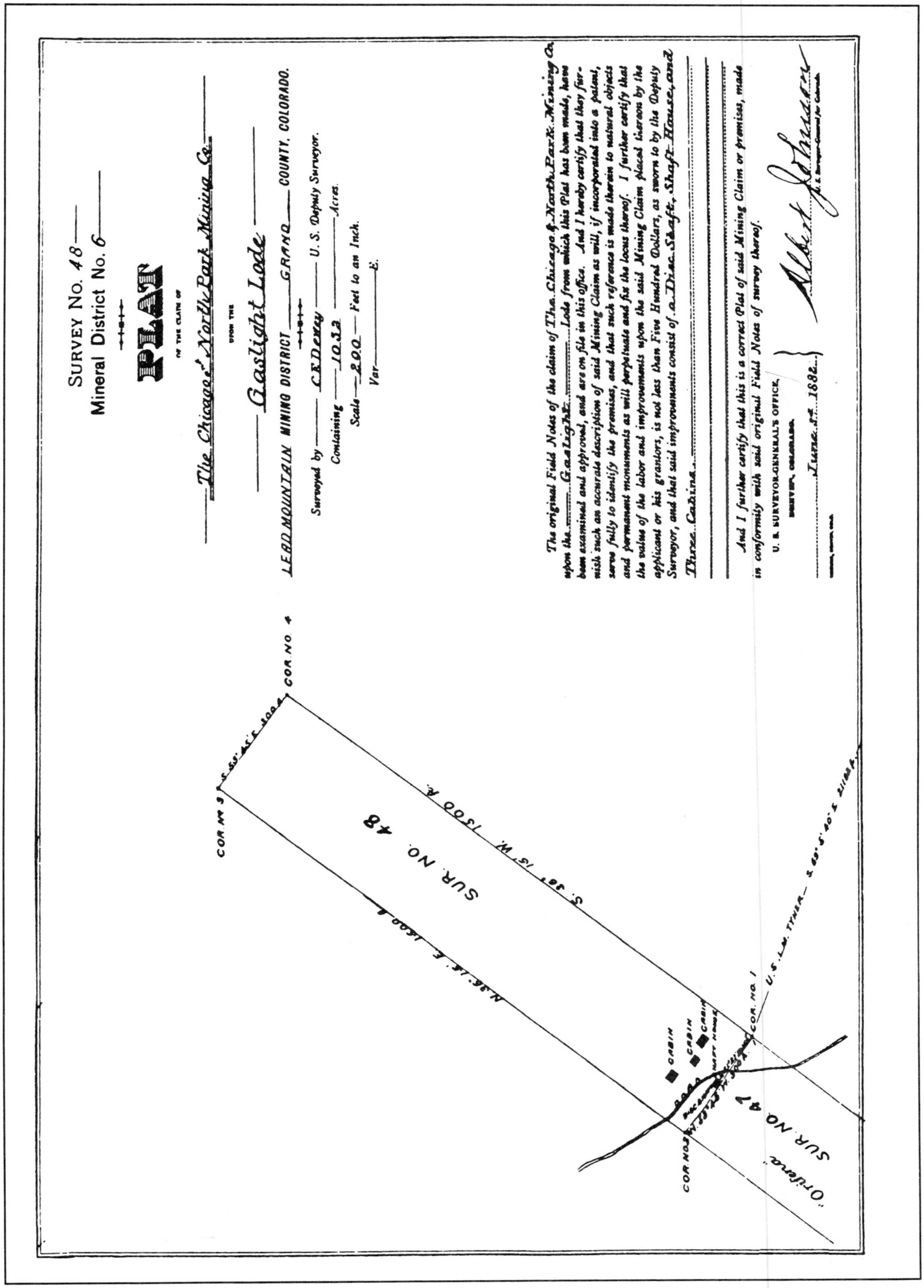
SURVEY No. 48
Mineral District No. 6
PLAT
OF THE CLAIM OF
The Chicago & North Park Mining Co.
UPON THE
Gaslight Lode
LEAD MOUNTAIN MINING DISTRICT GRAND COUNTY, COLORADO.
Surveyed by C.E. Dewey U. S. Deputy Surveyor.
Containing 10.32 Acres.
Scale 200 Feet to an Inch.
Var E.
The original Field Notes of the claim of The Chicago & North Park Mining Co. upon the Gaslight Lode from which this Plat has been made, have been examined and approved, and are on file in this office. And I hereby certify that they furnish such an accurate description of said Mining Claim as will, if incorporated into a patent, serve fully to identify the premises, and that such reference is made therein to natural objects and permanent monuments as will perpetuate and fix the locus thereof. I further certify that the value of the labor and improvements upon the said Mining Claim placed thereon by the applicant or his grantors, is not less than Five Hundred Dollars, as sworn to by the Deputy Surveyor, and that said improvements consist of a Disc. Shaft, Shaft House, and Three Cabins.
And I further certify that this is a correct Plat of said Mining Claim or premises, made in conformity with said original Field Notes of survey thereof.
U. S. SURVEYOR-GENERAL'S OFFICE,
DENVER, COLORADO.
June 1st 1882.
Albert Johnson
U. S. Surveyor-General for Colorado.
SUR. NO. 48
S. 38° 15' W. 1500 ft.
N. 38° 15' E. 1500 ft.
COR. NO. 1
COR. NO. 2
COR. NO. 3
COR. NO. 4
CABIN
CABIN
CABIN
SHAFT HOUSE
SUR. NO. 47
"Oriana"

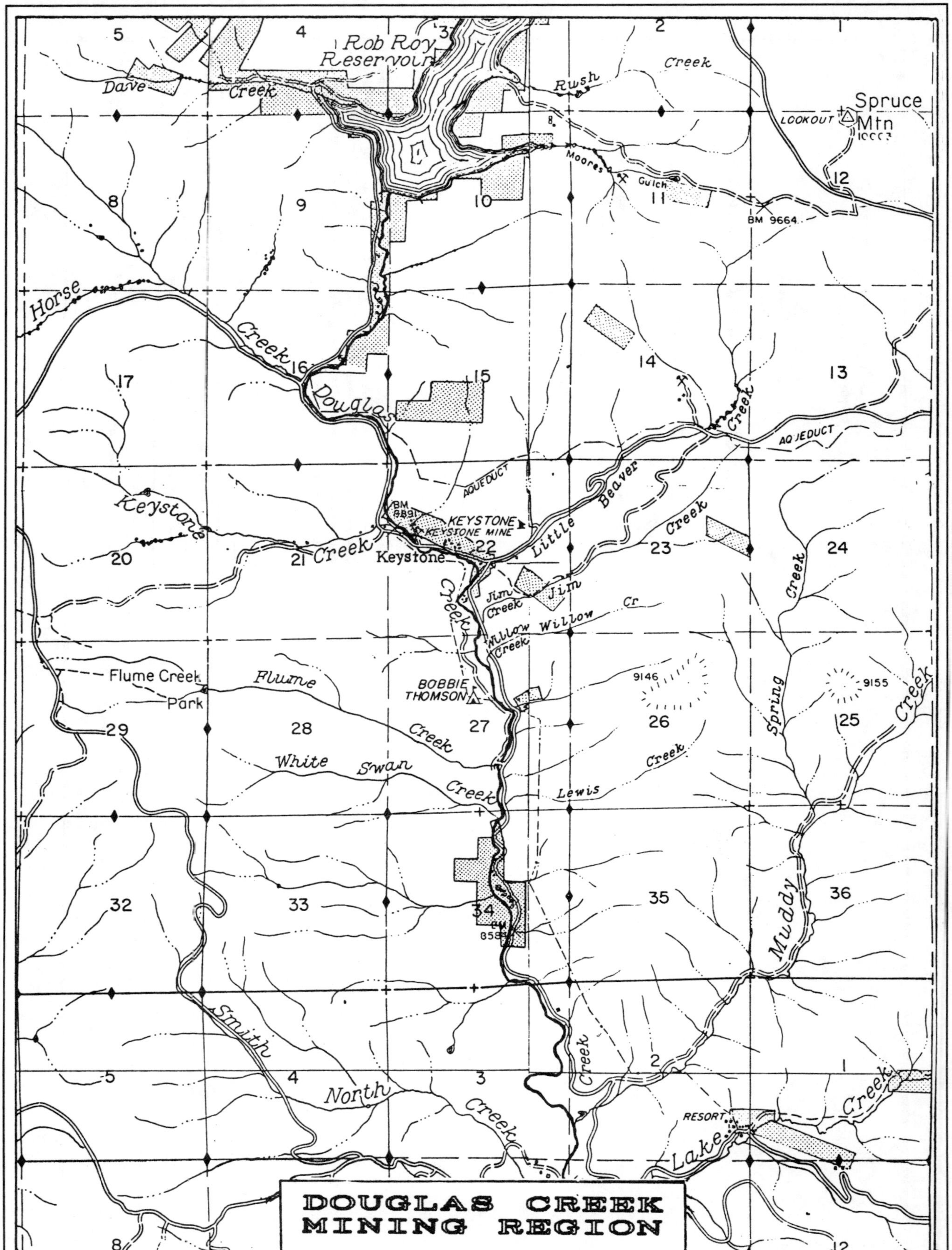
DOUGLAS CREEK
MINING REGION
Rob Roy Reservoir
Dave Creek
Rush Creek
Spruce Mtn
LOOKOUT
Moores Gulch
BM 9664
Horse Creek
Douglas
AQUEDUCT
Keystone Creek
KEYSTONE
KEYSTONE MINE
Keystone
Little Beaver Creek
Jim Creek
Willow Creek
Willow Cr
Flume Creek
Flume Creek Park
BOBBIE THOMSON
9146
9155
Spring Creek
White Swan Creek
Lewis Creek
Muddy Creek
Smith North Creek
RESORT
Lake Creek

VII THE DOUGLAS CREEK MINES VII

During all of the diverse mining activity of the 1870s, the old Last Chance District struggled on. New claims continued to be made up and down Douglas Creek and in the surrounding hills. Although Stephen Downey was caught up in the boom and bust of the Centennial Mine, he never lost faith in the old Last Chance District on Douglas Creek, and of greater significance, he maintained claim to extensive placer ground for several miles downstream from the old Last Chance claims.

The Mining Act of 1872 allowed placer claims of up to 160 acres for an association of eight persons. In 1873, Downey had the portion of Douglas Creek known as the Last Chance Creek and Park Gulch Placers, surveyed by R. W. Burkhardt. Downey and seven others then filed claim to this placer ground and several times in the following years he re-filed the claim, always with a total of eight persons so as to hold the maximum 160 acres.

Downey's group was not alone in their renewed interest along Douglas Creek and in September, 1876, J. H. Kelly claimed the "Mountain Lion" Lode located "300 yards west of Douglas Creek and about a mile below the Last Chance cabins." Also about the same time the "Great Republic" was also claimed by Kelly, G. B. Davis and Thomas Hodgson; the "Black Pine" by O. F. Elsworth and W. H. Lewis; and one of the few claims made by a woman, the "California Lode" by Pauline Lewis, located ". . .on the first left hand fork up Beaver Creek - about 3/4 of a mile from where the Beaver empties into the Douglas."[1]

> **ORGANIZATION OF A NEW MINING DISTRICT.**
>
> THE FOLEY DISTRICT—ITS BOUNDARIES—RECORDER, ETC.
>
> LAST CHANCE, Wyo. Ter.
> Sept. 16, 1876.

In conjunction with the filing of these new claims a meeting was held at the cabin of W. H. Lewis on September 16, 1876. Mr. T. Foley nominated W. H. Lewis as Chairman and the group decided upon the name of "Foley Mining District." Boundaries were established "Commencing at the Big Park on trail running from Last Chance to Centennial, thence west to divide between Douglas and French Creek, thence south on divide to the North Platte, thence east to divide between the Laramies and Douglas, thence northeasterly on divide between Laramies and Douglas to point of beginning." Election of a district recorder was postponed until September 18, when W. H. H. Brainerd was elected. Brainerd, a newspaperman from the Council Bluffs *Nonpariel*, apparently served as recorder for only a short time if at all.[2] Foley himself recorded no mineral claims but he did put up a stamp mill at his camp near the confluence of Beaver Creek and Douglas Creek.[3]

About the same time, September 18, 1876, the Douglas Mining District was formed. By-laws were drawn up and boundaries set, "Commencing at Big Park on trail running from Last Chance to Centennial, thence west to divide between Douglas and French Creek, thence South on divide to the North Platte, thence north-westerly on divide between the Laramie and Douglas Creek to point of beginning."[4] Exactly the same boundaries as the Foley Mining District. George W. McFadden was elected recorder and the by-laws were filed in both Albany and Carbon Counties. The Douglas and the Foley Districts were in conflict not only with each other but also with the old Last Chance District. The Foley District claims were being recorded at the former Last Chance recorders office of James Beagle and the Douglas District claims were recorded by George McFadden. James Beagle was still the Last Chance recorder, recording not only for that district but the newer Foley District as well. About twenty claims were filed in the Foley District before whatever differences the two districts had were resolved and all subsequent claims were filed in the Douglas District. The recorders would in turn make an occasional trip or mail the district claims to Rawlins and Laramie City to place them in

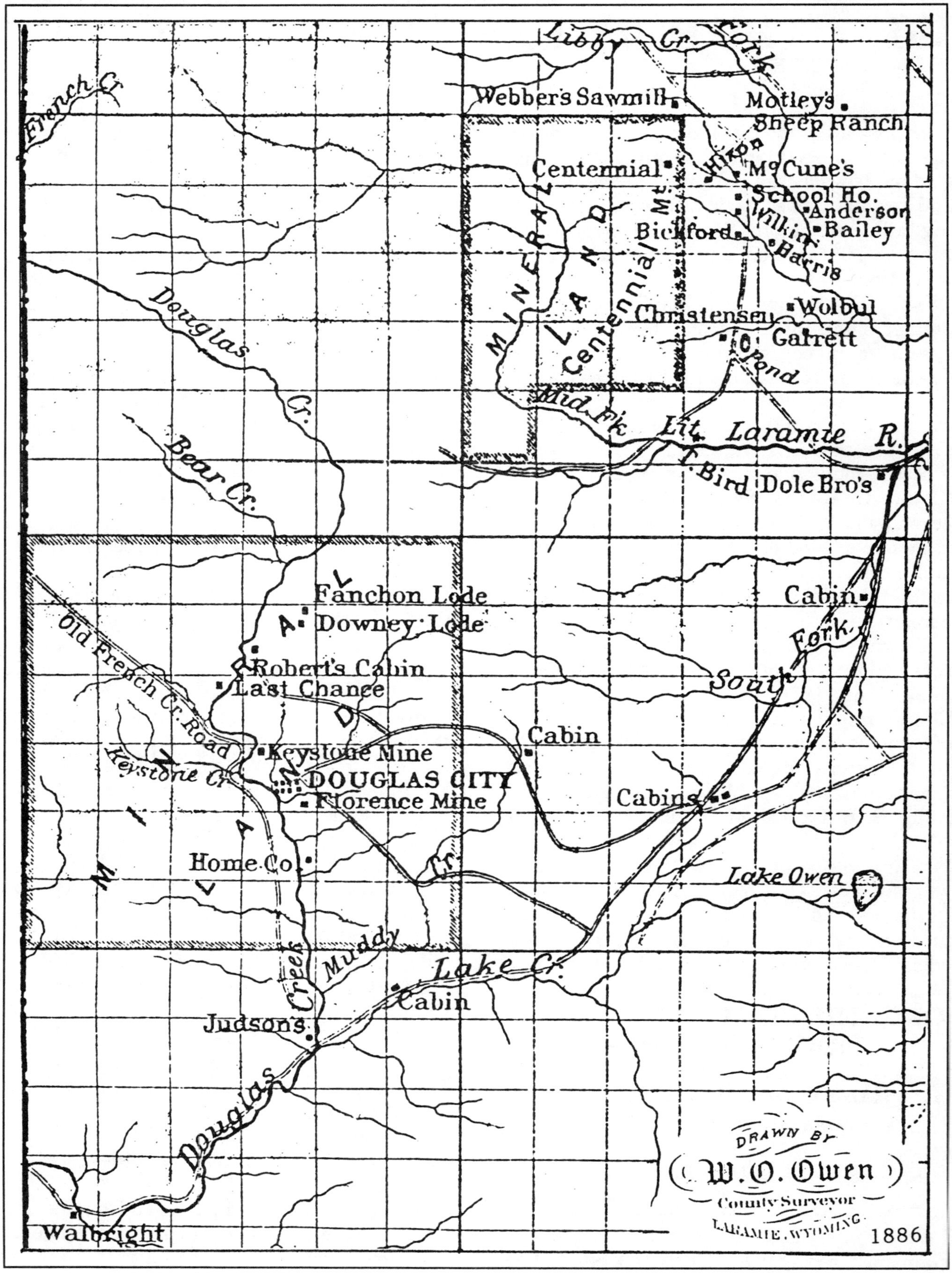
Libby Cr.
Fork
French Cr.
Webber's Sawmill
Motley's
Sheep Ranch
Centennial
Hilton
McCune's
School Ho.
Anderson
Bailey
Bickford
Wilkin
Harris
MINERAL LAND
Centennial Mt.
Christensen
Wolbul
Garrett
Pond
Douglas Cr.
Mid. Fk
Lit. Laramie R.
Bear Cr.
T. Bird
Dole Bro's
Fanchon Lode
Downey Lode
Cabin
Old French Cr. Road
Robert's Cabin
Last Chance
South Fork
MINERAL LAND
Keystone Mine
Keystone Cr.
DOUGLAS CITY
Florence Mine
Cabin
Cabins
Home Co.
Cr.
Lake Owen
Muddy
Lake Cr.
Cabin
Judson's
Crow Creek
Douglas
DRAWN BY
W. O. Owen
County Surveyor
LARAMIE, WYOMING.
1886
Walbright

the county records. Most of the miners realized that the Douglas Creek mines were at that time in Carbon County, but continued to file their claims in both Carbon and Albany Counties.

The forming of the new districts marked the beginning of a new era of activity along Douglas Creek. On October 31, 1876, the "Keystone" mine was claimed by Fletcher Dunham.[5]

• 1877

The increased mining activity along Douglas Creek began to attract not only prospectors but also drew the attention of the United States General Land Office. In mid-August, 1877, Surveyor General of Wyoming Territory, E. C. David, made a visit to the mines. David went first to Centennial where he inspected the failing mine and spent the night. The next day he rode to the Douglas Creek area where he inspected the most promising mines and visited with Charles Justus, J. T. Thompson, Jim Foley and others. He spent the night at the Metcalf cabin at Last Chance before returning to Centennial. His opinion was that the "Keystone" would become a major gold producer.

The "Keystone" and the later claimed "Blue Jay" and "Florence" were to become the predominant mines of the district and the entire Medicine Bow Mountains for years to come.

The "Florence" claim was not filed in the county records until February 22, 1882, noticeably after the Douglas Creek boom. However, according to Lalon Z. Farwell, it had been discovered some years earlier by David Rice and subsequently purchased by himself in January, 1878.[6]

The "Blue Jay" was originally located on December 2, 1876, surveyed during July, 1877, recorded in the Douglas Creek District record book in August, 1877, but not filed in the county records until April, 1878.[7]

In December, 1876, the Last Chance post office was moved downstream to the new mines and Beagle replaced W. H. Lewis as the postmaster. Beagle held the position for less than two months before being replaced on February 1, 1877, by William H. Bordway. Bordway in turn held the position until November 6, 1877, when Fletcher Dunham took over.[8]

In the closing weeks of 1877, the five stamp mill and steam engine from the Billie Waters' "Blind Lode" on Centennial Ridge was hauled to the Keystone where it was set up for a test run and eventually purchased by Fletcher Dunham.

• 1878

Early in January, 1878, encouraging reports of rich paying discoveries began coming into Laramie City from Douglas Creek. The *Daily Sentinel,* who promised to give the readers timely reports of the mining and outlook for the season, gave Colonel Stephen W. Downey ". . . great credit for his indefatigable labors in opening up of these treasures that lie at our very doors."[9]

During January, 1878, under the direction of Mortimer N. Grant, the small five stamp mill on the "Keystone" began crushing ore from the 60 foot deep shaft. The ore was assayed at 50 to 75 dollars per ton and by February 26, 20 tons had been processed. Grant took the first cleanup from the mill to Laramie City where it was melted down into a "brick" valued at $229.74 gold and $2.57 silver. Justification for the low yield was that the mill was only a five stamp affair and constructed of rough machinery parts. Great expectations were expressed for the new stamp mill which was on order and would increase the quantity of the ore processed and the percentage of gold produced.[10]

J. T. Thompson, one of the partners in the "Florence" mine, also ordered a ten stamp mill and planned to have it in operation within a month.[11]

In February, 1878, Fletcher Dunham the original owner of the "Keystone" mine sold four-fifths of his interest in the mine to Stephen W. Downey, Jacob B. Blair, Mortimer N. Grant and Thomas Feast. The Albany County Abstract Records show $1,500, however, the Keystone Ledgers show $200, as the price for the four-fifths interest in the mine and the five stamp mill and engine. Whatever the price, the agreement also stated that one-fifth of the yet undelivered ten stamp mill would be Dunham's.[12] (At that time, a new ten stamp mill sold for around $15,000.)

Adjoining the "Keystone" was the "Blue

Jay." Ownership of the "Blue Jay" had passed to "parties of the East" of whom it was reported would begin immediate development.

Nearby, the "Florence" mine, owned by Huse, Thompson and Farwell, was beginning a construction program to house their yet undelivered stamp mill and shelter their miners. All lumber for construction was being whip sawed and commanded a very high price.

In March, 1878, John Metcalf of the "Keystone" drove a team to Laramie City. Although the snowfall had been more than normal he experienced no difficulty and arrived in Laramie City by early evening. With him he brought $800 in gold from the little five stamp mill. He reported ten men were working on the mine and they were now 100 feet into the mountain.[13]

By March, the new machinery for the Keystone mill had arrived in Laramie City and on March 10, was loaded and headed for Douglas Creek. The mill, purchased from J. W. Jackson Company of Denver, weighed 18,000 pounds and required five teams of four horses each to convey it to the mountains. Mr. Blanchard, the wood contractor on the "Keystone" was in charge of transportation.[14] (It took three months to deliver and set up the new mill.)

Early spring, normally the best season for placer mining, brought no renewed placer mining activities, all efforts being in favor of lode mining, construction of the new mill, mill runs, boarding houses and cabins. Mr. Ah Say, of Laramie, offered to buy or lease the placer ground of Douglas Creek in order to put several hundred Chinese to work, an offer which apparently gained little or no support from the miners.

J. H. Hayford of the *Laramie Sentinel*, however, thought differently: ". . . We say turn the Chinese in. They will take out hundreds of thousands of dollars of gold there, and it will give impetus to the working of other claims in the neighborhood." [15]

In early June, H. S. Mills, recognizing the need for a sawmill, went to Laramie City where he struck a bargain with W. H. Holliday for sawmill machinery. Transportation by wagon was quickly arranged and although the road to Douglas Creek was still muddy, the equipment was soon delivered to Mill's site between the "Florence" and "Snowflake" mines. The water powered mill was quickly set up and began operating day and night to supply lumber to the mines and the new town which was being laid out. An advertisement was placed in the *Laramie Daily Sentinel*: "Wanted, four or five choppers at Douglas Creek mines within a week."

Mills, Donkersley and Moore were working the "Snowflake." They had dug a thirty foot tunnel on the mine and built a house near the sawmill.[16]

Under the guiding hand of Downey, who at the time was also serving as the Territorial Auditor, and encouraged by editor Hayford of the *Sentinel*, who enthusiastically printed all the mining news he could get his hands on, the mining season seemed bright indeed. Many thought this would at last be the great bonanza for which the miners and Laramie City had long hoped.

Almost every lode mine on Douglas Creek anticipated a stamp mill of their own or a custom mill to process on contract what they presumed to be valuable ore.

L. Z. Farwell, M. Farwell, S. E. Huse and E. M. Huse were working the "Florence" mine and finding enough encouraging ore to warrant ordering their own ten stamp mill.

M. Bruback was supervising the work on the "Blue Jay" mine and they too were planning a new ten stamp mill.

A. S. Blackbourne, who had been in the district for ten years, conducted Judge I. P. Caldwell, of the Territorial Legislature, a potential investor, on a tour of his "Morning Star" lode. He too anticipated a mill for his mine if he could raise sufficient capital.

However, the most prominent mine in the district was the "Keystone" which had its new ten stamp mill set up and in operation by June 10, 1878.

The new mill, under the supervision of M. N. Grant, became the timekeeper of the entire camp, sounding its steam whistle at 5, 6 and 7 in the morning, at 12 noon and 6 in the evening. The mill was running day and night and shutdown was planned only at 15 to 30 day intervals for "cleanup" of the gold accumulation.[17]

Anticipating increased travel to and from Douglas Creek, G. L. Russell, most recently of Red Buttes, built a "house" in Big Muddy Park, four miles east of the Keystone where he

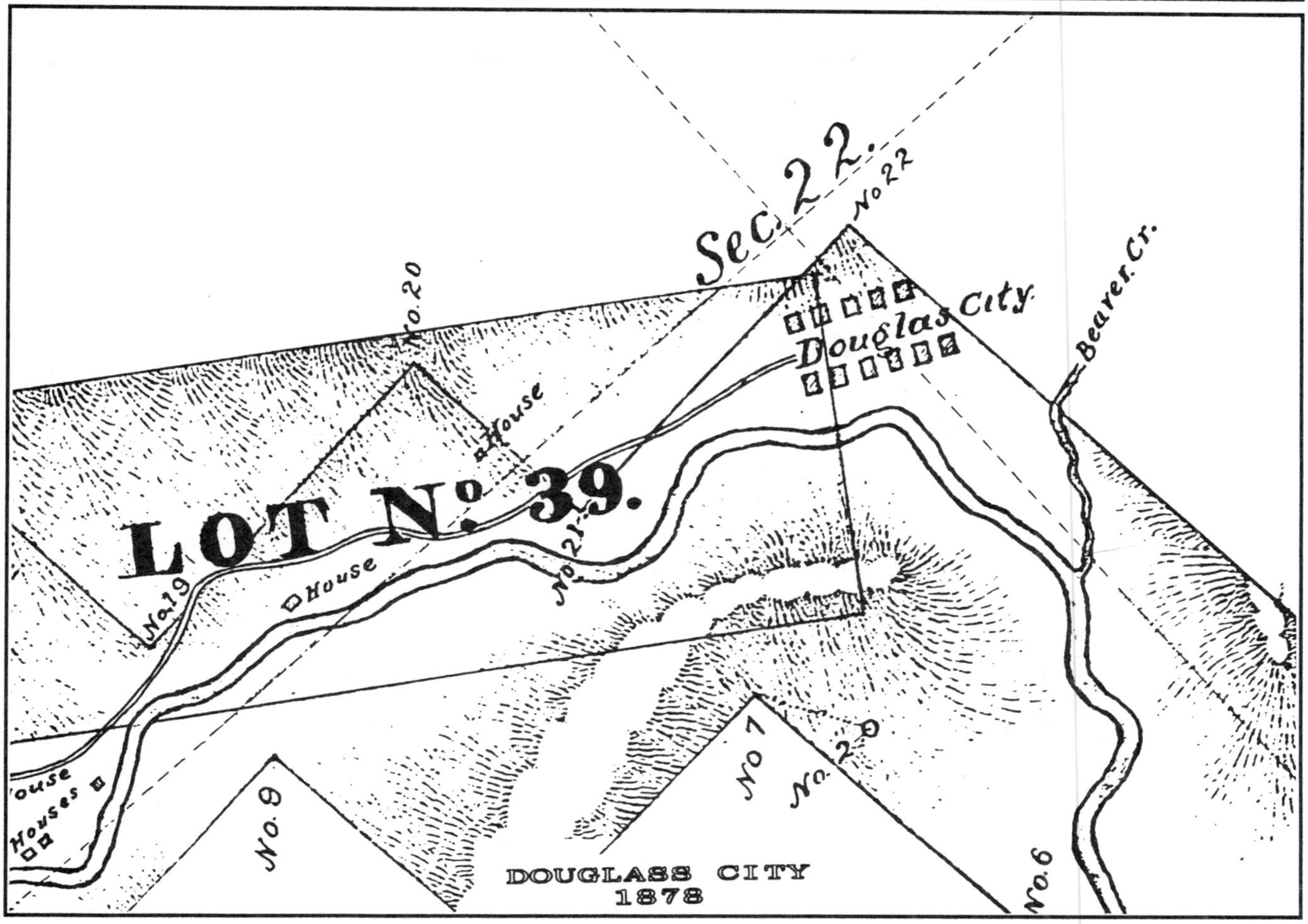

planned to open a restaurant to serve those using this route to the mines. He also planned to provide milk and other dairy products to the Douglas Creek miners.[18]

By the first of July, Charles T. Justus had completed his store on the main street of Douglass City and was open for business, selling provisions and miners supplies.

On July 9, Mills wrote the *Laramie Daily Sentinel,* describing the progress of the mining camp:

> Since my last letter we have laid out a town at the mouth of Beaver Creek. The settlers, calling a town meeting, gave the name of Douglass to the embryo city, and took measures to change the name of the postoffice from Last Chance to that of the new town. There are in course of construction twenty-one houses in our town, nearly all being erected for dwellings. Two are for boarding houses, and one for a doctors office, to be put up by Dr. Lee, from Kansas who will be our present physician.
>
> Among the new entries are five families. Most of them, of course, have children, and we shall have a schoolhouse and school soon. Your correspondent is also credibly informed that a shoemaker, blacksmith and wagonmaker are to be among the new arrivals at an early day.[19]

Mills had previously written;

> We have two women in camp, and although they do not often appear on the streets shopping, when they do they attract all eyes, as we are seldom blessed by the sight of the fair sex.[20]

The new town, looking forward to a rapid buildup, adopted a system of disposing of building lots:

> Anyone who wants a lot is entitled to take one, without money and without price, subject to the following conditions: Within twenty days from the date of location he must have the foundation of a house laid, and in sixty days a house built on the ground. Failing to comply with these requirements, the lot is subject to be jumped by a more enterprising miner.[21]

Jacob Fein made a new discovery on the "Snowflake" mine and promptly declared this was the richest district in the world. One might question just how serious he was in this claim as his first thought was not of mining but of putting on an express wagon between Laramie City and the mines.[22] On July 18,

1878, Fein began a "once a week" express and freight line, leaving Laramie City on Thursdays at 6 a.m. and arriving at Douglas Creek at 6 p.m. Return trips were made on Fridays.[23]

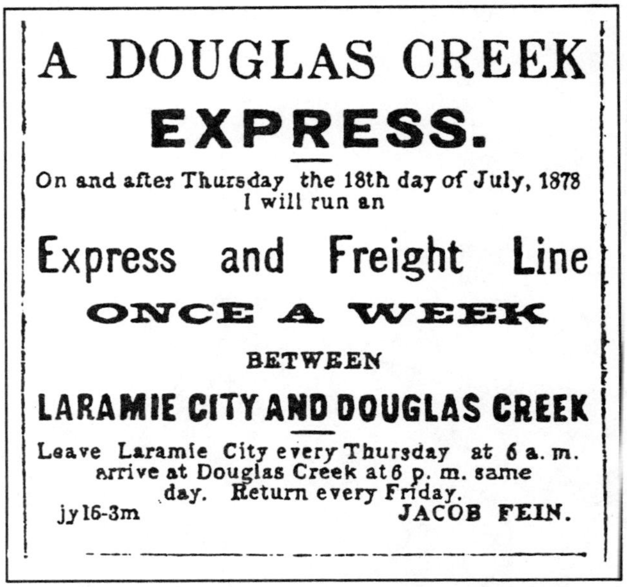
A DOUGLAS CREEK
EXPRESS.
On and after Thursday the 18th day of July, 1878 I will run an
Express and Freight Line
ONCE A WEEK
BETWEEN
LARAMIE CITY AND DOUGLAS CREEK
Leave Laramie City every Thursday at 6 a. m. arrive at Douglas Creek at 6 p. m. same day. Return every Friday.
jy16-3m JACOB FEIN.

William O. Downey surveyed the township in July and described the area in his field notes:

> The town of Douglass was laid out by M.N. Grant . . . on the left bank of Beaver Creek, near the confluence of the two streams. Douglass . . . containing at present a population of about one hundred inhabitants, occupied in mining and prospecting the country in the vicinity.

Downey also described the nearby "Florence" mine and settlement:

> The inhabitants number about 30, who are engaged in mining. . . .The Florence Mine . . . is located upon a gold bearing ledge and lode. The owners of the mine have made improvements valued at $15,000 consisting of a ten stamp quartz mill; a sawmill; a dwelling house and a shaft and tunnel.

William Downey had more than a casual interest in the progress of the region and described the area with somewhat more enthusiasm than usually shown in the dry and brief field notes.

> The Keystone has extensive improvements . . . mill building, containing a ten stamp quartz mill and a fifteen horsepower engine; a large boarding house, and other extensive improvements consisting of tunnels, shafts and underground works. The estimated value of these improvements is $25,000.
>
> The Blue Jay mining claim . . .owners . . . have improvements . . . consisting of a dwelling house, a shaft and a tunnel. Valuation of improvements, $3,500.
>
> The Fanchon Lode mining claim and the Downey Lode mining claim are adjacent and upon a gold bearing ledge.
>
> The Grant Lode and the Metcalf Lode mining claims . . . are also adjacent and upon a gold bearing ledge.
>
> The Home Placer Mining Claim is located upon Douglas Creek . . . The owners of this mine have made extensive improvements . . . consisting of a bed rock flume 560 feet long, requiring an expenditure of $10,000. They also have four dwelling houses . . .
>
> The New York and Wyoming Placer Mining Company upon Douglas Creek . . . has expended a considerable sum of money on their claim in improving the channel of the creek, preparatory to putting down a bed rock flume, the material for which is at the present time scattered about the Co's claim . . . They also have two dwelling houses.
>
> There are 5 cabins situated upon the central part of NW 1/4 of Sec. 10.

The 5 cabins in section 10 were the remainder of the Last Chance settlement first known as Cinnabar City, the others having been torn down, some to be rebuilt in Douglass City.

Downey ended his field notes with an observation on the production of the area, information of which he was well apprised, being brother to the principal claim holder of the area, Stephen W. Downey:

> The production of gold bullion in the Township up to the present time has been about $30,000 and in connection with the fact that gold bearing quartz has been found all over the Township, I therefore return the entire Township as valuable mineral land.[24]

On July 4, 1878, the William O. Downey survey party hosted a celebration at Lake Owen. The people from Laramie City as well as miners and prospectors from the surrounding mountains were invited. The number that attended is in question but the event was termed "a grand success."

On July 21, 1878, Grant, Campbell and Donkersley arrived in Laramie City with $3,200 in gold, the result of the first cleanup from the new ten stamp mill on the Keystone. This was from about one month of operation and claimed as ". . . not a fair test of what may be done, as the mill is entirely new, and a large amount of gold of course is left in the timbers and on the new copper plates."[25]

CIVIL ENGINEER

WM. O. DOWNEY,

DEPUTY UNITED STATES MINERAL SURVEYOR,

COUNTY SURVEYOR FOR ALBANY COUNTY,

MINING ENGINEER AND

Practical Draughtsman.

On August 23, M.N. Grant was again in Laramie City. This time he brought in $2,120 in gold from the Keystone Mine, the product of twenty days of operation.[26]

In September the people of Douglass City reported their first death:

> Last Friday morning [Sept. 6, 1878] our camp was somewhat startled by the sad news of the death of Mrs. Estus, after an illness of only one week, with some disease of the throat. Funeral services were held Saturday at 2 o'clock P.M., very largely attended by the citizens of this place.

Mrs. Estus, a widow, had arrived about one year prior to her death. She had traveled from Arkansas with Doctor Lee and party in an ox-drawn wagon. Apparently, Doctor Lee, her son-in-law, could provide neither a cure nor determine the cause of her death. She was buried about a mile below the camp on the southwest side of Douglas Creek.[27] (Present site of Bobbie Thomson Campground.)

Early on the morning of September 19, 1878, a rather prestigious group left Laramie City on a four day tour of the mines. Newly appointed Territorial Assayer, John G. Murphy provided the expertise for the group which consisted of Governor John Hoyt; Professor James Baylor of the United States Coast and Magnetic Survey; "General" Noah C. Worth; James H. Hayford, editor of the *Laramie Sentinel* and Stephen W. Downey.[28]

Two days later, on September 21, another group consisting of Judge Jacob Blair, Judge M. C. Brown, Judge "Bill" Nye, W. E. Wheeler of the *Evanston Age* and Sheriff N. K. Boswell also left for a tour of the Douglas Creek, Centennial and Snowy Range mining districts.[29]

The Governor's group made their base of operations Charlie Bussard's ranch in the Centennial Valley. From Bussard's they traveled to Douglas Creek on an overnight trip and then after another night at Bussard's, rode to the La Plata District and back in one day. On his return to Laramie City, Hayford wrote a full column article displaying his usual fervor for the mines:

> The first thing that strikes one who is at all familiar with geology and mineralogy is the universal indication of rich deposits of gold and silver all through that belt. The ground is everywhere covered with rich blossom rock, and all the gulches prospect good pay.
>
> Another thing that impresses one is the great amount of work that has been done by a few men in a short time. The Keystone Company, at Douglas Creek, under the supervision of M. N. Grant, began operations in the dead of winter, less than a year ago. They got their mill, material and provisions in there under circumstances which called forth more energy and heroism than the famous historical incident of Bonaparte crossing the Alps. This company have built a mill, built cabins and boarding houses, driven tunnels and sunk shafts to the amount of about 400 feet, have their mine better opened than anyone we ever saw, have half a million dollars worth of ore in sight and with their little five stamp mill have made their lode pay all the expenses from the start. We don't think the whole history of Rocky Mountain mining furnishes any parallel to this.
>
> Two other lodes there, the Florence and Blue Jay are not yet extensively opened but there is quite a quantity of quartz out and looks richer and assays higher than the ore from the Keystone. A very fine mill is being put up on the Florence by Thompson and Huse, and it will soon be running. Quite a thriving town is growing up in the gulch. A row of nice hewed log cabins runs up the gulch for a half mile long. They are built with much neatness and taste. There are now about 200 men at work in the gulch, and several families have moved in there.

> They depend for their lumber upon a little sawmill, owned by Mr. Mills. We wish we could describe and do justice to that mill but we give it up. It saws about 1,000 feet a day of real good lumber, and the growth of the town is only limited by the capacity of this mill.[30]

By the end of September the Florence Mine had its own sawmill set up and ready to begin operation.[31]

Albany County Sheriff, N. K. Boswell, found time from his official duties to do some prospecting and in October he brought ore from his "Martha Bedford" claim into Laramie City where he crushed it by hand and then panned out the gold. On one portion he claimed a yield of $1,000 per ton.[32]

On October 14, 1878, the post office was officially changed from Last Chance to "Douglass." The Laramie newspapers and the miners apparently preferred "Douglas" but the official post office name was now "Douglass." Fletcher Dunham was the first postmaster under the new name.[33]

Storekeeper Charlie Justus took on Isaac G. Skiles as a partner in his Douglass City enterprise which was apparently doing enough business to require repeated trips to Laramie City to replenish the stock.[34] Also, during October, Bickford and Rush began operating a tri-weekly stage from Laramie City to Centennial and Douglass City. Leaving from Bakers Store at the corner of Second and South B Street the stage made the run to Douglass in about seven and one-half hours, terminating at Charlie Justus' Store. Three changes of horses were made on the fifty mile trip. George Hughes was also running a "Fast Express Stage" from Laramie City to Douglass City, for a fare of $3.50. Return trip was $3.00. A new road was inaugurated shortening the distance between Laramie City and Douglass City needing only a bridge over the South Fork of the Little Laramie River and some work on the last mile into Douglass City to be complete.[35]

When the 1878 election for Delegate to Congress was held, the Douglas Creek District cast 97 votes for their favorite son, Stephen W. Downey and 4 for E. L. Pease. This total of 101 votes was the third largest turnout in Carbon County, being exceeded only by Rawlins with 387 votes and Carbon with 170 votes.[36]

DOUGLAS CREEK
Passenger and Express Line.
The undersigned wish to notify the public that they are running a
TRI-WEEKLY STAGE LINE
BETWEEN
LARAMIE CITY AND DOUGLAS CREEK
Leaving Laramie City every Monday, Wednesday and Saturday, at 7 a. m.
Returning, leave Douglas Creek on Alternate Days.
RATES REASONABLE – For further particulars, call on BICKFORD & RUSK. Laramie City.
[illegible]

On November 20, 1878, the "Keystone" superintendent, M. N. Grant, arrived in Laramie City with about $1,600 in gold dust, the results of 15 days of operation of the Keystone mill.[37] It appears the "Keystone" was producing about $3,000 a month from day and night operation of the mill. Enough to keep their operation going, but hardly a bonanza.

In December, Mrs. W. H. (Pauline) Lewis visited Laramie City and on her return to Douglass City, which required two days of cold, hard travel, wrote a letter to the *Laramie Sentinel*. In her letter, she said, the Florence mill had started operation on December 16, and crushed its first ore at 10:30 a.m. Mr. Arista Percival, formerly of Finfrock's drugstore was acting as engineer, Mr. Rogers as head machinist, Mr. Huse as superintendent, and W. H. Lewis as foreman. The mill, she said, was now operating day and night on eight hour shifts. She also mentioned the Keystone Mill was grinding away as usual.

The weather at Douglass City, she said, was cold with "plenty of the beautiful, rendering the roads heavy and the travel light."[38]

By years end the residents on Douglas Creek had settled into their winter doldrums. Christmas morning dawned cold and cloudy with a temperature of 24 degrees below zero. Travel and mail, however, continued to and from Laramie City and the *Laramie Daily Sentinel* printed a letter from the mining camp:

> We have no little stockings to fill at our house, but

we have thankful hearts, and everyone seems to be happy and to enjoy their Christmas turkeys with as much relish as though housed in the city with all the refinements of modern life.

On Christmas night Mr. and Mrs. Reese Davis gave a rousing house warming, they have just moved into their new boarding house, or more properly, what is to be called the Miners Hotel, situated near the Florence Mine.

A school meeting was called on the 27th and three directors elected - Grant, Robinson and Lewis. A school house is now on the tapis.[39]

An earlier article had stated that they had eighteen children of school age living within a half mile of the proposed school site.

In 1878, there were many rumors and proposals for a railroad from Laramie City to North Park. Such a road would not only serve the gold mines but would also tap the soda lake deposits on the Laramie Plains and the timber in the Medicine Bow Mountains. The first company to file for incorporation was the North Park and Wyoming Railroad and Telegraph Company, on November 26, 1878. This company failed to materialize. On May 24, 1880, another company, the Laramie, North Park and Pacific Railroad and Telegraph Company filed for incorporation. Construction of this line was begun and for several years was used to aid in mining the soda lakes, but never extended as far as the mountains. It would be many years before a railroad would actually reach the Medicine Bow Mountains.

On December 30, 1878, the *Laramie Daily Sentinel* ceased publication. The *Weekly Sentinel* continued to publish but mining news became infrequent and meager.

• 1879

With the single short lived exception of the original Centennial discovery, the Medicine Bow Mountains had yet to yield a paying mine. Still, the miners, developers, investors and most certainly J. H. Hayford were convinced that the only thing the Medicine Bow Mines lacked was greater investment to develop the prospects into bonanzas. Hayford, editor of the *Laramie Sentinel*, used his first edition of 1879 to lament the shortcomings of the mines and sound his latest vindication:

It seems strange that lying close to the railroad and to our town, so convenient of access and so safe from Indian depredations, these mines have not been extensively opened and worked before.

Had the Centennial, La Plata and Douglas Creek mining districts, which are all in sight of Laramie, been situated five hundred miles away from civilization and the railroad in the heart of hostile Indian country, there would long ago have been thousands of people there, hundreds of rich quartz and placer mines opened and scores of mills pounding out the precious metals.[40]

Hayford's implications may not have been obvious but it would take more than oratory to turn the mines into paying propositions. Development of the area mines followed a pattern that was to become all too common in the Medicine Bow Mountains. However, the Douglas Creek mines were showing they might just be the first to produce precious metal on a continuing basis. In mid-January, M. N. Grant was in Laramie with an undisclosed amount of gold from the "Keystone." The shaft, he said was now down 110 feet.

The last week in January, 1879, C. T. "Charlie" Justus, "the merchant prince of Douglas Creek," traveled to Laramie City. He informed the *Sentinel* that the road had never been in better condition, that the stage company used four changes of horses and the trip could be made in eight hours. Also, he said, the snow was about two feet deep with a good hard track and sleighs were used on eight miles of the mountain road.[41]

In February, 1879, the *Cheyenne Sun* published the long awaited report of Professor J. G. Murphy, Territorial Assayer. Following his lengthy description of the geological formations of the Medicine Bow Mountains, Murphy described the Douglas Creek Mines and the mining camp of Douglass City.

A tri-weekly stage coach runs between Laramie City and the town of Douglas. The latter is a mining town of about 200 inhabitants. The road runs over the Laramie Plains for a distance of thirty-three miles then ascending the eastern slope it passes over the most level portion of the range for a distance of twelve miles, when Douglas is reached, the entire distance being about forty-five miles. The road has been graded where it was found necessary so that there are good facilities for travel and transportation of heavy machinery between the range and the Union Pacific railroad.[42]

Apparently work in the Douglas Creek District was slow during the first months of 1879.

However, the *Laramie Sentinel*, at the time being published weekly, continued to enhance and print the rather meager news from the mines.

February 14, 1879:

A party of capitalists from Galena, Illinois, having purchased the Eureka Mine, on Douglas Creek, will commence developing it in the spring.

A public hall at Douglas will soon be completed and then, 'like St. Paul's Cathedral, be doomed to have a ball.'

Peabody loaded one or two large outfits with supplies for Keystone on Wednesday.

The business of Messrs. Bickford and Rush, the Douglas Creek stage men, is increasing so rapidly that they are having two four-horse coaches built to be put on their route with the opening of spring. Next season will witness lively times in Laramie and her adjacent mining camps.[43]

Isaac G. Skiles became postmaster on February 14, 1879, relieving Fletcher Dunham.[44]

In February, James H. Hayford of the *Sentinel,* in his ceaseless efforts to promote the mines, proposed a telephone line to connect Laramie City with the Douglas Creek mines. Hayford proclaimed and apparently believed ". . . wire could be procured and laid down here for less than $500. The mines and business men out there, and the teamsters on the road would probably be willing to contribute the poles and put them on the ground. The whole expense would be less than $1,000, probably. There will be a big business between the mines and this city in a few months." [45] Hayford seemed enthralled by the fledgling telephone, only two years since invented by Bell, and had one installed in his home which was connected to Doctor Finfrock's office in the post office building. He was a participant in one of the earliest of long distance telephone conversations which occurred between Cheyenne and Laramie City in February, 1878. (It might be noted that the first long distance line for public use was in January, 1881, between Boston and Rhode Island.)

Hayford, it seems liked to speculate about the lively times for big business and the grand development yet to come and seldom mentioned any downbeat news. It is surprising he let even a small mention of slack times into his paper even though he himself had been forced to forego his daily paper for a weekly edition, but on February 18, 1879, his weekly *Sentinel* noted:

Mr. S. P. Holtburg, well known in this city, and a miner of eight years experience in Wyoming who has been spending a month at Douglas arrived in our city last Friday evening. He reports that camp as in a flourishing condition with every prospect for lively times this summer. Work just now is somewhat slack and there are plenty of men in the camp to supply all demands for labor, but when spring opens there will be employment for quite a number of men.[46]

In March, 1879, Professor Murphy, Territorial Assayer, submitted his resignation. He gave his reason as insufficient salary, as the office was located at too great a distance (Rawlins) from principal mining districts and also the fact he could find no other work to augment his salary, ($600.00 a year plus $1.00 per assay). The *Sentinel* added, "Had the office been located at Laramie adjacent to the mines and where he could have opened a school of mines, chemistry and mineralogy, he would have stayed with us. It is a great loss to have him go away."[48]

The same week, Stephen W. Downey resigned his seat on the Territorial Council and his office of Auditor of the Territory to fulfill his duties as Delegate to Congress, a $5,000 a year position. Downey departed for Washington, D.C. and temporarily left his mining ventures to his faithful friends of which he apparently had many. There were those in Laramie City who proclaimed Downey as the best legal talent in the Territory. His friends in Douglass City also thought highly of Downey and his penchant for mining, saying ". . . with a few more like Colonel Downey, than whom none better live, . . . we shall develop into a live mining town."

By the first week in March, the "Keystone" was reported to be producing $100.00 of gold per day. On March 1 they reported a $1,800.00 shipment "to the East."

The "Keystone" continued digging into the mountain and the mill continued to average around $100.00 a day. Employing 27 men at $3.00 a day plus superintendent Grant, the mine was still not breaking even. By April the mine was reported to be 160 feet deep with a 55 foot cross-shaft where they reported a new vein of rich ore. Although troubled by water from the melting snow, a "big cleanup" was expected soon.[49]

Winter travel apparently posed only a small problem, but springtime proved more troublesome. In April the Douglas stage which had been operating without problems for most of the winter was stuck in a snowdrift about six miles west of the Little Laramie and abandoned, the driver continuing on to Laramie City on horseback. The westbound stage was lost for two days in a spring blizzard on the Laramie Plains. Mail service was resumed on horseback but was taking three days instead of the usual one from Laramie City to Douglass City.[50]

M. N. Grant visited Laramie City in mid-May and reported they were still having water problems but things were improving. He brought with him a $2,500 gold "nugget," the results of a reported week's cleanup.[51]

The Keystone mill continued operating day and night with improving results. The output, it was said, being limited only by the capacity of the mill. Several other mines in the area were lying idle for lack of a mill and money to proceed.

Placer mining which could be pursued with a limited amount of money was again drawing some attention. On Jim Creek, a tributary to Douglas Creek, an old prospector named Miller was reportedly sluicing $10 to $15 a day. Again, Hayford printed his opinion that if mined properly, anyone could pan $5 to $10 a day from Douglas Creek and its tributaries. At that time, hard rock miners were earning $3.00 a day, if they could find work.

The mining season of 1879 at the Douglas Creek mines was drawing little attention, even Hayford of the *Sentinel* failed to print any noteworthy stories. The mining camp of Douglass City, however, maintained an only slightly diminished population due probably in part to diversity of the miners and their ability to become lumberjacks and tie hacks when the time and season demanded. When the election of 1879 was held for Legislative members, Douglass City was able to amass a total of 83 votes. I. G. Skiles, the Douglass City merchant and postmaster, garnered 81 of these votes which helped elect him as the Carbon County member of the Sixth Legislative Assembly which convened on November 4, 1879, in Cheyenne.[52]

• 1880

By 1880, Douglass City was beginning to show a decline in both reported mining and population. When the 1880 census was conducted on June 7, J. F. Crawford enumerated 43 people in the town of Douglass. Six were listed as quartz miners and nine as placer miners. Undoubtedly some were missed but this nevertheless indicates a decline in the local populace. The November elections drew 29 voters, a noticeable decline from 83 the previous year.[53]

In November, 1880, a Mr. Hines took over M. N. Grant's job as superintendent of the "Keystone."

By the end of 1880, Wyoming Territory had produced $20,000 in gold for the year. The majority of this amount was from the Douglas Creek Mining District and most of that was from the Keystone mine.[54] (In contrast, Colorado produced $17,500,000 in gold for the year 1880).

The 1880s brought wider diversity to the Medicine Bow Mines. Cummins City was demanding much attention, Centennial was still drawing interest, La Plata was drawing attention to its promise of paying silver mines, and North Park was attracting many Wyoming miners to the Teller mines. The Douglas Creek District and the old Last Chance placer mines were seldom heard from. The Keystone mill, however, continued to pound away at the quartz from the Keystone group of mines, producing only a modest amount of gold but still the major producer in the entire Territory. Interest in the "Keystone" and surrounding mines, however, was beginning to wane.

• 1881

On January 3, 1881, Fletcher Dunham sold his remaining one-fifth interest in the "Keystone." This included not only his share in the mine but also the mill, mill-house, and water rights. The reported price was $1,000.

Although on the decline, Douglass City endured, and on April 11, 1881, Jay H. Donkersley became postmaster, relieving Isaac Skiles. (Donkersley was in later years, 1888, postmaster of Laramie).[56]

On October 18, 1881, the old Keystone Mining Company was reformed and incorporated

Gus Hurley on his Lincoln Gulch Claim.
Wyoming State Museum

under the name of The Wyoming Keystone Gold Mining and Milling Company. Three of the original company's co-owners, Blair, Feast and Stephen Downey joined with John J. Strode, Isaac P. Caldwell, Richard Butterand and William Downey to form the trustees of the new company.[57]

Some trading in mines and shares continued and during the year, Francis E. Warren, at the time Territorial Treasurer, purchased 250 shares of the Keystone Company.[58]

One of the old company's members, John Metcalf, died in 1880. During probate of his estate, in November, 1881, it was stated that of the $60,000 that had been expended on the "Keystone" development, $30,000 was still outstanding. The deceased owed one-tenth or $3,000 of this amount. In addition, Metcalf had owned one-eighth of the New York and Wyoming Mining and Milling Company. It was also stated that the New York and Wyoming company property could not be worked without a large expenditure of money and therefore his shares should be sold to satisfy debts of the estate.[59]

• 1882

In 1882, two prospectors, Gus Hurley and John Walbright located a claim on Lincoln Gulch, (the eastern limit of the Douglas Creek drainage). A third prospector joined them in the fall of that year.[60]

Again, the year produced only limited activity on Douglas Creek and for the entire Medicine Bow Mountain Range.

Water in the lower levels of the "Keystone" continued to be a problem and new machinery was put to work pumping out the shaft. Plans were made to drive a 100 foot drift on the vein of gold producing quartz but only a limited amount of work was accomplished.

A railroad to the mines continued to be one of the hopes of the mine owners, and although plans were being laid to extend the Laramie Plains Line into the mountains, the outcome was still uncertain. In an effort to provide a better transportation system, the Keystone

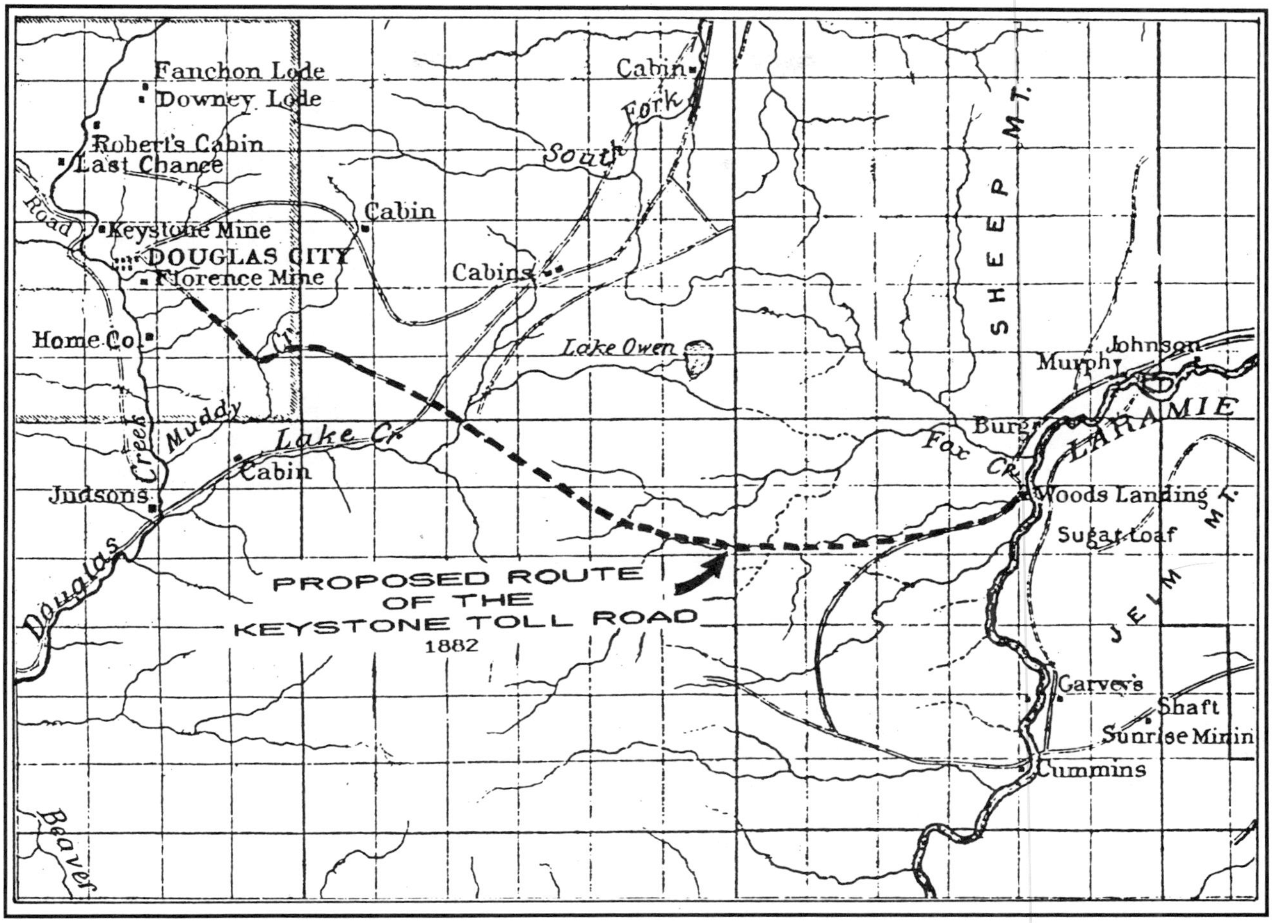

Toll Road Company was formed and incorporated in July, 1882. As could be expected Stephen W. Downey was the instigator, director and trustee of the company. John H. Neal, Charles H. Bussard and his son Charles W. Bussard were the remaining trustees. The senior Bussard was very familiar with the country, having lived in the area and done considerable logging during the previous ten years. The road was to begin at the county bridge over the Big Laramie River one mile south of Burg's ranch and just north of Sugar Loaf Mountain, (the present site of Woods Landing). The route was to extend to the North Platte River somewhere between Douglas Creek and Big Creek. "Beginning at the said county bridge, thence southerly up the west side of said Big Laramie river to the mouth of a Dry Gulch or ravine about one-half mile from said county bridge, thence westerly up said Dry Gulch about four and one-half miles to its head, thence in a northwesterly direction crossing Little Fox Creek, thence westerly to the south fork of Big Fox Creek, thence in a continued general westerly course by the most feasible route to or near the 'Keystone' mine on Douglas Creek in Douglas Creek mining district in said Carbon County. . . ." The road was then to continue on to the North Platte River. Only a portion of the road was completed, and apparently no tolls were ever collected.[61]

During the election of 1882, postmaster Donkersley was elected Justice of the Peace and George Glass was elected Constable of Douglass City.[62]

The output of gold for 1882 was estimated by the U.S. Geological Survey at $5,000 for the entire territory.[63]

• 1883

In May 1883, Frank Wilkeson took over the postal duties of Donkersley and became the last postmaster of the fading Douglass City.[64]

On May 10, 1883, the Hurley Mining District was formed. Named after Gus Hurley, its headquarters was located in Lincoln Gulch, about seven miles southeast of the Keystone. George W. Moore (most recently of Cummins City) was elected as recorder. Boundaries

Remains of John Walbright's cabin, 1990.

were established as the Carbon County line on the East, the Platte River on the West, the Colorado line on the South and 11 miles north of the Colorado line on the North.[65] These boundaries were in conflict with the older Douglas Creek District which was by this time practically defunct.

On Lincoln Gulch, Burham, Leavens and Ramey were working their "Cora" claim and reported they had taken out eight ounces of gold (probably a true but rather humble claim). (Gold was still officially priced at $20.67 an ounce.)

During the 1883 season, the miners on Lincoln Gulch, were reported to be mining $300.00 a season, mostly from their claim the "Maggie May." They constructed a simple hydraulic apparatus and planned to do hydraulic mining in the spring of 1884.

Walbright was also planning on using the hydraulic method to mine his "Walbright" claim. Moore and Ingle were also planning to mine hydraulically. Not to be left out, Hugh Kelly was also planning to hydraulic his claim, the "Memphis." These claims used up the entire three mile length of Lincoln Gulch.

The placer gravels of Lincoln Gulch were assayed to contain $3.75 gold per cubic yard. With a simple hydraulic, three men could wash 50 yards per day. It was stated at the time, "as anyone can see, it will pay." [66] However, as with the Douglas placers, although the placers could sustain a few isolated prospectors with their gold pan or simple sluice, it would not sustain extended and expanded mechanized operations.

During the remainder of 1883, the only other claims noted were the group made by Peter M. Jepson near the Mountain Home Ranch on the new Laramie City to North Park road.

The 1883 gold production for Wyoming Territory was estimated to be $4,000. [67]

• 1884

Again, 1884 revealed little mining news.

On the first day of January, 1884, Stephen W. Downey filed papers to relocate the apparently abandoned "Keystone." [68]

In August, 1884, the *Weekly Sentinel*, reported Asbury Johnson, an old California "forty-niner" who was working placer ground

John Walbright grave site, 1990.

in Lincoln Gulch, was in Laramie. Johnson apparently had $13.00 in gold which he had washed out in two and one-half days. He declared that with a pick, shovel and a simple sluice box, anyone could earn $5.00 a day.

Gus Hurley working alone on the adjoining claim was operating a "cheap and primitive hydraulic apparatus" and was reportedly taking out $8.00 to $20.00 a day.[69]

These modest returns were certainly no bonanza, but for the lone prospector, the individual return was just as great as the miner working for wages in the rich mining districts that surrounded the territory.

The 1884 gold production for Wyoming Territory was estimated at $6,000.[70]

• 1886

A few days before Christmas, 1886, John Walbright, one of the first prospectors and miners of the Medicine Bow Mountains was found dead in his cabin. Walbright, who had not been seen for two weeks, was found lying on the floor of his cabin by Orin Porter. Frank Leavins made a trip to Laramie to notify Albany County Coroner Sam Barnes, who instructed Leavins to return and bury the body near the cabin as no indication of foul play was evident. Today, only an unmarked depression indicates the site of his grave near the deteriorated remains of his cabin. His name is perpetuated in Walbright Creek, some miles away from his final resting place.

Placer claims continued to be made up and down Douglas Creek. Around 1886, William "Billy" Roper came into the area and filed several claims, among them, the Big Bend Placer near the mouth of Lake Creek. Roper built a cabin on his claim where he ran a hunting lodge for Laramie sportsmen. He died in 1932, at the age of 80 and was buried near his cabin.

• 1887

Douglass City continued it's gradual decline with only a sporadic population. When Barnes and Jerry Quigley visited the town in February, 1887, they found it deserted. However, they prospected in the area for several days and reported good placer discoveries above the Keystone mine on Douglas Creek.

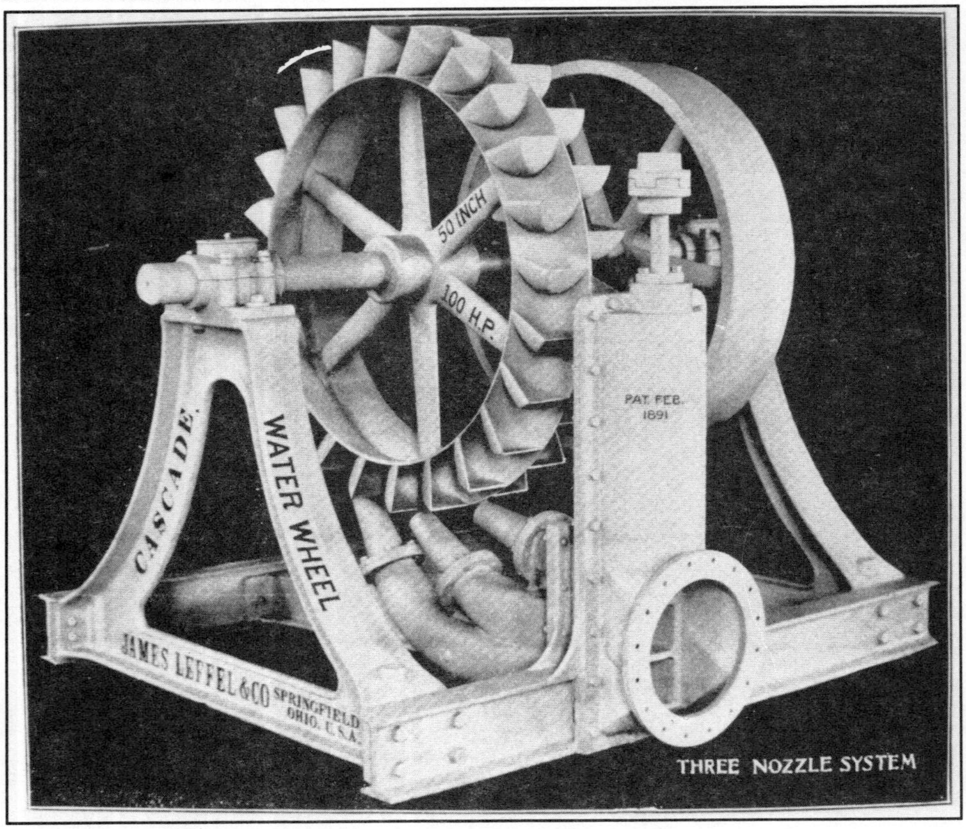

They thought enough of their discovery to organize the Douglas Creek Mining Company.

During November, 1887, Stephen W. Downey again put the Keystone mill in operation with 40 men under the supervision of M. N. Grant. Twenty-five wagon loads of supplies were sent to the mine which included $4,000 in provisions and 2,000 pounds of dynamite. Production from the mine was said to be $150 to $200 a day.[71]

• 1889

In July, 1889, Stephen Downey sold the "Keystone" to Lionel Sartoris for a reported $30,000. At the time of the purchase it was reported that the mine had produced, in it's thirteen year history, a total of $70,000 in gold. Under the supervision of Mat Nichols, a new building was constructed to house the new twenty-stamp Fraser-Chalmers mill. The building, was 60 by 100 feet on the ground floor, 40 by 60 feet on the second floor, 20 by 60 feet on the third floor and the head frame making up the fourth and top floor. The entire frame structure was roofed with corrugated iron. To support the twenty stamps of 850 pounds each, a solid timber support 12 feet square made up of 16 by 16 inch timbers set 12 feet into the ground was constructed. The mill was supplied with Challenge ore feeders, and four Frue Vanner concentrators. Motive power was supplied by an 85 horsepower Leffel water turbine. Cost of the mill was said to be $27,000 and of the water turbine, $2,000.[72]

The old ore stamp mill was converted to a sawmill run by Gus Reed and Jack Martin.

For a short time, the old district was to realize a superficial comeback. Under superintendent Wilbur C. Knight, the "Keystone" was again producing a nominal amount of gold. More ore than ever before was said to be "in sight" and the mine was developed to a total of 1,280 feet of drifts of which, 80 feet were at the bottom of the 230 foot shaft.[73] At the time they speculated that an average of $20.00 per ton would provide sufficient profit for continued operation.

On October 16, 1889, Knight was married to Emma Howell in Omaha, Nebraska. The newlyweds traveled through the early heavy snow to their new home near the Keystone mine by bobsled. Here they set up housekeeping in the comfortable log building which also housed the mine office. Sixteen brides were in the camp that winter and 16 babies were born the following summer and fall.

In November, 1889, the old Douglass City post office was reopened under the name of "Keystone" and David B. Dole became postmaster.[74] Delivery of the mail however became a problem, with the postal service wanting to deliver weekly by way of the Wood's post office and the residents of Keystone wanting a tri-weekly stage using the more direct Laramie to Douglas Creek road. Claiming 150 residents, 20 new houses, a school, and officers to run their little community including two justices of the peace, a constable and a recorder, the miners apparently won out.[75]

The 1889 Keystone Stamp Mill building.
Courtesy of Lynn Benson

Keystone Post Office, ca. 1890.
Courtesy of Lynn Benson

By 1889, it was becoming apparent that Wyoming was not going to be a great gold and silver producer. According to the 1890 Mineral Census, gold production for the Territory of Wyoming for 1889 was $14,512 or 711 ounces. (Added to this was an estimated 375 ounces of unrefined gold from assay offices and mints). The grand total of expenditures for this production was $80,712. Of the $14,512, $6,172 was from Albany County and the remainder from Fremont County. Although there were fifty-six gold mines within the territory at the time, only eleven were considered producing mines. The census showed 226 people employed in metal mining within the territory, 120 in Albany County, 10 in Carbon County and the remainder mostly in Fremont County. Miners wages at the time averaged $3.05 per day for underground work and $2.36 for above ground labor. Hardly the viable industry that Downey and his supporters had anticipated.

However, this elusive promise of rich gold and silver mines had played a part in attracting hopeful prospectors and miners to Wyoming and many stayed on, adding to the rather meager population of the territory. With the entry of the Dakotas and Montana into the Union in 1889, Wyoming was anticipating statehood. A sustained population had become the crux of the presentation for statehood before Congress and Wyoming needed every man, woman and child possible.

An election of Constitutional Convention delegates was held in July and by the end of September they began gathering in Cheyenne. Of the eight delegates from Albany County, four had considerable interest in the Medicine Bow mines, S. W. Downey, M. N. Grant, M. C. Brown and G. W. Fox. One other delegate, ex-governor John Hoyt had frequently shown interest in the mines and was part owner in several non-metallic mines in Albany County. M. C. Brown was elected president of the convention.

Among the articles incorporated into the Constitution was Article IX, Mines and Mining, directed primarily at coal mining which was already the largest industry within the territory. The article also provided for mining and metallurgy to be taught in one of the state institutions.

• 1890

After holding the Keystone mining property for little more than a year, Sartoris sold out to the Otras Mining Company for a reported, although doubtful, $750,000. (Although the sale included the "Blue Jay," the "Home Placers," the "New York and Wyoming Placers" as well as the Keystone mine, the price was probably to be paid only if the mine reached it's greatest potential). The Otras Company took out promissory note for over $65,000 to finance their initial operation. It should be noted that Sartoris and Knight were both on the board of directors of the Otras Mining Company.[76]

THE

DOUGLAS CONSOLIDATED

PLACER MINING

... AND ...

MILLING COMPANY.

ORGANIZED AUGUST 5TH, 1892.

Its property consists of 2,100 acres of placer ground located on the Muddy, Lake, Beaver and Douglas creeks, 40 miles southwest from Laramie, Wyoming. Water rights secured on all property. Abundant supply of water and timber. Expert examination shows the ground to be worth from 30 cents to $1 per cubic yard. Stocked at 3,000,000 shares of $1 each and owned by nine men. No stock on the market.

OFFICERS:

C. W. BRAMEL, President. M. A. HANCE, Treasurer.
H. D. BEEMER, Vice President. C. S. DAWSON, Secretary.

• 1891

During 1891, Wilbur Knight moved to Laramie where he opened an assay office and laboratory.

In March of 1891, Stephen Downey applied for patent to several thousand acres of placer ground, which, even after exclusion of all valid previous claims, still amounted to over 2,800 acres.[77] Downey was challenged in court and spent years trying to perfect the patent.

C. W. Bramel, A. A. Johnson, A. Trabing, Geo. W. Fox, M. N. Grant, Chas. Culross, M. A. Hance, A. C. Jones,

BOARD OF TRUSTEES OF THE LARAMIE MINING AND STOCK EXCHANGE.

Wyoming State Museum

In December, 1891, Keystone postmaster Dole was relieved by Peter Pfandler.[78]

• 1892

Although it is difficult to track the decline of the mining camps, election returns give one comparison of the population of the mining camps. It was said at the time that one-third of the population would vote in the general elections. Others said one-half would vote and still another said only one-fifth. Most likely it depended on several factors including weather, the candidates and the issues. Nevertheless, elections give a comparison, sometimes the only comparison, of the population of the camps. In the election of 1892, the old Douglass Precinct now known as Keystone cast 8 votes, indicating a further reduction in the population.[79] Polls were in the school house and Gus Hurley, David Myer and Peter Pfandler were the election judges.[80]

With lode mining failing to produce the expected return, attention again shifted to placer mining. On August 5, 1892, The Douglas Consolidated Placer Mining and Milling Company was organized. Company officers were: C. W. Bramel, President; H. D. Beemer, Vice President; M. A. Hance, Treasurer and C. S. Dawson, Secretary.[81]

• 1893

The Keystone post office ceased operation on February 15, 1893.

Just a few miles to the north of the Keystone a new mine called the "Rambler" opened up. The Holmes District was formed and a post office established. John E. Holmes became the first Holmes postmaster on March 3, 1893.[82] The promise of a producing mine was still a dream of the miners and Holmes became one of the longer lived camps of the mountains. This was not entirely due to mining as Holmes was also a center of timber cutting activities.

In July, 1893, The Douglas Consolidated Placer Mining and Milling Company was incorporated. Capital stock consisted of 3,000,000 shares at one dollar each, but no stock was placed on the market. Seven directors were named; M. N. Grant, M. A. Hance, C. B. Richie, C. W. Bramel, W. H. Nelson, J. R. Root and Edwin S. Raymond. The main office location was designated as Laramie. The company planned not only to conduct mining but also to engage in construction of railroads, tramways, wagon roads, electric power plants, irrigating canals and telephone lines.[83]

In August, 1893, the Otras Mining Company defaulted on their promissory note of

$65,432 plus three years interest and the property was ordered to be sold at public auction. The year 1893 was the beginning of a general depression across the entire country and property was not selling well. The sole bid for the property was $3,600 by the Pacific Investment Company of Salt Lake City, Utah, who held the note.[84]

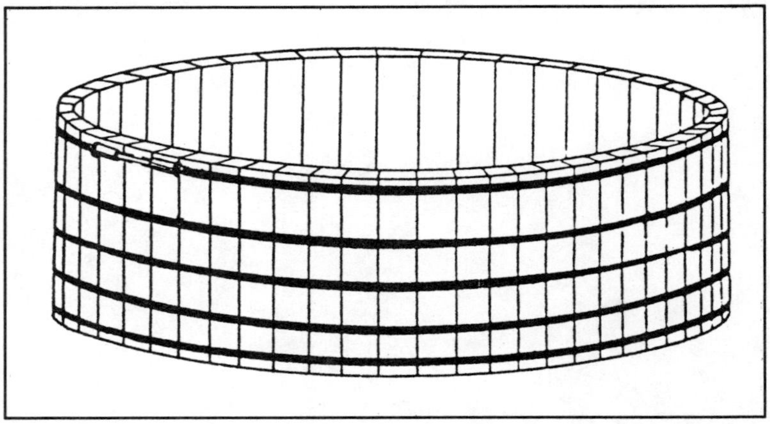

Cyanide leaching tank.
Courtesy of Bud Owens

• 1895

In December, 1895, Mortimer Grant, who was at the time serving as Albany County Sheriff, collected four pounds of black sand from the Douglas Creek placers and sent the sample to New York to be analyzed and assayed. The assay indicated $10.75 of gold and $2.01 of silver per ton. Wilbur Knight, working at the University of Wyoming, had also run an assay on this black sand and estimated the gold value at $9.40 per ton. ("Black Sand," by its very nature, is the heavier concentrate of placer gravels and is limited in amount.) The *Laramie Boomerang*, who had taken over the praising of the Medicine Bow Mines from the defunct *Sentinel*, printed their optimistic view, "There is millions of tons of this sand in the placers of Douglas Creek." [85]

• 1896

The Laramie Mining and Stock Exchange was organized February 1, 1896, with a membership of 64 members consisting of "the best business men of Laramie." Their admitted purpose was to influence moneyed men of the country to invest in the mineral development of Albany County. Their first project was a pamphlet extolling the virtues of Laramie and the Medicine Bow Mountains.[86]

In 1896, Charles S. Crysler purchased the Keystone mill, the Keystone mine and the adjoining Blue Jay for the Breitung-Crysler syndicate. The purchase price, an estimated $12,000, included the mill, which in reality was two mills, one a twenty-stamp and the other the old ten-stamp mill, all the related machinery and 150,000 tons of ore on the dump.[87]

North of the Keystone, the Rambler mine failed to meet expectations and the Holmes post office was closed September 15, 1896.[88]

• 1900

The turn of the century brought a copper boom to nearly all the Medicine Bow Mines. Every mine in the old Douglass District, except one, the "Ida May," switched to copper mining.

The "Rambler" mine in the Holmes District was again active and at the time was considered to be the best hope for a paying copper mine. Water, however, was a problem. The *Laramie Republican*, November 21, 1900, reported:

> A pump with a capacity of 25 gallons a minute is being unshipped from a car this afternoon and will start for the Rambler Mine tomorrow. When the good strike of 25 percent copper was made a few days ago, manager A. T. Holmes started at once for Denver to buy the pump, the water having begun to flow into the mine too rapidly to be controlled by ordinary methods of bailing. The pump arrived from Denver this morning. It will be installed at the mine within the next three or four days.[89]

Although the quality of the copper ore was good, (a sample resides in the Smithsonian mineral collection) the quantity could not compete with the giant open pit mines of Montana, Utah and the Great Lakes regions. Transportation of the ore added to the problems and again paying mines were not to be.

• 1901

By 1901, old Douglass City was again, practically deserted. Only three families remained, "holding the fort." These families were apparently living in the double row of cabins on the west side of Douglas Creek. The

Drawing by Dan Swift

The $11,000 was to be paid in installments and a royalty was added to the price should any valuable ore be found. The company began new operations on the Keystone and Florence mines. A contract for six cyanide vats, 24 feet in diameter and 5 feet deep, was let to a Pueblo firm. At the time it was reported that the old Keystone mill of 1878 was rusted and worn to the point of retirement. The log building housing the old mill had decayed and the roof had caved in. The later mill of 1890, however, was reported in perfect shape an had available both water and steam power. A contract for several hundred cords of wood was let to insure continued operation when the stream flow diminished. Steam was to be piped to the cyanide vats for cold weather operation. Plans were made to treat not only the newly mined ore but also the mine tailings on the dumps which still assayed $7.50 a ton. However, a reasonable profit still could not be realized and the operation closed down still owing the miners and contractors for their work. The Pacific Investment Company was again in possession of the Keystone group of mines.[93]

settlement was by that time generally known as "Keystone."[90]

The Holmes post office re-opened on May 9, 1901, with John R. Cordiner as postmaster.[91]

• 1903

In 1903, several of the old mining entrepreneurs made an attempt to revive the old district with the forming of a new company, "The Medicine Bow Mines."[92]

• 1904

For a short time in 1904, it appeared there would be a revival of the district. The Ideal Mining and Investment Company was formed and incorporated and under the direction of C. W. Bramel bought the Keystone mine from the Pacific Investment Company for a reported $11,000. The sale included not only the Keystone but the adjacent Blue Jay mine and all the associated equipment and rights.

Several of the old time prospectors were still poking around in their never ending search for their personal bonanza. One of these was an old prospector known as Pete Kennesaw. Pete was about seventy years of age and had a small ranch near Sheep Mountain. He also had a cabin at a place called Podunk near Keystone. In early September, 1904, Pete was traveling from Sheep Mountain to Keystone with a string of five burros and his dog. The burros alone showed up at Keystone which led to a search and the eventual discovery of his body at the side of the trail where he had apparently sat down to rest. His faithful dog had stayed with him. The cause of his death was speculated as heart disease and he was buried on the spot by John Hughes and Gus Reed. His grave was marked with a crib of lodgepole pine logs.[94]

Pete was considered as a peculiar character by those who knew him. Some believed him to be a pensioned sailor. When questioned on where he was from he invariably answered, "I came from over the mountain." Locally he was known variously as "Dirty Pete," "Placer Pete" and "Burro Pete." His last name had been spelled Kennesaw, Kennison, Kenisaw and Kennisaw. However, shortly after his death, the *Laramie Boomerang* revealed that his true identity was Egbert Singleton, a former colonel in the Federal Army. He had gained his nickname during the battle of Kennesaw Mountain during the civil war and was apparently known as "Kennesaw Pete" for a time. He and his partner had come to the Medicine Bow Mountains on the strength of some ore shown to him when he was in Montana. They supposedly staked out a claim and after his partners death, "Pete" stayed on guarding the alleged valuable mine. (A thorough search of existing mining records revealed only one claim made by "Pete Kenisaw," a placer claim on the east fork of Beaver Creek.) [95]

On July 21, 1904, the Keystone post office was re-opened, under postmaster Charles H. Marsh. It remained open until November 30, 1910. The Holmes post office continued in operation until 1950.[96]

• 1906

In 1906, Stephen Corlett Downey, son of Stephen W. Downey who had died in 1903, bought the Keystone mines from the Pacific Investment Company for a reported $12,000. Again the note was to be paid in installments and again a royalty payment was added.[97]

• 1907

In 1907, S. C. Downey assigned control of the mines to James Negus of the North Park Copper Syndicate Limited, an organization of "London Capitalists." Once more the "Keystone" began mining "promising ore."[98]

• 1908

In 1908, the "Famous Keystone Mine" was again abandoned. The superintendent, F. Percy Rolfe, planned to sell all the valuable machinery and return to England. The machinery was said to consist of a 30 horsepower engine with a 100 horsepower boiler. The complete cyanide plant was also included. It was estimated that up to that time, the "Keystone" mine had produced $75,000 in gold.[99]

• LATER

The Keystone property sat idle for many years and when the property taxes for 1916 and 1917 became overdue, a public sale was held on June 29, 1918. There were no bids received and the "Keystone" and "Blue Jay" were sold to the county for $11.57. In September, 1919, E. E. Smith of Kansas City, Missouri, paid the county for all the back taxes and within the month, assigned title to the property to Victor Carlin of Wyoming. The required three years elapsed and on March 24, 1923, Carlin became the sole owner of the mines.[100] In 1939, Carlin began selling portions of the old Keystone thus beginning the era of summer cabins and recreational use. For a time in 1939, there was some anticipation of the Keystone again resuming operation under L. J. Landrith. At that time it was said the mill was in good condition but the building was in need of repair. A trial run of the mill made in July, but as in the past no profitable results could be realized.[101]

The Keystone shafts were sealed in the 1950s. The stamp mill building was torn down in 1956.

Hydraulic mining on Douglas Creek.
Courtesy of Lynn Benson

Hydraulic mining, Douglas Mining District.
Courtesy of Lynn Benson

Florence Mine, ca. 1890.
Wyoming State Museum

Town of Keystone, ca. 1890.
Wyoming State Museum

Keystone Mill.
University of Wyoming American Heritage Center

Keystone Mill.
University of Wyoming American Heritage Center

Holmes New Rambler mine.
Wyoming State Museum

Home Placer steam shovel.
University of Wyoming American Heritage Center

The Albany Placer Mining Company

OF LARAMIE, WYOMING.

Officers: C. E. Denig, President; Chas. Culross, Vice President; A. C. Jones, Treasurer; A. S. Harmon, Secretary

Directors: M. N. Grant, J. N. Marks, E. T. Duffey, J. E. Porter, A. H. King, F. B. Cramer.

This company was organized at Laramie, August, 1895, and owns 1,500 acres of valuable placer land in Albany county, Wyoming, on Douglas Creek and tributaries, including Moore's Gulch and Dave creek, Gold Run, Elk creek, Willow creek and others. Over 1,200 acres of this ground is pay gravel running from five to ten feet in depth to bed rock.

The following water rights have been acquired:

DITCH NO. 1.

Is six miles and a half in length and takes water from Douglas creek, and runs so as to work the following creeks and gulches: Mark's creek, Willow creek, Dunn's gulch, Evangeline, Rush creek and Moore's gulch, which it reaches at an elevation of 208 feet above Douglas creek; estimated cost, $1,800.

DITCH NO. 2.

Takes the water from Elk creek and conveys it to the head of Dave creek at an elevation of 140 feet; length of ditch, one-half mile.

DITCH NO. 3.

Takes the water from Douglas creek and Bear creek; length, three and one-half miles, two and one-half of which is now completed to the mouth of Dave creek, where it reaches an elevation of 72 feet. The water passes through a bulk-head, to which is connected an iron pipe 800 feet long, eighteen inches at the head, tapering to six inches at the discharge. The hydraulic plant was not completed in time last season to admit of any extended washing of the ground, but a small test run was made Oct. 12th of twenty-one yards of ground which netted $29.45, and from numerous tests on different parts of the property by panning and ground sluicing, a result of over $3 per yard was shown. The gold is coarse and can easily be saved, samples of which can be shown at the company's office in Laramie.

The company is clear of debt and has 300,000 shares of the capital stock in the treasury, 10,000 shares of which are now offered for sale for development purposes.

For further information address—

A. S. HARMON, Secretary, Laramie, Wyo.

WM. O. DOWNEY. *M. N. GRANT.*

Downey & Grant,

SURVEYORS & MINING

ENGINEERS.

MINING CLAIMS SURVEYED

AND

PATENTS SECURED.

THE BEAVER CREEK

PLACER MINING CO.

CAPITAL STOCK, $250,000. ACREAGE MINING LAND, 960

This company is preparing for active work as soon as the mining season opens.

Persons desiring profitable investments can find none safer than the Beaver Creek Placer Mining company.

A. B. HAMILTON, Secretary.

Reference: Laramie Mining and Stock Exchange.

THE NEW RAMBLER DISTRICT

Buy "Medicine Bow's Mines Company" Treasury Stock. They have the ore. They have the properties. : : : : : : : : : : :

This Stock is Guaranteed to Pay Dividends in 1904.

Will be worth lots more money soon. Send drafts to the Albany County National Bank. : : : :

STOCK $10.00 PER SHARE. CAPITAL 10,000 SHARES.

WM. BENTON CO., HOLMES, WYO.

THE McKINLEY MINING AND MILLING COMPANY

Incorporated Under the Laws of the State of Wyoming.

The property of this company consists of the McKinley claim (which is the western extension of the famous Keystone mine, from which $96,000 has been taken), located in the Douglas Creek district, Albany county, Wyoming.

CHARACTER OF VEIN.—The claim is 1,500 feet long and 600 feet in width, along the strike of the vein. The vein is a true fissure of decomposed quartz, averaging three feet in width, between well defined walls of gneiss, and assays taken from the surface run from $12.50 to $16.50 per ton.

CAPITALIZATION.—The capitalization is $500,000, divided into 500,000 shares of the par value of $1 each, fully paid and non-assessable.

TREASURY STOCK.—The company has donated to the treasury 100,000 shares of the capital stock to be sold for development purposes.

Directors: C. E. Denig, Pres't; W. E. Finfrock, Vice Pres't; L. Tyvold, Sec'y and Treas.; J. E. Porter, Sup't; J. R. Scandlon.

For further information address

L. TYVOLD, Secretary.

Reference: Laramie Mining & Stock Exchange.

The Snowy Range Placer Mining and Milling Company

Its property consists of 610 acres of placer ground located on the headwaters of Douglas creek.

Stocked at 250,000 shares at $1 each.

A. A. JOHNSON, President. L. C. COLBURN, Vice President.
H. D. BEEMER, Treasurer. P. J. GOSDEN, Secretary.

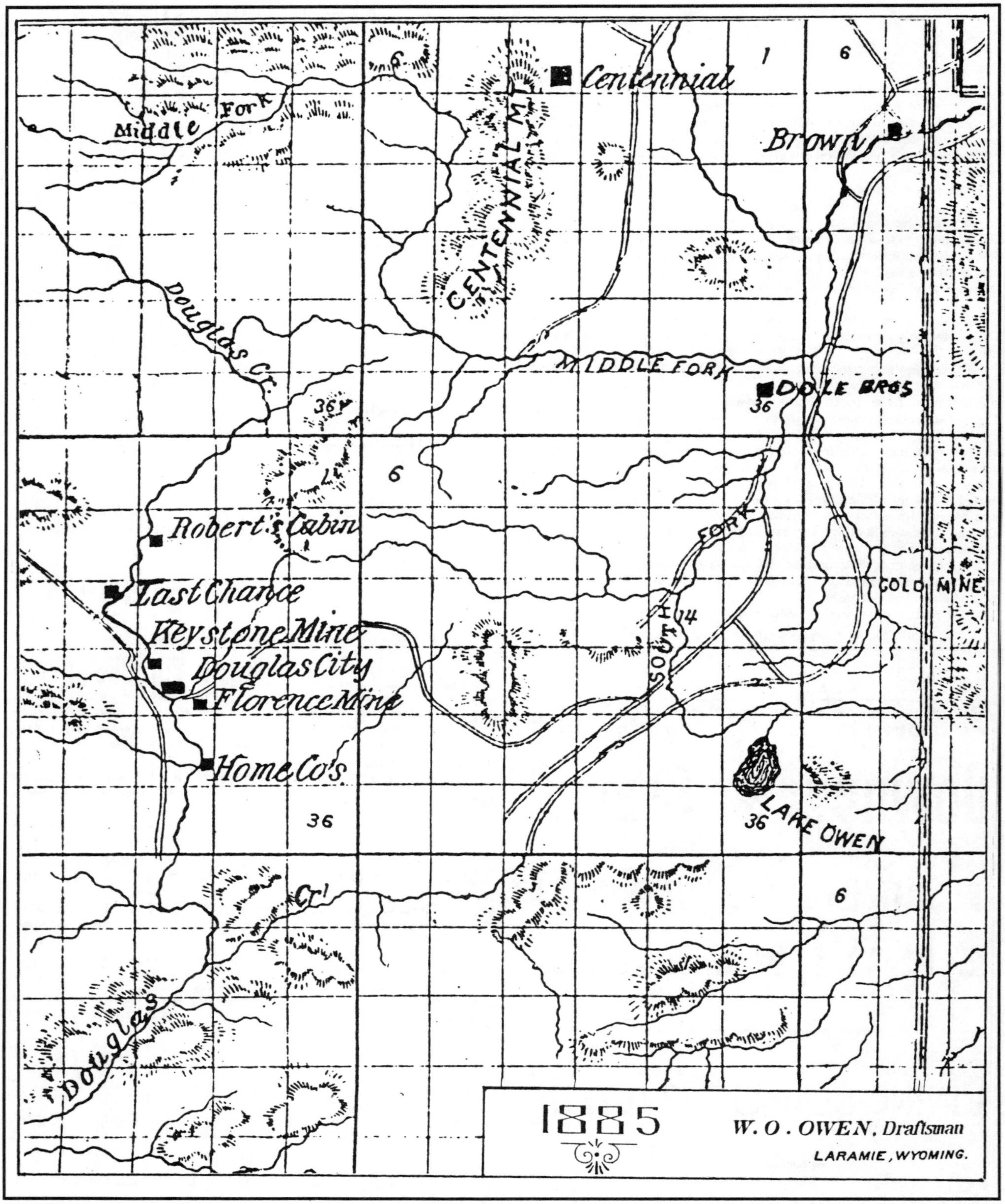
Middle Fork
Centennial
CENTENNIAL MT.
Brown
Douglas Cr.
MIDDLE FORK
DOLE BROS
Robert's Cabin
Last Chance
Keystone Mine
Douglas City
Florence Mine
Home Co's
SOUTH FORK
GOLD MINE
LAKE OWEN
Cr.
Douglas
1885
W. O. OWEN, Draftsman
LARAMIE, WYOMING.

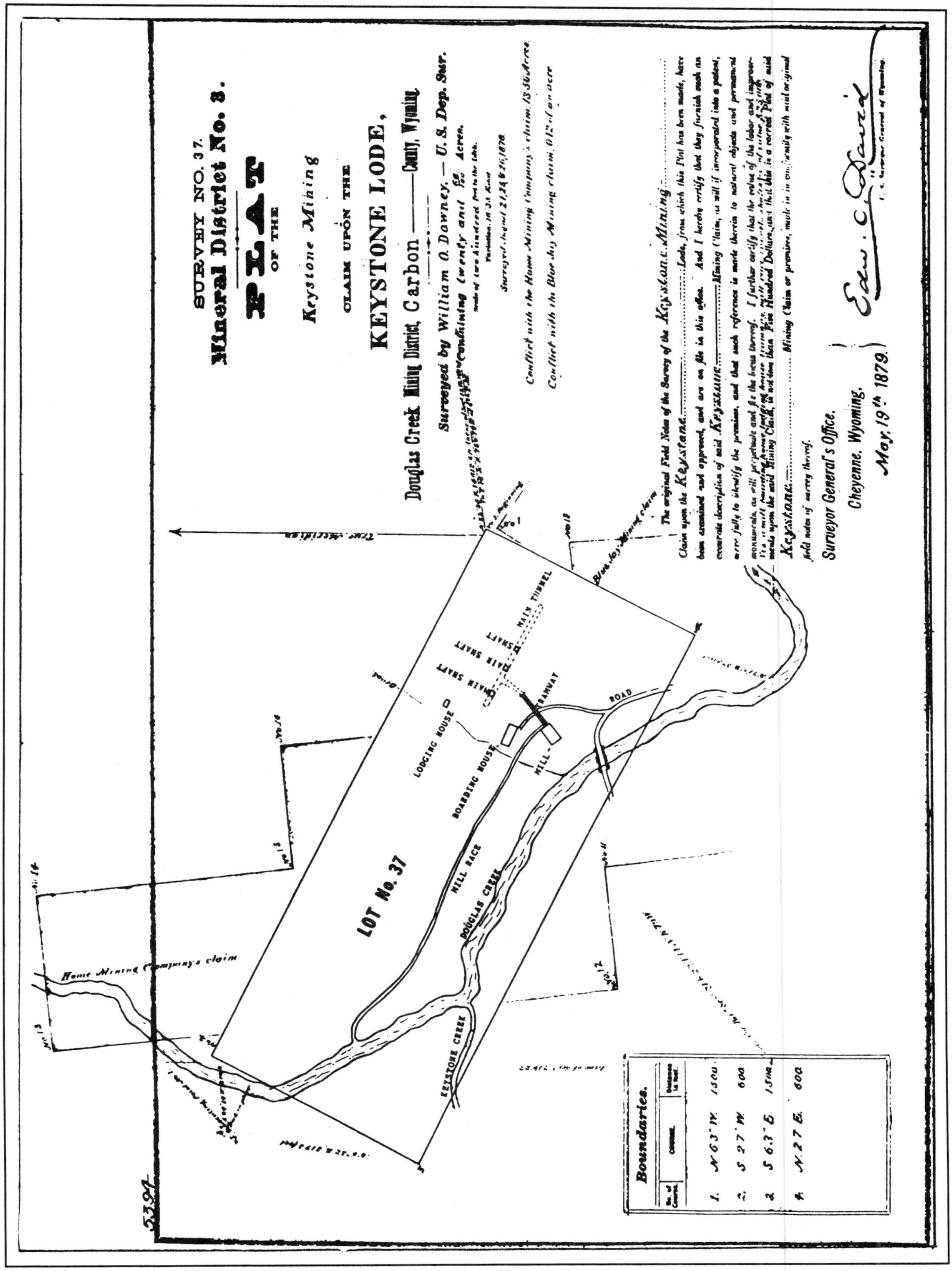
SURVEY NO. 37.
Mineral District No. 3.
PLAT
OF THE
Keystone Mining
CLAIM UPON THE
KEYSTONE LODE,
Douglas Creek Mining District, Carbon County, Wyoming
Surveyed by William O. Downey, — U.S. Dep. Sur.
Surveyor General's Office.
Cheyenne, Wyoming.
May 19th 1879.
LOT No. 37
LODGING HOUSE
BOARDING HOUSE
MILL
MILL RACE
DOUGLAS CREEK
KEYSTONE CREEK
ROAD
TRAMWAY
MAIN SHAFT
MAIN TUNNEL
Home Mining Company's claim
Boundaries.

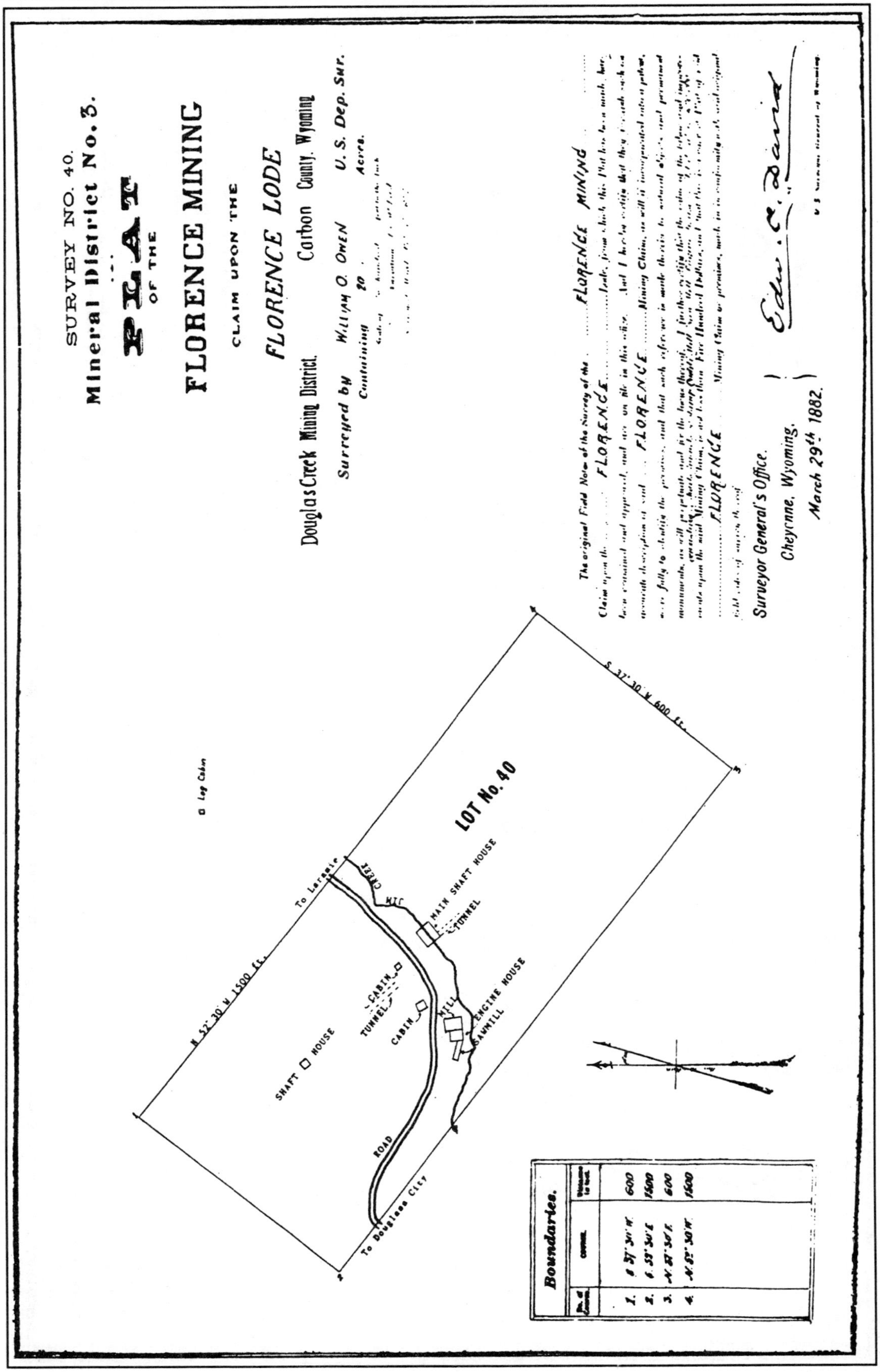
SURVEY NO. 40.
Mineral District No. 3.
PLAT
OF THE
FLORENCE MINING
CLAIM UPON THE
FLORENCE LODE
Douglas Creek Mining District. Carbon County, Wyoming
Surveyed by William O. Owen U. S. Dep. Sur.
Containing 20 . Acres.
FLORENCE MINING
FLORENCE
FLORENCE
FLORENCE
Surveyor General's Office.
Cheyenne, Wyoming,
March 29th 1882.
LOT No. 40
MAIN SHAFT HOUSE
TUNNEL
CABIN
ENGINE HOUSE
MILL
SAWMILL
SHAFT HOUSE
ROAD
To Laramie
To Douglass City
Log Cabin
N 52° 30' W 1500 ft.
S 37° 30' W 600 ft.
Boundaries.
1. S 37° 30' W 600
2. S 52° 30' E 1500
3. N 37° 30' E 600
4. N 52° 30' W 1500

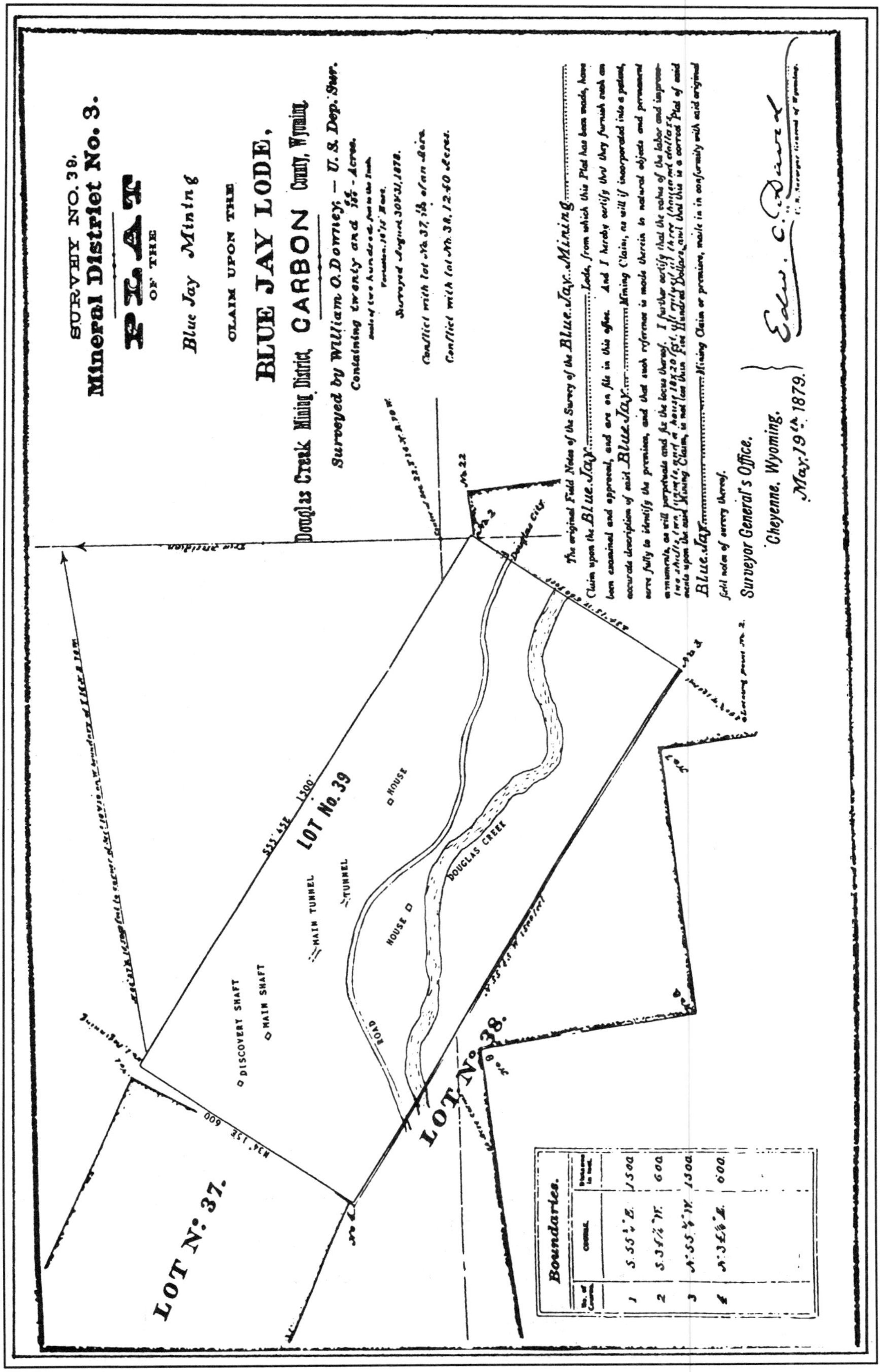
SURVEY NO. 39.
Mineral District No. 3.
PLAT
OF THE
Blue Jay Mining
CLAIM UPON THE
BLUE JAY LODE,
Douglas Creek Mining District, CARBON County, Wyoming,
Surveyed by William O. Downey, — U. S. Dep. Sur.
Surveyor General's Office.
Cheyenne, Wyoming,
May, 19th, 1879.
LOT No. 39
LOT No. 37.
LOT No. 38.
DISCOVERY SHAFT
MAIN SHAFT
MAIN TUNNEL
TUNNEL
HOUSE
ROAD
DOUGLAS CREEK
Boundaries.
S. 55¼° E. 1500
S. 35¼° W. 600
N. 55¼° W. 1500
N. 35¼° E. 600

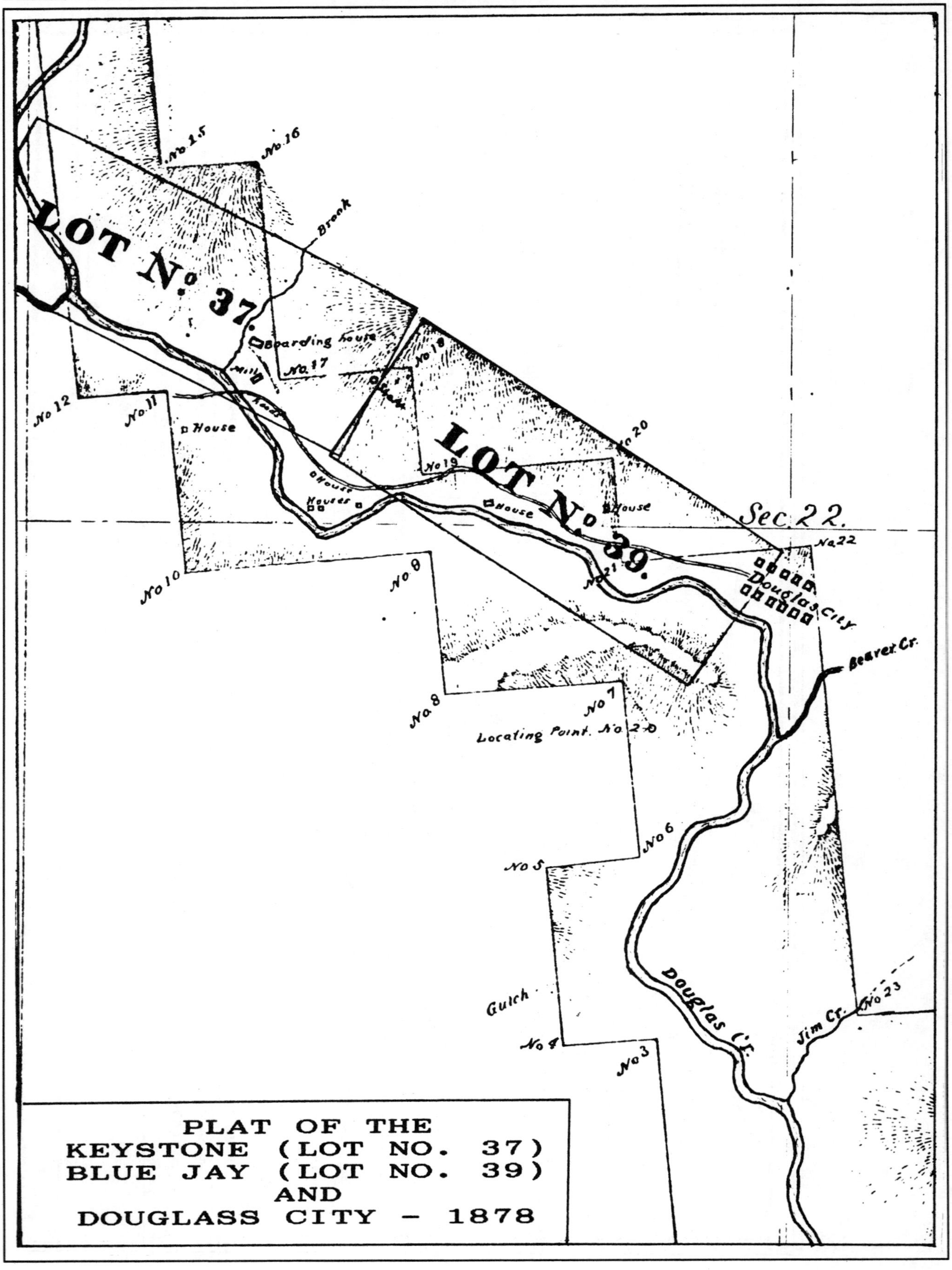

PLAT OF THE
KEYSTONE (LOT NO. 37)
BLUE JAY (LOT NO. 39)
AND
DOUGLASS CITY - 1878

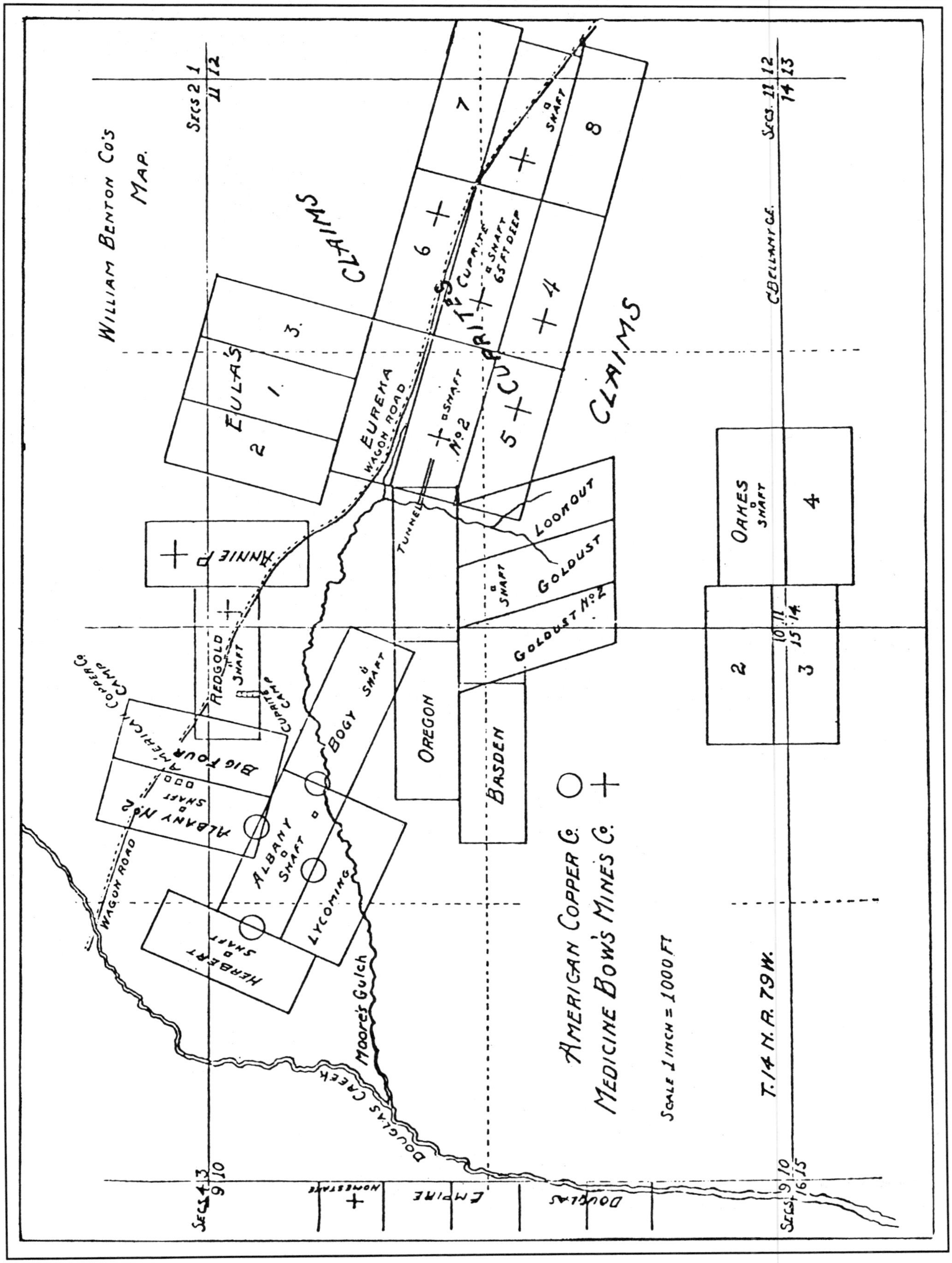

William Benton Co's Map.
American Copper Co.
Medicine Bow's Mines Co.
Scale 1 inch = 1000 ft.
T.14 N. R. 79 W.
Eula's Claims
Cuprite Claims
Eureka
Wagon Road
Annie P.
Redgold
Big Four
Albany No. 2
Albany
Herbert
Lycoming
Bogy
Oregon
Basden
Lookout
Goldust
Goldust No. 2
Tunnel
Oakes
Moore's Gulch
Douglas Creek
Empire
Douglas
Homestead
Copper Co. Camp
Cuprite Camp
Cuprite Shaft 65 ft deep
C. Bellamy Cl.
Shaft

THE OTTUMWA COPPER COMPANY

Capitalization, $100,000

Par Value; $1.00. Non-Assessable.

Claims: SARAH A, TWO MACKS, IOWA, POCAHONTAS.

Located on Lake Creek, 48 miles west of Laramie.

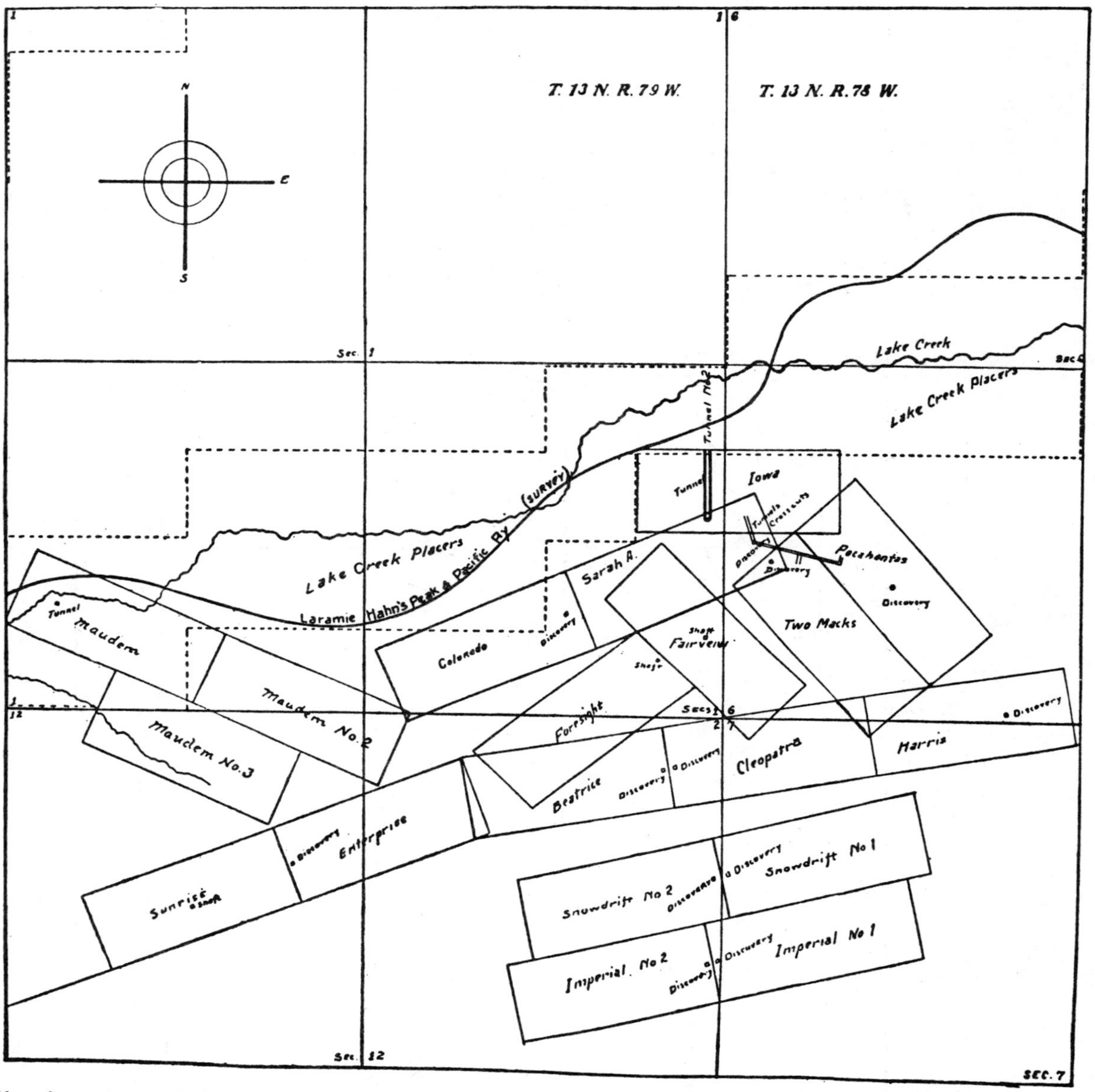

Showing surveyed line of The L. H. P. and P. Ry., crossing the dump of the Iowa, upon which work is now being done.

For all information, prices of stock, etc., address

JOHN MacINTYRE, SEC'Y AND TREAS.

LARAMIE, WYO.

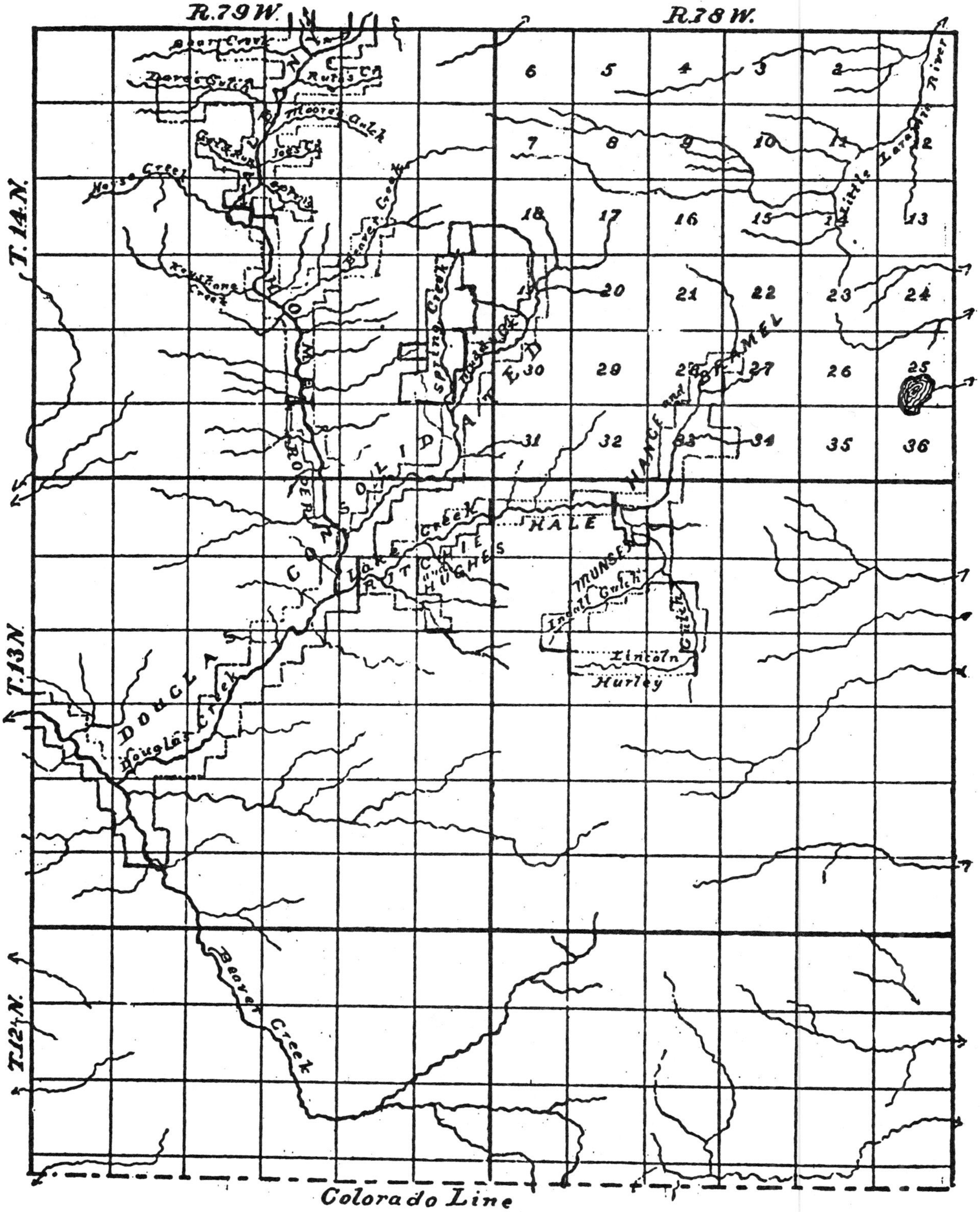

Map of the Douglas Placer District.

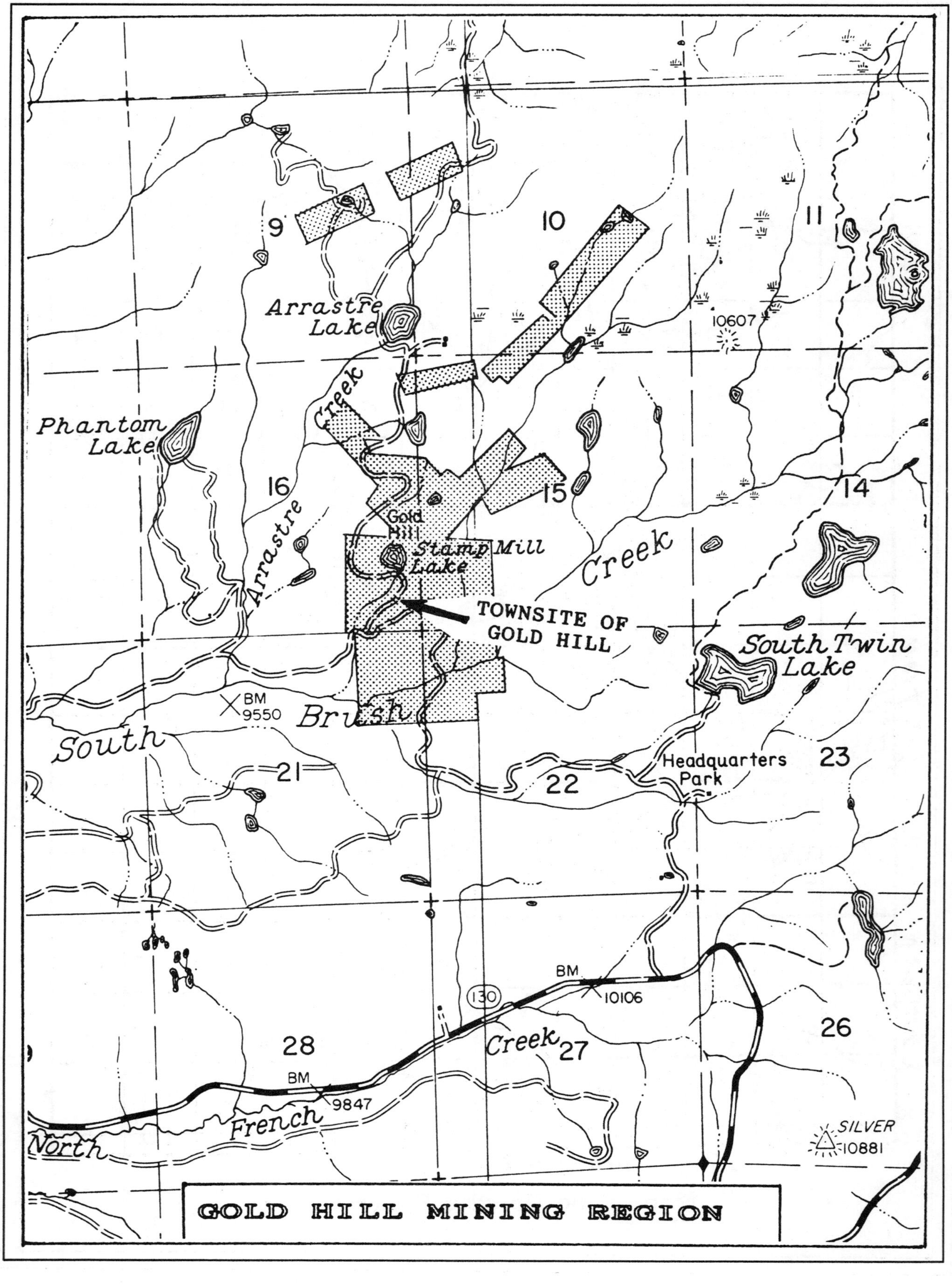

GOLD HILL MINING REGION

VIII GOLD HILL VIII

From as early as 1858, the Medicine Bow Mountains had been considered a likely area for gold prospecting. The very first issue of the Denver *Rocky Mountain News*, April 23, 1859, published a map of the gold regions and indicated access to the Medicine Bow Mountains and North Park from Fort Laramie along the Overland Trail system. However, the region remained largely unpopulated until the Overland Trail was moved south in 1862, and a ferry and small settlement was established at the Platte River crossing. These were probably the first settlers in the area.

In 1868, Fort Fred Steele was built near the railroad crossing of the Platte River. A post office was established and the North Platte River Valley to the south of the railroad crossing began to draw more settlers.

As early as 1868, gold had been reported in the ridges along the North Platte River. In November of that year Captain F. Trench Townshend, traveling with an English hunting party wrote, . . . "On one of the peaks overhanging the Platte gorge we picked up quartz with gold scattered through it. Gold is supposed to exist in considerable quantities on some of the mountains of the Medicine Bow Range, though no mines are yet worked in that part of the Rocky Mountains. . . ." Townshend camped that night on Brush Creek, which was apparently already well known by name.[1]

From about 1849, migrants had used the Cherokee Trail in their travels west. From northern Colorado, the trail diverged, circumnavigating the Medicine Bow Mountains, one branch running from North Park, down the North Platte River Valley to converge with the northern trail before crossing the Sierra Madre Mountains. It is not unreasonable to assume that some rudimentary prospecting had been done along the route.

In 1875, the Brush Creek and Cedar Creek area began attracting prospectors. In June of that year, George Ritter, D. D. Wolff and Robert Burkhardt traveled from Laramie City to the Brush Creek region to investigate reports that a man named Mullison had found indications of silver and copper in the area. A group of Laramie City citizens attempted to follow them but were outwitted when Ritter and his party took the train to Fort Steele, leaving the horses and wagons of the group far behind. Several claims were made before the party returned to Laramie City. The *Laramie Daily Sentinel*, however, declared that these were not new discoveries at all but old discoveries which had been relocated.[2] Existing records do not support the *Sentinel*, but there may have been some unrecorded claims.

Prospecting for the time was restricted to placer mining but soon expanded into the gulleys and streams of the foothills above the North Platte River Valley. During October, 1876, a group of eight prospectors, including Stephen W. Downey, working along South Brush Creek claimed 160 acres at the mouth of Lincoln Creek. They also claimed the water rights for mining.[3]

On the North Platte River a small settlement grew up around the warm springs and on October 4, 1878, the Warm Springs post office was established. Walter French served as postmaster until February 4, 1884, when it was renamed Saratoga with Wilbur B. Hugus as postmaster.[4]

Apparently a few prospectors were searching the placer gravel of the North Platte River and in May, 1883, the 160 acre Bennett Peak Placer on the North Platte River was claimed by a group of eight men, each claiming 20 acres.[5]

In Saratoga, the *Platte Valley Lyre* made its initial appearance on June 7, 1888, and under editor George R. Caldwell, began printing area news.

By the summer of 1888, the Brush Creek Placers, discovered by James Cusick, were drawing attention to the North Platte Valley east of Saratoga. Twenty five men were said to be in one camp working the gravels of Brush Creek. Beverly W. Henry and rancher John B. Hunter were the owners of a number of claims. According to the *Platte Valley Lyre*,

"... gold fever has gradually but surely pervaded the entire valley." "Everyone talks of gold. Old timers tell over and over of the glittering stories of early days of the gravel placers of the Platte. . . ." Stories were told of "... a now prominent ranchman of the Platte Valley who washed gold in the Brush Creek region twenty years ago." [6] This again was apparently John Henry Mullison. Another man, Joe Cassidy, a timberman, had also found gold in the Brush Creek region some years before.

John Henry Mullison, a Civil War Veteran, had come to the area in 1867 and prospected, hunted, cut timber and ranched in and around the Medicine Bow Mountains from that time. Mullison had prospected in North Park and was given credit for the naming of Independence Mountain. In 1900, he joined the Forest Service and served 11 years. He helped organize a school district and became a school board member. He was apparently well thought of in Saratoga and Laramie and was said to be the most energetic man in the valley. Known locally as "Hank Mullens," both the Mullens Creeks in the Medicine Bow Mountains were said to be named after him. For sure, Mullison Creek and Mullison Park carry his name.[7]

In the Fall of 1888, so called "rich placer claims" were made in the Gross Gulch area between Cedar and Brush Creeks. On October 25, 1888, the "Henry Placer Claim" was made by Beverly W. Henry, James Cusick and Josiah M. Morrison. This was the first claim of record in the newly formed Cusick Mining District and was duly recorded by Alfred Heath, the Cusick District recorder.[8] Shortly thereafter, on November 8, the "Elk Placer Claim" was made on Buffalo Gulch, by John Henry.[9]

Stories of lost mines and early mining in the valley were being told and retold. One story was of a prospector named Bradfield who had come into the region around 1864. Bradfield, apparently prospecting by himself, came upon the warm springs along the North Platte River. From here he prospected up a tributary to a deep gulch where he, according to the story, found quantities of coarse gold. In two days work, Bradfield mined $80.00 (about 4 ounces) of gold. Indians surprised him and he was forced to flee leaving all his belongings, including his pick and shovel, behind. Bradfield made his way to the Virginia Dale stage stop and then to Denver. The following year Bradfield organized a party of well armed men and returned to the area in search of the gold find. After a season of searching he was unable to find either the gold deposit or his belongings. The lost mine became known as the "Lost Pick and Shovel" mine. The story was widely circulated and by 1888 speculation placed it in the Brush Creek region.[10]

In the wake of the Lost Pick and Shovel mine story, the "Lost Pick Placer" mining claim was made on Tip Top Gulch on November 12, by Joseph B. Hassett[11] and the "Lost Shovel" claim on November 17, 1888, by H. M. Bennett,[12] both in the Cusick District. Numerous claims followed, most recorded as being located on Buffalo Gulch or Tip Top Gulch. None of these claims extended into the mountains, but were made in the placer gravels where the streams ran out to the foothills and plains. The prospectors reported "good colors" in all the panning and sluicing being carried out and much speculation was made as to the source of this gold.

• 1889

Cusick spent the winter in Casper but was back on the Brush Creek Placers by late March, 1889. He brought with him not only the financial backing from Casper but also several new mining partners. Also on the Brush Creek placer claims were H. M. Bennett, C. T. Waldron and Fenimore Chatterton who were installing a hydraulic sluice to mine the placers.[13]

In April, 1889, the Cusick District was reorganized and new district boundaries were established; the Medicine Bow Mountains on the east, Pass Creek on the north, the North Platte River on the west and French Creek on the south. Al Heath was reelected recorder. The *Platte Valley Lyre* reported, "Saratoga is fast putting on all the mining activity of a mining town. Mining implements and supplies are in brisk demand in Saratoga and our merchants are laying in a new stock."[14] H. W. Maghee brought down enough lumber from Ryan's Cedar Creek sawmill to manufacture 21 sluice boxes for the placer miners. Some eighty claims were made up and down Brush

Creek and it's tributaries. Forty-one men were at work on the placer claims.

On April 4, forty cents in gold was washed from the stream, on April 18, sixty cents was obtained and on April 25, a gold wafer as large as a man's thumbnail was mined. Minuscule amounts to be sure but enough to encourage the miners to continue their search.

During April, the Platte Valley Hydraulic Mining Company was formed and incorporated to work Buffalo Gulch. J. G. Rankin was president and Fenimore Chatterton was treasurer of the new company which was capitalized at $40,000.[15]

By May 2, fifty or sixty men were said to be working the placers and finding traces of gold in all the gravels of Brush Creek and its tributaries.[16]

Evidence of gold in the placers of the lower creeks led many of the prospectors higher and higher into the mountains searching for the source. One of these prospectors, Benjamin P. Arendell, "A native of Texas and a prospector by profession . . . , had prospected from the Mexican border to Canada." Arendell and several other prospectors spent the summer of 1889 searching the upper reaches of Brush Creek. Finally on September 5, 1889, they located a claim which they called the ". . .Patent Right lode or ledge which is located in Carbon County. . .about 3 miles from the base of Snowy Range, 3/4 miles from the summit of the low range between Brush Creek and tributaries of the Medicine Bow; Township, Range and Section unknown." The location notice was filed by Arendell, Charles M. Drake, S. D. Phillips and V. H. Patterson on November 6, 1889.[17]

These same four prospectors also located the "Passtime Lode" on upper French Creek.[18] Arendell and Charles Drake then located the "Little Giant" on a tributary of Brush Creek.[19] The "Little Giant" was to become the best known claim of 1889 and was instrumental in attracting more miners the following year.

Nearby, another prospector, W. O. Drake (probably a brother to Charles Drake) located the "Centinal" (sic) lode.[20] In the same vicinity, Henry M. Johnson located the aptly named "Henry M. Johnson Lode."[21] All of these claims were located in September and filed in the Carbon County Courthouse on November 6, 1889.

After the November filing there is no evidence to indicate that any of the prospectors returned to the upper Brush Creek area until late summer of 1890. However, there were two other prospectors who apparently traveled into the high country in the dead of winter. Fred Barclay and S. A. Lawson stated that they had located the "Home Patent Mining Claim" on January 2, 1890, which they filed in Rawlins on January 14, 1890.[22]

• 1890

Due to the deep snow and the late runoff on the western slope of the Medicine Bow Mountains, it was July and August of 1890 before prospectors were again active in the upper Brush Creek area. By late summer and early fall, reports began coming down from the high country, of new discoveries, a new district and new hopes for a rich gold camp.

The *Laramie Boomerang* reported that Jack Sharp of the Keystone mine was in Laramie with ore samples from the head of Brush Creek. The white quartz samples "showing free gold" stirred considerable interest in the men at the Douglas Creek camp and Sharp said that many of them had departed for Brush Creek, among them, David B. Dole, the Keystone postmaster. With him went Benjamin W. Towner, Thomas Hale and Henry P. Dahl, all Douglas Creek miners.[23]

The *Laramie Boomerang* also reported that three of Ford Barkley's men came in from Brush Creek with stories of a rich new discovery on Barkley's Brush Creek claim. "These facts becoming known a good rush is being made to the Brush Creek region and there is already quite a large settlement in there."

The *Cheyenne Daily Leader* reported:

> A new mining district has been organized at the head of Brush Creek in the Snowy Range. It has been called the Chase Mining District, after L. D. Chase, an old Colorado miner, who made the first important discovery in the district. Charles Benton was elected recorder. D. B. Dole, B. W. Towner, Charles Benton, Thomas Hale and others were the organizers. The district is in Carbon County. It contains six or seven fine prospects. The leads are large and the ore rich. There will be considerable development work done this fall.[24]

In little more than a week the regional newspapers were calling the new district

"Gold Hill." Stories of new gold discoveries were always newsworthy and the *Denver Republican*, the *Rocky Mountain News* and the *Salt Lake Tribune* soon followed with stories of the new district. Shortly thereafter the more distant *Omaha Bee*, *New York World* and *St. Louis Globe Democrat* also picked up the news. By the end of October, Gold Hill in the new State of Wyoming was known, at least to the readers of mining news, across the United States. For the next two years the struggling miners at Gold Hill enjoyed favorable, although sometimes overly enhanced, news coverage.

By October 4, 1890, the *Laramie Sentinel* estimated there were 100 miners at work at Gold Hill, probably an exaggerated guess at best, but indicating nevertheless a robust interest in the new area. Ben Towner, who made his second trip from Keystone in early October estimated there were about 30 men working at Gold Hill.[25] Whatever the number, the Laramie Board of Trade speculated there would be 5,000 to 10,000 people going into Gold Hill in the spring and summer and proceeded to organize a company to build a road and operate a stage.[26]

The Union Pacific coal town of Carbon quickly recognized that they were the closest railroad town to the mines and they too began to organize a group to build a road from Carbon to Gold Hill.

Saratoga, although without a railroad, was still the closest settlement and had, already in use, a road into the sawmill on lower Brush Creek. A number of citizens began working on this road to extend it to Gold Hill.

The *Wyoming Commonwealth* remarked that if the camp should prosper, all the roads would be used and the entire county would benefit.

By mid-October, the winter weather that had effectively stopped the miners the previous year, began anew. The first major snowfall of the season stopped all travel into Gold Hill. Charles Benton and several others left Laramie for Gold Hill by way of the old La Plata road. They got as far as Gressley and Robin's cabin where they were stopped by snow which they reported to be ten feet deep. Ben Towner and several other miners attempted to travel from Keystone to Gold Hill. They too were turned back by heavy snow.[27]

THE ONLY

Practicable Route to

GOLD HILL

IS VIA

SARATOGA

The Center and Supply Point of

The Upper Platte Valley.

Some concern was expressed for the now estimated 30 or 40 men stranded at Gold Hill. Vinet Ferguson, of the Saratoga Stage and Stables, however, made it into camp with two wagon-loads of supplies, estimated to weigh 4,000 pounds.[28]

A brief respite in the winter weather apparently encouraged the miners to stay and make ready for the impending weather.

It appears the miners thought more of the mining opportunity than they did of personal shelter. When Ben Towner finally made it into camp, he reported only three cabins were constructed for living quarters but several shaft houses and tunnels were complete.[29]

In Rawlins, the county seat, the miners began filing claims for their recent discoveries. The "State of Wyoming" lode claim filed on October 13, 1890, was said to have been located July 10, 1890, the day Wyoming became a state. The "Crawford Lode," "Smuggler," "Conflict," "Bill Nye," "Challenge," "Podunk," "Leviathon No. 1," "Leviathon No. 2," "Omaha Lode" and the "Tin Horn" were all located and filed before the end of October, signaling what the miners believed would be the end of the mining season.[30]

A meeting was held November 1, 1890, at the cabin of Ben Arendell and Pat Woods for the purpose of forming a new mining district. Twenty-eight miners attended and proceeded to elect J. F. Crawford, Saratoga attorney, as chairman, James J. Hopkins as secretary and William H. Greene as recorder. Greene proposed naming the new district "Gold Hill," to which the majority agreed. Boundaries were specified as those of Townships 16 and 17, Ranges 80, 81 and 82. (The boundaries of the new district were in conflict with the Chase District and also the older Cusick District. Notably absent from the meeting were any of the Chase District organizers).[31]

Anticipating a booming gold camp, the miners drew up a petition for a post office to be called "Gold Hill" and proposed that Alexander Lute become the postmaster. They further proposed that Joseph M. Cassidy become the constable and W. H. Greene, Justice of the Peace.[32] Within a few days following the meeting a group of nine miners located and filed on 160 acres of land, called the "Gold Hill Placers," to the northwest of Gold Hill, which they admittedly planned to hold for a townsite rather than a placer mine.[33]

About the same time, L. D. Chase refiled on the "Leviathon No. 1" and "No. 2," amending his location description to indicate their location in the new district.[34]

Although the northern boundary of the Gold Hill District had been surveyed in 1870 and the township boundaries in 1877, the surveyors and miners experienced difficulty in precisely locating the claims. To alleviate this, a United States Locating Monument (U.S.L.M. Gold Hill No. 1) was constructed in November, 1890. All of the subsequent surveys utilized this monument as a reference to more accurately locate the claims.[35]

Colonel Stephen W. Downey, unsuccessful at Last Chance, Centennial and most recently Douglas Creek, took a not unexpected interest in the new district. During October, 1890, he was in Denver assaying and testing ore from several of the Gold Hill mines. Apparently convinced that the ore was valuable, he returned to Wyoming and began negotiations for a stamp mill for the camp.[36]

Meanwhile the Laramie road building company realizing the futility of attempting road construction across the front of the Snowy Range in the dead of winter, contemplated a different route which would utilize the old Mill Creek tie road to Rock Creek Park then through a pass with an easy grade across the "backside" of Snowy Range and into the camp from the north.

N. F. SPICER in the 1880's

Nicholas Spicer, in Laramie after three weeks at Gold Hill, said he believed the road could be built for $500. Lee Hughes of Laramie guaranteed he could put in a road that would enable one team to haul 2,500 pounds.[37] Hughes and Arendell traveled to the John Webb ranch and after investigating the alternatives they too decided the Mill Creek road would be the best choice. Arendell and Charles Drake had already started a road from the Gold Hill side and the Mill Creek road was planned to connect with that road. Drake was chosen to do the timber cutting work and Hughes the necessary grading.[38]

The number of miners expected to "winter over" at Gold Hill steadily increased. Stephen Downey negotiated a share in several mines. Lionel Sartoris (of the Douglas Willan Sartoris Land and Cattle Company) entered into a contract with Pat Woods, Arendell and Franks to work four claims in which he owned

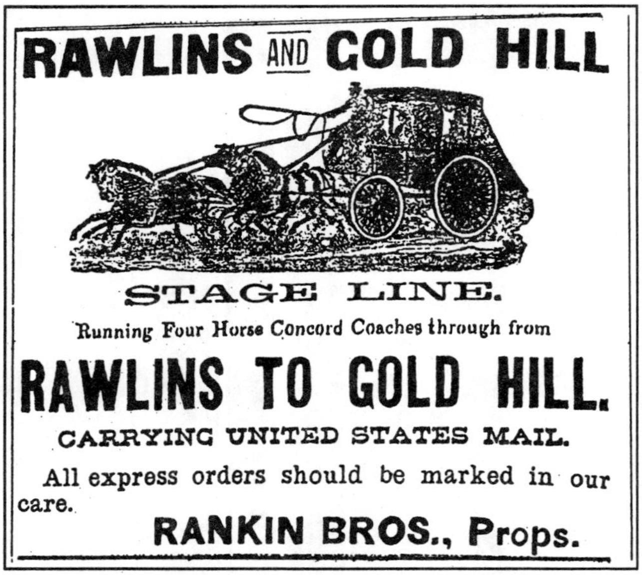

an interest. Mr. Louis Johnson, a "mining expert" from Park City, Utah, closed a deal with the Greene brothers, Dorr Keech and Arendell to dig a 325 foot tunnel on the "Acme," "Lakeside," "Enterprise" and "Prospect" claims.[39]

"Postmaster" (not yet official) Alexander Lute visited Laramie and reported he was building a post office building which was to be 16 by 32 feet.

The day before Christmas, 1890, Ben Arendell visited Saratoga bringing with him about 25 pounds of ore from the "Wyoming" mine and about the same amount from the "Leviathon." All of the ore was reported to show free gold and was sent to Cheyenne for the impending Mineral Convention scheduled for the first week in January, 1891.[40]

According to various sources, by the end of 1890, some 70 to 175 claims had been made and between 30 and 50 men were working the most promising. Shaft houses had been built over eleven of the most active mines and numerous cabins, lean-tos and tents were erected for shelter from the onslaught of winter.[41] Two camps were established, the upper near the working mines and the lower below Gold Hill on Gold Run Creek, a tributary of South Brush Creek.

• THE MINING CONVENTION

On December 30, 1890, the Wyoming Mining Convention convened in Cheyenne. Both houses of the first state legislature, which was in session at the time, attended the opening ceremonies as a group. Stephen W. Downey was elected to preside over the convention. In his opening address, Downey compared Wyoming's Medicine Bow Mountains to Colorado's mountains which had produced many millions of dollars in precious metals. "There is no doubt," he said, "that we have a second Leadville in the Medicine Bow Mountains in Carbon and Albany Counties, as we have seen strong geological formation and find the same mineral. . . ."

During the course of the convention, each of the representatives from the 13 attending counties, presented a paper on their county's mineral wealth. Representing Albany County, Professor Wilbur C. Knight presented the first paper extolling the mineral wealth of Albany County including the Douglas Creek mines.

John F. Crawford of Carbon County presented his paper on the "Precious Minerals of Carbon County" which included a description of Gold Hill and the progress achieved.

> . . . It is reached by a good wagon road from the town of Saratoga, over which road mail, express and freight are being regularly carried. . . . There are, here and there, within a radius of a mile, about a dozen mining cabins, erected within the last ninety days. In that altitude the snow falls early, and lies late, giving not more than four months of bare ground to prospect over.

Crawford, an attorney from Saratoga who acted as agent for many of the miners, was chairman of the Gold Hill Mining District, owned shares in several mines, and could hardly have been expected to present a modest report on the district, continued with a description of the mines:

> Gold is found in all in varying quantities from $10.00 to $1,000.00 per ton . . . the Leviathon No. 1, $125.00 per ton . . . the Little Giant from $300.00 to $2,800.00 . . . The Wyoming from $47.00 to $4,500.00 and the Laramie and Little Bonanza $125.00 per ton.

(At the time, it was estimated that values above $2.82 per ton were profitable when processed with a twenty-stamp mill).

During the course of the convention, Alex Lute and Wilbur Knight presented a gold panning demonstration using ore from Gold Hill. The four day session concluded with the forming of a permanent organization calling themselves "The State Board of Mines." A trustee from each attending county was named, among them, Downey from Albany and Crawford from Carbon county.[42]

• 1891

At Gold Hill, snow fell every day of the first week of 1891. Strong winds on January 5, 6, and 7, piled the snow in great drifts. The onset of winter was creating new problems with transportation to and from the camp.

One traveler, a Mr. Griffin and his son reached Gold Hill from Carbon on January 6, packing in an iron cook stove and supplies. His route of travel from Carbon was through Cedar Pass to South Brush Creek and then joining and traveling up the Saratoga route to Gold Hill. The original route from Carbon was completely snowed in.[43]

The Laramie to Gold Hill road was little better, however, the initial road did finally reach Gold Hill on January 20, after 18 days of tough work, only to be snowed in again almost immediately. To relieve the anticipated communication problem, C. F. Annett, of the Rocky Mountain Bell Telephone Company, proposed a branch telephone line from Laramie to Gold Hill.[44]

Although the Saratoga stage was reported to be running daily, the teams could not pull the sleds all the way to camp and necessities were packed in on "snowshoes." The "snowshoes" were described as being 8 to 12 feet long, 3 to 4 inches wide with a leather binding and propelled by a single or double pole, usually single. Often referred to as Norwegian snowshoes, they were obviously what we now call nordic or cross-country skis.[45] Some were ordered from as far away as Norway and some were crudely crafted by the miners. A Laramie newspaper noted, "A Laramie carpenter made two pair of snowshoes of native pine eight feet long, four inches wide and one and one-half inches thick for the Gold Hill men." These "snowshoes" were a new mode of travel for most of the miners and several quickly adapted to them and traveled at will throughout the winter.

Using the "snowshoes," the miners found they could not only get around in the deep snow but travel was actually easier and faster than in the spring and summer. No longer restricted to slogging from cabin to mine shaft, these residents expanded their travels to the entire Medicine Bow Range.

In early January, Charles Benton and Sam Love "snowshoed" from Gold Hill to the Gressley-Robins ranch in the Centennial Valley in one day, spent the night at the ranch and then traveled on to Laramie by team the next day.[46] Benton became quite proficient with his "snowshoes" and journeyed, mostly alone, across the Snowy Range seemingly at will. Only once did he admit to having difficulty, a difficulty which almost cost him his life.

Benton had made a trip to Laramie and was returning alone. He had left Laramie early in the morning and caught a ride to the Ed Nellis ranch and then walked about a mile to the Peter Johnson's ranch where he spent the night. The next morning, declining a lunch offered by Mrs. Johnson, Benton walked across four miles of bare ground to where he put on his "snowshoes." The day was bright and warm and the snow stuck to his "snowshoes," making slow and tough going. By 11 o'clock he stopped long enough to eat the meager lunch he had brought from Laramie, already wishing he had accepted Mrs. Johnson's offer of a more substantial lunch. By 2 o'clock, he reached the head of Rock Creek, by his own calculation already four hours late. By 4 p.m. he cleared the timber and found himself confronted with wind and blowing snow so strong he could scarcely stand upright. At nightfall he was directly north of the highest peak where he elected to camp for the night. Using a small hatchet he had brought along, he built a fire and kept it burning all night. Snow continued falling throughout the night. With no supper and no breakfast, Benton left his meager camp at 6 a.m., heading for Gold Hill. Snow was still falling, cutting visibility to 20 yards. Sinking to his knees in the soft snow which stuck to his "snowshoes" he continued all day. Compelled to stop every ten feet to rest, he continued until darkness again overtook him. The area looked familiar to him but he could see no signs of the camp at Gold Hill. Using his last bit of energy he "hallowed" as loud as possible. To his relief he was answered by one of the miners, Mr. Griffin, camped in a tent above the main camp. He was soon eating his first meal in 32 hours. Suffering from frostbite of the face, nose, ears, feet and fingers he fully expected to be completely recovered in a few days.[47]

Drawing by Dan Swift

The post office at Gold Hill received its commission on January 9, 1891. As earlier recommended, Alexander Lute was designated postmaster. The new post office was given the official name of "Gold," as the name "Gold Hill" was already in use.[48] As in most of the camps, the post office became a social and political gathering place. Visitors to the camp were finding the only overnight accommodations at the post office located at the upper camp in Lute's cabin. From all reports it appears that from 25 to 40 men spent the winter at Gold Hill. For the average miner the isolation and loneliness was difficult enough, but for Alex Lute, with his fourteen year old son, the remote camp life apparently became a little too much. On January 25, 1891, after a trip to Rawlins to obtain the necessary license, Alexander Lute and Mollie Berry were married at the Hot Springs Hotel in Saratoga.[49] Mollie became literally and figuratively the first lady of Gold Hill.

Even though Gold Hill was still locked in winter weather, predictions and anticipation of a great mining camp were being printed in the regional newspapers. The *Laramie Republican* published an optimistic prediction that by the end of July, Gold Hill would be a great mining camp with Laramie as the center of all the activity with 12,000 strangers in the city. They further forecast that a $500,000 depot and hotel would be built and Gold Hill would be making monthly shipments of $175,000 in gold bullion. They could also foresee three, 150 stamp-mills in operation and a 220 stamp-mill on the way.[50]

The *New York Times* January 1, 1891, published a story of a rich silver strike in the Gold Hill District which they said contained 2,000 ounces to the ton. The story seems to have originated with the McAllister Brothers Podunk Mine located a mile and a quarter from the Greenville townsite, from which an assay of 1,886 ounces of silver was obtained.

Competition among the various towns for a serviceable all-weather road into the camp now included Rawlins, Fort Steele, Carbon, Rockdale, Laramie and Saratoga. Rawlins and Fort Steele were vying for a branch railroad line to Saratoga.

Lee Hughes and Jake Farr had been working on a road from Laramie to Gold Hill and during the last few days of January were in

Acme Mine, Gold Hill, Wyo. ca. 1892
Wyoming State Museum

Laramie. The *Laramie Republican* January 28, 1891, reported:

> . . . Mr. Hughes reports that the road he built in from Laramie across the north end of the Snowy Range is complete and well broken. He had no trouble whatever in getting through with his load. There are now about thirty-five men in the camp and most of them are working like beavers on various locations.

Laramie began pressing for a mail route directly to Gold Hill, anticipating using the Hughes' road. The road, however, was quickly closed again by drifts of snow and most of the travelers continued to use the Saratoga route.

Travel from the railroad terminals was primarily from Rawlins and Fort Steele with Rawlins getting the edge over Fort Steele by introduction of a six-horse Concord stage coach, the first of which arrived in Saratoga on April 15, 1891.[51]

When the Laramie Board of Trade met in early March, they were visited by Stephen W. Downey, George Morgan and Wilbur Knight. Morgan proposed a stage line and steps to keep the Laramie to Gold Hill road open for all season travel.[52]

Even Cheyenne was taking an interest in the budding gold camp and when a staff correspondent of the *Cheyenne Daily Leader* visited the camp in March, he counted about fifty men and four women, Mrs. J. M. Morrison, Mrs. Harbison, Mrs. Currigan and the recently married Mrs. Lute.[53]

The miners held a meeting at the Gold post office on April 1, 1891, and whether to fulfill an immediate or perceived future need, elected Grant Young as Constable and Alexander Lute as Justice of the Peace.[54]

By mid-March, about 125 claims had been made of which only eleven were being worked. The "Leviathon" was said to be the most developed and was down 75 feet. The "Acme" was down 20 feet, the "Wyoming" was said to be the most valuable, and the "Enterprise," the "Laramie," the "Conflict No. 2," and "Lone Jack" all reported very promising veins. One thousand tons of "pay rock" was claimed to be on the dumps ready to process.[55]

Financial backing, including plain old "grub staking," was being provided by business men of Laramie, Saratoga, Rawlins and Carbon.

Pat Woods, one of the more practical miners, apparently recognized the uncertainty of his venture at Gold Hill and wrote to George Morgan, manager of a large ranch on

Rawlins &

Gold Hill

STAGE LINE

RANKIN BROS. -:- Propriet or

Six-horse Concord **Coach leaves**

RAWLINS DAILY

-o-o-for-o-o-

GOLD HILL.

VIA SARATOGA.

First-class Accomodations

In Every Respect

Address Rankin Bros. Rawlins.

the Laramie Plains; ". . . if the lead holds out, I will walk out of here on silk hats, and if it does not I will be out in July to get a job of putting up hay."[56]

The mid-March snow depth at Gold Hill was ten feet and snow was still falling.[57]

On the Saratoga to Gold Hill road, John C. Brewer and Joe Lucien built a new stage station about ten miles from Gold Hill which they appropriately named "Ten Mile House." A daily stage line was in operation between Saratoga and Gold Hill, although there is occasional reference that the "daily stage" was running three times a week.[58]

The U.S. Mail arrived at Gold Hill in an official mail sack for the first time in mid-March. It was carried by "snowshoe" from Greene's in the lower camp to the post office in the upper camp.[59]

E. J. Lewis reported that a petition had been forwarded for an additional mail route between Laramie and Gold Hill. About the same time, a telegraph line from Saratoga was proposed.[60]

Gold Hill at this time had no store but W.B. Hugus & Company of Saratoga owned a building which they planned to stock with goods for the camp.

By April 1, the snow on Brush Creek was beginning to melt causing new problems. The stage from Saratoga was not yet running the promised Concord coach. The stage was a "four-in-hand bus" capable of carrying 12 passengers. The stage made the 20 mile run to Headquarters where the passengers changed to a horse drawn sled which took them to the cabin of W. H. Greene in the lower camp. From here the traveler was required to don snowshoes for the final one-half mile to the post office and hotel in the upper camp. Cost for the trip from Fort Steele, the nearest railhead was:

Bed and breakfast at Fort Steele	$1.00
Stage fare to Saratoga	2.50
Supper, bed and breakfast	1.50
Stage to Gold Hill	2.50
Meal at Headquarters	.50
Total	$8.00 [61]

In Laramie, the Gold Hill Investment Company was organized and incorporated on April 2, 1891. A modest company with capital stock of $1,200, the five members of the company planned to deal in real estate and mining properties at Gold Hill.[62]

George Canis, most recently a correspondent of the *Omaha Bee*, was in the camp examining the mines. He apparently saw enough in one day to convince him the mines were valuable. Canis, soon after, announced that the Union Pacific would build a branch line from Fort Steele through Saratoga "almost" to Gold Hill. He did not disclose the source of his seemingly exclusive information. However, at least he was convinced, and filed claim to 480 acres of land about a mile below Headquarters for a townsite and railroad station.[63]

With the promise of Spring and a booming camp, sales and trades were taking place and in April, 1891, the Carbon *Black Diamond* reported that "Old Man" [L.D.] Chase had sold his "Hackney" claim for $5,000 and was no longer a prospector but a capitalist loaning money to hard up prospectors.[64]

Although the closest railroad station was Fort Steele, the preferred route seemed to be from Rawlins. The Rankin Brothers were running a four or six-horse Concord Coach from

GOLD HILL

HOUSE,

GOLD HILL, WYOMING.

Meals 50 cts. Beds 50 cts.

BOARD By The WEEK, $10.00

Lewis, Lute & Poutre,

Proprietors.

Rawlins to Saratoga, advertising daily service but apparently running only three times a week. From Saratoga, Vinet Ferguson ran a connecting stage to Gold Hill on alternate days. In Saratoga, the Rawlins, Saratoga and Gold Hill Road Company was formed.[65]

As the spring thaw began in earnest, travel became increasingly difficult. The preferred manner of travel was to take the evening stage from Saratoga to Headquarters for an overnight stay. Then at 5 o'clock the following morning the drive was completed to Greene's cabin at the south end of the camp, and as quickly as possible the stage would start back to Headquarters in order to complete the trip before the snow began to soften. If the travelers wished to continue to the higher portions of camp they were required to use "snowshoes."[66]

At his boarding house, Greene was said to be feeding 30 to 40 men at every meal.[67] Reports of rich discoveries at Gold Hill continued to come into Saratoga, Laramie and Rawlins, each one better than the last.

Charles Bellamy returned to Laramie after seven weeks of survey work at Gold Hill. Bellamy was apparently regarded as a reliable man and considerable faith was placed on his reports. He reported that by the first of May there were between 75 and 100 men in camp and many more in Saratoga awaiting the arrival of spring. The snow, he said was settling but still quite deep. Travel by "snowshoe" was possible in the morning but by noon travel by any means was impossible. The "Leviathon," he reported, was the most developed mine and was down over 100 feet. He also reported there were 35 to 40 cabins in camp.[68]

Ben Arendell, visiting Rawlins about the same time, gave a slightly cautious but nevertheless glowing account of the camp saying, he did not allow enthusiasm to overshadow the difficulties of the camp. Gold Hill, he thought, was properly named although he and several others were already prospecting the range east of the camp on ledges of what he called excellent indications of silver.[69]

With spring impending, Carbon began a push to open their road to a Concord coach. By the first week in May, they were claiming that seventeen passengers rode the Carbon stage into camp. However, one of the men on the journey said that he paid $5.00 for the privilege of walking to Gold Hill behind the Concord. Over one portion he said they walked in front of the coach stamping down the snow. Furthermore, he said the seventeen "passengers" were not passengers at all, but were men hired to push the road into the camp.[70]

Stephen W. Downey, still the outstanding proponent of the Medicine Bow Mines, had taken an early interest in the Gold Hill mines and now took measures to promote his interests.

On May 9, 1891, Downey, who was again in possession of the old Centennial Mill which had been at Cummins City since 1880, had the mill moved to the Soda Lakes railroad spur. Here he had the mill loaded and shipped to Carbon. At Carbon, against advice of those who knew the condition of the Carbon to Gold Hill road, he hired the Beckwith Commercial Company to transport it to Gold Hill.[71] Under the direction of Thomas O. Minta, Mayor of Carbon and manager of the Beckwith Commercial Company, the mill was started toward Gold Hill on May 12. Ten or twelve days were anticipated for the 35 mile journey. Within a week the mill was hopelessly bogged down a little more than half way to the camp.

Downey, not to be discouraged, promised the miners he would transport a second mill to the camp. He also contracted to have the Carbon road completed into camp for $100.00 a mile.

Another stamp mill was proposed by W. S. Tolls, a Park City, Utah miner working at Gold Hill.[72]

The little settlement of Rockdale, (now Arlington) recognizing they were the closest

GOLD HILL

EXCURSION!

Will leave Denver at 10 o'clock A. M.,

AUGUST 10, 1891.

Round Trip Tickets $22.

This excursion will be the event of the season. It will afford an opportunity for all to visit Wyoming's famous gold camp. Gold Hill is without doubt the most wonderful gold strike made during the last quarter of a century. The camp consists of rich quartz and placer claims, and there are better opportunities for parties to make money than has been known for twenty years. Not one-tenth of the claims are filed upon.

The Excursion will be accompanied by a large number of Colorado mining men who will pass judgment on the camp. This is an opportunity of a life time. If you desire to go register your name at the city passenger office of the Union Pacific Railroad, 1703 Larimer Street, corner Seventeenth, Denver.

---

town to Gold Hill, began construction of a road from the Rock Creek crossing to Gold Hill. Their road was to run up the ridge west of Rock Creek, cross over to the Medicine Bow drainage, and into camp from the north using the same last two miles as the Carbon road.

The *Platte Valley Lyre*, at the time being published by Gertrude and Laura Huntington, faithfully printed all the mining news that came to them. Many of their articles were picked up by *The Wyoming Commonwealth*, published in Cheyenne.

George F. Canis, who had come into the Platte Valley and the Gold Hill mining district as a correspondent of the *Omaha Bee*, saw not only the mining opportunity but also the possibility of a growing community around Saratoga. In July, 1891, he printed a few preliminary issues of what he called the *Saratoga Daily Sun*. Apparently meeting with favorable response, on July 14, 1891, he began printing the *Saratoga Sun*, a weekly newspaper, "Devoted to the interests of the Upper Platte Valley and Wyoming." Vowing to print only reliable information regarding the development of the mineral resources and proclaiming "the Sun shines for all . . . and has nothing to boom for." Canis, it should be remembered, owned an interest in several lode and placer claims. His first issue was predominately devoted to the mining activity in the hills and mountains rising from the North Platte River Valley. Several columns were devoted to the story of the Lost Pick and Shovel Mine and several more to the strange discoveries made by John H. Mullison on his Brush Creek claims. In the months ahead, the *Saratoga Sun* was the prime source for stories of the mines.

Although the *Sun*, and the *Platte Valley Lyre* were the prime boosters of the Gold Hill District, the Rawlins *Carbon County Journal*. the *Carbon Black Diamond* and the *Laramie Republican*, *Boomerang*, and *Sentinel* all anticipated a gold rush and printed all the encouraging mining news available. The *Wyoming Commonwealth* re-printed many of the stories and by early July were planning to sponsor an excursion trip from Denver to Gold Hill. The fare was set at $22.00 which included round trip rail fare from Denver to Rawlins, then by six horse Concord coaches from Rawlins to Gold Hill and included all meals and overnight accommodations. Only the date of travel remained to be set.[73]

Failure of Downey's stamp mill to make it to Gold Hill and an increasing amount of ore on the dumps prompted Ben Arendell to erect an arrastra near his mine, the "Little Giant." A ditch, 300 feet long was constructed to provide water and "eastern cement" was used to provide a solid grinding base, 14 feet in diameter. Arendell with his primitive arrastra was the first to process the ore that promised to be a great bonanza.[74]

By mid-July, "Ingot," the correspondent for the *Wyoming Commonwealth* reported the boom was on at Gold Hill. "There are between 350 and 400 miners prospectors at work here and in the near vicinity. New buildings are rising in everyone of the three townsites which are located here." [75] (Depending on the source, three, four or five townsites were being laid out. Actually, three of the sites were connected or very nearly so). "The Gold Hill townsite lies south of the Leviathon hill and cuts into the claim of that name. It adjoins the

Acme Hotel
Wyoming State Museum

Gold City townsite on the northwest. There are two other townsites in the camp besides those mentioned, the Golden Courier, that joins on to Gold City and Greenville . . . ."

Greenville (sometimes called Greensboro) named after William H. Greene, was the head of the stage road from Saratoga. The townsite had in July, 1891, ten cabins and about twenty-five tents. Included were two hotels and one saloon. The Acme House was considered to be a "first class hotel" (two story log, 14 rooms, board on the first floor, sleep on the second, one dollar per day).[76] A. W. Ainsworth, proprietor of the Acme Hotel, had come to Gold Hill from Percy, a small siding on the Union Pacific Railroad. He took Greene's 20 by 27 foot cabin, added a second story and made continued improvements. When he received word of the impending Denver to Gold Hill Excursion, he added a 20 by 22 foot addition to the hotel allowing him to accommodate forty people.

One other townsite called Summit was laid out near the hydrographic divide of Brush Creek and the Medicine Bow River. The site was claimed by people from Carbon. Apparently the site was never developed.[77]

The combined Gold Hill sites had three hotels, three saloons, a blacksmith shop and a post office by August 1, 1891.

Ben Towner, still living on Douglas Creek, visited the camp in early August and was quoted as saying there were two or three hundred people at Gold Hill.

Stephen W. Downey went to Gold Hill to investigate the delay in moving his stamp mill into camp. He was apparently pleased with the progress the camp was making but not pleased in the progress his mill was making. The Beckwith Company had failed to fulfill their contract and Downey looked elsewhere for a freighter with the means to bring his mill into Gold Hill. Sam Wood, from Cummins City and Woods Landing, who at that time was running a sawmill on Brush Creek, was hired by Downey to complete the transportation.

Apparently the Denver to Gold Hill Excursion went off as planned on August 10, 1891, with no notable remarks being made for or against the camp.

Gold Hill, ca. 1892
Wyoming State Museum

By September, Arendell and Woods had processed enough ore in their crude arrastra to produce a small button of gold. The *Lyre* estimated the weight at four ounces while George Canis of the *Sun* estimated twelve ounces.[78] Whatever the weight, this was the first gold produced at Gold Hill and plans were made to send it to the Mineral Convention in Cheyenne. The production sparked new interest in the arrastra process and Miller and Streight put in a second arrastra on Gold Run Creek.

The gold production also added impetus to the erection of a stamp mill at Gold Hill. At that time, plans were being made for the erection of three separate mills. Downey's mill, however, was the only one that had gone beyond the planning stage.

During the last days of September, Wyoming's Senator, Francis E. Warren visited Gold Hill, "for the purpose of satisfying himself regarding the camp."

On the last day of September and the first few days of October, 1891, Stephen W. Downey purchased the "Leviathon No. 1," "Leviathon No. 2," the "Little Giant" and the "State of Wyoming" claims from the original locators and became the sole owner of the most promising mines in the camp.[79]

On Monday, October 5, 1891, work was begun on a building to house Downey's stamp mill. The site chosen for the mill was on the north side of Arrastra Lake. Ground breaking was started in a blinding snow storm. The main building was to be 34 by 55 feet. Adjoining the mill a blacksmith shop and a storeroom were also planned. Downey's mill under the direction of Sam Wood had finally been moved into the vicinity of the camp after spending several months on the Carbon to Gold Hill road.[80]

The State Mining Convention was held in Cheyenne from September 7 through 12, 1891. At the request of the convention committee, George Canis, editor of the *Saratoga Sun*, addressed the convention on "Gold Hill and the Medicine Bow Mountain Range." Canis spoke on the recent history of the district and the hardships of the miners. He also addressed the ancient history of the gold fields

W. O. JONES,

Official Recorder

GOLD HILL MINING DISTRICT

Business of non-residents promptly attended to. Correspondence solicited.

GOLD POST OFFICE, WYOMING.

GOLD HILL

MINING AND MILLING COMPANY,

Incorporated Under the Laws of Wyoming.

—OFFICERS—

THOS. O. MINTA,	President,
H. B. ALLEN,	Vice President,
J. S. ALLEN,	Secretary,
MANDER PETERSON,	Treasurer,
GEORGE H. FALES,	Superintendent.

Capital Stock, - - $60,000,

In Shares of One Dollar Each.

Fifty Per Cent. Paid Up Capital.

Principal Offices, Carbon and Gold, Wyo.

and expressed his personal belief that the area had been mined "by the Aztecs or by Coronado or other adventurous Spaniards." Canis added, "In the output of Wyoming for the year 1891, Gold Hill will not figure very extensively," and concluded that "Gold Hill could use more miners and fewer prospectors."[81]

With the impending onset of winter, many of the miners who had left the district, returned to spend the winter at Gold Hill. Apparently these miners had left the mines to earn a grub stake to carry them through the winter. Mining, they found was eased by the lack of flooding they had encountered during the summer. Prospecting and some mining had extended in all directions from Gold Hill, to the old La Plata District on the south and east, to Rockdale on the northeast and to Lower Brush Creek on the west.

Some friendly and some not so friendly rivalry arose between some of the miners. In Laramie, Henry Dahl from Gold Hill and Sam Parker from La Plata got into a controversy as to the merits of the two camps. The argument came to a conclusion that evening in a local hotel dining room. Dahl irritated Parker to the point that Parker drew a pistol and fired at Dahl, hitting not Dahl but an innocent bystander, Jack McLennan. McLennan was not seriously wounded and refused to press charges against Parker. The case was dismissed and Parker later apologized.[82]

On October 19, 1891, a post office was designated for Headquarters and James D. Minshall was appointed postmaster.[83]

In mid-October, the Gold post office was moved from Lute and Poutre's cabin in the upper camp to W. O. Jones' in the Greenville townsite.[84]

Jones, who was also the district recorder, took over as acting postmaster until his official appointment was made on November 11, 1891. About the same time Jones took over management of the Tilton Mining and Development Company store. The Tilton Company was formed by investors from Holyoke, Massachusetts, and was incorporated in the

Elk Head Saloon

DOGGETT & ANDERSON.

Keep Constantly on hand The Best Quality and Largest Stock of

WINES, LIQUORS AND CIGARS

In Saratoga and Gold Hill.

MILWAUKEE AND DENVER BEER.

State of Wyoming on October 31, 1891. Twenty-nine lode and placer claims were included in the company's property. Capital stock in the company was set at $1,500,000, divided into 300,000 shares at $5.00 each. George Nightingale of Holyoke was designated president and six trustees were named.[85]

On October 22, 1891, the Gold Hill Mining and Milling Company was formed and incorporated. The amount of capital stock was set at $60,000 at a par value of $1.00 a share. Thomas O. Minta, mayor of Carbon was appointed president and six trustees were named.[86]

During the closing months of 1891, Gold Hill was settling down to the lifestyle of a small but promising mining camp. Stearley's Hotel (Louie and wife) was undergoing repair. Ainsworth's hotel, the Acme, was building a new stable and was talking of enlarging the hotel again. The Acme Hotel had become

Fast Time and Comfort

—Guaranteed By—

RAWLINS AND GOLD HILL

STAGE LINE.

Six-Horse Concord Coaches Leave Rawlins and Saratoga Daily to and from

GOLD HILL.

EXPRESS -:- AND -:- BAGGAGE

For All Points in the Upper Platte Valley and Gold Hill Should be Sent via Rawlins. Prompt Attention and Rapid Transit Insured to all who Send Goods Over this Line.

J. W. HUGUS & CO., Saratoga Agents,

From Whom Information Can be Obtained, or by Addressing

RANKIN BROTHERS,

RAWLINS, WYO.

New Brush Creek Mines

ARE LOCATED IN VICINITY OF

GORE'S TEN MILE HOUSE,

On the GOLD HILL Road.

First Class Accommodations for Man and Beast.

MEALS, - - 35 Cents.
LODGING, - - 35 Cents.

P. F. GORE, : : Proprietor.

popular center of entertainment for the camp and was the designated polling place for the Gold Hill Precinct. The Elkhead Saloon of Doggett & Anderson was operating and Riggs and Company had their saloon stocked and were fixing up for the winter. F. O. Sawin was building a new residence. Seemingly well set for winter, the miners declared what they needed most was a washer woman.

Downey's mill was beset by problems in getting set up and operating. The mill had remained idle at Cummins City for several years and with the additional year on the road many parts were lost, broken or rusted.

On the Saratoga to Gold Hill road, the Headquarters road house had become the preferred stop rather than Ten Mile House. The first official U.S. Mail was received at the Headquarters post office on December 2, 1891. The Headquarters Hotel was advertising "Road House and Eating Station on the Gold Hill Road, Eight Miles from the camp. All Stages Stop Here and Relays are made. Private Stage to Gold Hill. J. D. Minshall, Proprietor." In direct competition was P. F. Gower, proprietor of the Ten Mile House, was advertising "First Class Accommodations for Man or Beast, Meals 35 cents, Lodging 35 cents." [87]

• 1892

On January 3, 1892, Downey's mill began its long awaited first day of operation.[88] After only thirty-six hours of crushing ore the mill had a major breakdown. A key had come loose, striking the cylinder cross head and breaking it into five pieces.[89] After four days of repair the mill was again in operation and ran day and night for thirteen days, processing ore from the Wyoming mine, but was again forced to shutdown, this time for lack of ore.[90] It was determined however, that the ore could be processed for $5.00 a ton. The mill was said to be able to crush fifteen tons of ore daily.

Reports from Gold Hill proclaimed the "Wyoming," "Little Giant," and the "Acme" were "showing up better all the time." Free gold was said to be visible in the Acme tunnel with the dim light of a candle. "Every man who owns a prospect in Gold Hill is now using every effort in his power to develop it, and the more development work is done, the more certain becomes the future of the camp as the greatest gold producer ever discovered in the western mountain ranges."

A second run was started through the mill on January 22, this time with ore from the Acme Mine. A "cleanup" was planned in about 15 days. Two pieces of the Acme ore were sent to Denver for the exhibition.

During the month, the Hampden Mining and Development Company was formed and incorporated by a group of Holyoke, Hampden County, Massachusetts investors. Their plan was to buy and develop a group of mines at Gold Hill.[91]

Six new claims were made during the last week of January.

On January 24, 1892, Gold Hill recorded its first birth. The A. W. Ainsworths of the Acme Hotel had a baby boy. Great preparations were made for the christening which they planned to coincide with the first cleanup from Downey's mill. Such were the great expectations from the cleanup that it was suggested a souvenir should be made from the first bullion shipment to present to the young Mr. Ainsworth.[92]

At Gold Hill, snow fell twenty-three out of the thirty-one days of January. In one day Gold Hill recorded four feet of snow. The total snowfall for the season had been more than twenty feet. At the upper camp, near the mill, the snow was twelve feet seven inches deep. For four days, four teams were kept traveling back and forth over the ore road just to keep it open. A foot trail was kept open between the post office and the lower camp. The road to Headquarters was kept open except for a few days after each heavy storm. Sleds were being used from Ashley's Cedar Creek ranch, six miles from Saratoga. Sam Wood was using his teams to pack down the snow for easier passage. George Lathrop, the stage driver, was forced to carry the mail the last one-half mile into camp on "snowshoes." While Gold Hill was laboring under the extraordinary snowfall, Saratoga was enjoying a mild winter.[93]

The snow did little to dampen the spirits and recreation in camp carried on at a seemingly lively rate. A masquerade ball was planned at the Acme Hotel on the evening of Saint Valentines Day. The Acme Hotel was doing a lively business, advertising, "Square Meals and Good Beds, Meals — 50 cents, Lodging — 50 cents a night, Board and Lodging — $10.00 per week, Table Board — $7.00 per week."[94]

The ball and the christening of baby Ainsworth were delayed due to conflicting entertainment at the Saratoga School.

Entertainment in camp took more than one direction and on January 30, a fight was staged between the camp pugilist, Corrigan,

PEN SKETCH OF GOLD HILL, OCTOBER, 1900.

University of Wyoming, American Heritage Center

and Al Roberts. Roberts won the knock-down fight, administering a thorough pounding to Corrigan.[95]

The "Wyoming" mine work was being pushed by Jack Martin under contract from Downey. The "Acme" mine was producing ore for the Downey mill but the owners were planning a mill of their own to be built on the "Patriot Placer" claim at the lower camp on South Brush Creek.

The efficiency of the old Downey mill was in doubt and 1,950 pounds of ore from the Acme was sent to the Black Hawk, Colorado mill for tests. This ore produced about $24.00 in gold. Estimating they could mine, transport and mill the gold for $5.50, the tests were a much needed encouragement to the miners.[96]

At the post office, a dance was given by postmaster W. O. Jones. The event was held in honor of visiting ladies, Mrs. Miller, the Misses Marie and Bertha Miller (Doebler) and Mary Stearly. The christening of baby Ainsworth was again set for February 22.[97]

Other forms of entertainment were not forgotten and to provide local amusement, several of the miners formed the Gold Hill Comedy Company to perform at the Acme Hotel.[98]

Near Gower's Ten Mile House a claim which was made by "Old Joe" Lucin in 1891 was by early 1892 attracting attention due partly to the assays and partly to the year around access to the mine. Lucin's mine, the "Fairview," had assayed as high as $364.00 to the ton. By February, 1892, "Old Joe," in spite of his infirmities caused by rheumatism, had tunneled more than one hundred feet into the hill.[99]

At the upper Gold Hill camp, a new post office named Altamont, was established on February 5, 1892. John A. Burman was designated postmaster.[100] Burman's saloon was the most popular spot in the upper camp, containing not only the new post office and saloon but also a barber shop run by Albert Sloan.

On February 25, 1892, Gold Hill made its first shipment of gold bullion from the Downey mill. This was the product of more than a month of crushing about one hundred tons of ore.[101] The size of the gold "brick" was not disclosed by Downey or his associates. Since his Centennial enterprise, Downey had been quite secretive about his mining endeavors. Downey at this time had thirty five men on his payroll at Gold Hill.

On March 3, 1892, a group of miners met at Gower's Ten Mile House and voted to "secede" from the Gold Hill District. At the meeting, presided over by one of the owners of the Fairview Mine, E. W. Holland, the miners named their new district "Villette." Seventeen people attended the meeting including George Canis and his wife, Villette. P. F. Gower was elected recorder for the new district. Boundaries were set which extended to the Platte River and included the claims on Bennett Peak and Corral Creek. Plans were made for a newspaper, a waterworks, a short-order house, a large stable, a smelter and at least a dozen new houses. Postmaster Minshall of Headquarters petitioned the post office department to change the location and name of the Headquarters post office to the new townsite of Villette.[102]

Meanwhile, Gold Hill was beginning to show the distress of the Medicine Bow mines seen earlier. Downey's mill was to run only as long as the "Leviathon" ore held out. Work on the "Acme" was suspended. Thomas O. Minta resigned as president of the Gold Hill Mining and Milling Company and Hiram B. Allen was elected as the new president.[103]

On March 20, 1892, Gold Hill experienced an unrelated (to mining) tragedy. Marie Doebler committed suicide by taking strychnine at the stamp mill office. Twenty year old Marie and her sister, Bertha, had been living in the isolated camp with their mother and step-father C. Miller who ran the boarding house for the men employed by Downey. Marie had been in low spirits about a love affair with a Colorado man. She had gone to the mill superintendents office to write letters. The letters it transpired, were suicide notes to her mother and fiance. She was found by Charlie Richie, in convulsions and died shortly afterwards. She was buried in the "new" cemetery at Saratoga.[104]

Although prospects for the future were not bright at Gold Hill, apparently no plans were made to abandon the camp. On March 31, Grout and Hopkins of Saratoga sent 3,000 pounds of potatoes to Stearly and Miller at Gold Hill.

Snow continued in unprecedented amounts and on March 28, three feet fell at Gold Hill. The storm and winds continued unabated for three days and was said to be the worst since the storm of 1878. The mail was delayed twenty-four hours due to snow drifting so badly that Lathrop did not dare to venture out for the trip. During this storm, eighteen inches of snow fell at Villette.

However, during this time, several new prospects were claimed and word was spread that power machinery was in the valley ready to be transported to Villette when weather permitted. As late as 1892, power equipment was quite rare in the Medicine Bow Mining camps and procurement of the Ingersoll steam drill and boiler were signs of an optimistic camp.[105]

Several of the new claims were said to assay higher in silver than gold. One silver lead ran from Brush Creek Canyon over the ridge to Buffalo Gulch a distance of 4,500 feet.

At Gold Hill the Downey stamp mill was shut down the first week in April. The supply of wood for the mill boiler was exhausted. Snow was said to be twelve to fifteen feet deep on the level and for the time, the cutting of a new supply was out of the question. Apparently only one mine was working, the "Acme Fraction" owned by Clark, Ashley and Forrest.[106]

The only other news from the camp was that Bertha Doebler, sister of Marie who died at the camp a month earlier, was in Saratoga under going treatment by Doctor Bennett. A. W. Ainsworth's three month old baby was reported seriously ill. (Apparently the much awaited christening never took place.)

Saratoga, although not dependent on the mining activities, had nevertheless prospered in their wake. Several new businesses had begun with service to the miners their primary purpose. The population of the town was said to have doubled in a year. Aside from the Gold Hill and Villette districts, a new mining district was established on Jack Creek, west of Saratoga. Miners from as far away as Savery and the Little Snake River were using Saratoga as a supply point. A $15,000 iron bridge spanned the North Platte River providing not only access to the mountains to the east of Saratoga but also the opportunity to expand the settlement to both sides of the river.

On North Brush Creek, J. H. Mullison was again planning to work the placers. Theories of ancient workings in the area persisted, and were repeatedly printed in the *Saratoga Sun* by George Canis.

At Villette, the snow was almost gone by the first week in May and work was progressing on the mines. The "Fairview" was considered the greatest hope and silver leads were now being pursued along with the gold. The discovery of a gold nugget weighing

Pen Sketch of saratoga, ca. 1900.
University of Wyoming, American Heritage Center

"twenty-five cents" was considered a newsworthy occurrence.

Spring and summer of 1892 brought little new activity to the Gold Hill mines. Carbon County granted business licenses to: Sam Stearly, hotel; Doggett, Anderson and Smith, saloon; J. M. Riggs and Company, saloon and John A. Burman, saloon.[107] Ainsworth was still at the Acme Hotel. Ben Arendell, founder of the district, was camped on French Creek, prospecting. The hired miners of the camp had almost all departed. Downey's mill was sitting idle. In August, Barton T. "Tom" Ryan made an agreement to start up the mill and treat ore from the nearby mines. Also in August, Ainsworth left the Acme Hotel and moved up to the Downey camp to run the hotel there.

On August 9, 1892, Gold Hill experienced another suicide. Franklin O. Sawin, former recorder and postmaster at La Plata, miner and prospector at Gold Hill, Keystone and La Plata left Gold Hill where he had been working and living alone. His wife and daughter were staying at the A. G. Bullock ranch on Lake Creek where Mrs. Sawin had been teaching. Sawin rode to the ranch and after a brief conversation with his daughter walked to the barn and shot himself. The coroners inquest revealed that Sawin had suffered "business troubles and financial reverses." His claims had not "panned out" and to add to his dismay, his wife had preferred to take a teaching position at Fort Steele rather than go to Gold Hill with her husband. Franklin O. Sawin had been mining in the Medicine Bow Mountains since at least as early as July 1872 when he made his first claim the "Albany County Lode." He had claimed, either by himself or with partners, fifty-one claims in Albany County and several more in Carbon County where Gold Hill was located. Had he only known, some of the claims he made in northern Albany and Carbon County were located on what later proved to be valuable oil property, and would have eventually made him a wealthy man. Sawin was given Masonic rites and buried at Saratoga.[108]

At Gold Hill, after about two weeks of mill operation, a cleanup was made yielding approximately $10.00 of gold per ton.[109] The mill

Acme Mine, ca 1900
Wyoming State Museum

was said to do very poor work, saving only half of the gold. Gold Hill was slipping but not yet dead. The Acme mine, under several miners liens, was ordered appraised for sale by a Rawlins court. Ryan, Ferguson and Meason appraised the mine at $9,666.66. [110]

On November 15, 1892, the post office at Headquarters was discontinued. On November 30, the Gold and Altamont post offices were also closed down.[111]

Gold Hill had followed the course of the Medicine Bow Mines before. One of discovery, great hopes, primitive development to the point of erecting a stamp mill, followed by the reality of limited profits from the mill and the reluctant death of another camp. As with most of the camps, Gold Hill would experience repeated attempts at revival but none to equal her little boom of 1891.

• 1893

During the first days of January, 1893, Charles Miller crossed the range between Camp Herman and Saratoga on "snowshoes" and reported the camp was deserted.[112] However, Saratoga newspapers were talking of a revival of the camp.

For a brief time in January, 1893, there was a glimmer of hope that the "Altamont Star" would produce the amount of gold to again boom the camp.[113] Shortly thereafter, Ben Hoffman and Dick Christenson made their way to Gold Hill and began to work the "Altamont Star." Apparently they were the only miners working in the camp.[114]

Early in March, Alex Lute made a trip to the camp to relocate his old claim. He found the camp covered with five and one-half feet of snow but spent several days panning two small leads which he hoped to develop. On returning to Saratoga, Lute spoke highly of the "Altamont Star" which he reported could be panned for 600 to 1,000 "colors" per pan. At the time, the "Altamont Star" was the only mine working and miners were attempting to tunnel into the mountain to intercept the lead 160 feet under the vertical shaft.[115] By midsummer, the owners of the "Altamont Star" were offering one-half interest to anyone who would develop the mine.

During the summer of 1893, B. T. "Tom" Ryan leased Downey's mill and ran mill tests on several groups of ore. Rumors were that "Eastern Capitalists" were interested but apparently there were no takers.

Several mines were being worked on a small scale and on a visit to the camp, J. R. Smith reported five different groups of miners in camp.[116]

In November, Ryan was hired by L. G. Davis, Rawlins based owner of some of the predominant mines, to do some assessment work and plans were made to continue operating the mill as long as it paid the wages of the men. No further reference was made as to the results or shutdown date.[117]

Still there were those who kept the faith, vowing it was only a question of time when Gold Hill would again come to the front.

• 1894

In 1894, W. H. Greene became interested in the McArthur-Forrest or cyanide process and formed a new company to work the old Acme mine. A small company was formed but again failed to produce any notable amount of gold.[118]

• 1896

By 1896, the Gold Hill road was said to be sadly in need of repairs. In March, Isaac Van-Horn, (most recently of Lincoln, Nebraska), relocated thirty-three claims at Gold Hill, the former owners having forfeited their rights to the claims by failing to do the required annual assessment work. R. A. George and M. B. Hassett, both of Saratoga, accompanied VanHorn and also relocated some claims.

• LATER

Gold Hill experienced a short rebirth along with the other Medicine Bow Mines around the turn of the century only to confirm the findings of those before them.

The Gold post office was reopened on February 21, 1901, and again closed on July 31, 1902.[119]

The gold camp began it's inevitable decay and by 1931, a Laramie newspaper reported: "There are 25 to 30 cabins standing as reminders of the mining settlement. Snow has caved in the roof of the old hotel but the post office, two saloon buildings and other souvenirs of the old days are still standing."[120]

Throughout the years there has been hardly a time when the Medicine Bow Mountains were not being prospected. To this day prospectors continue to search the area during the summer months, hoping to find their personal "gold hill."

Remains of concentrator buried at Gold Hill.

Remains of Gold Hill Boiler.

STANDARD RETURN TUBULAR PORTABLE BOILER.

Fig. 82.

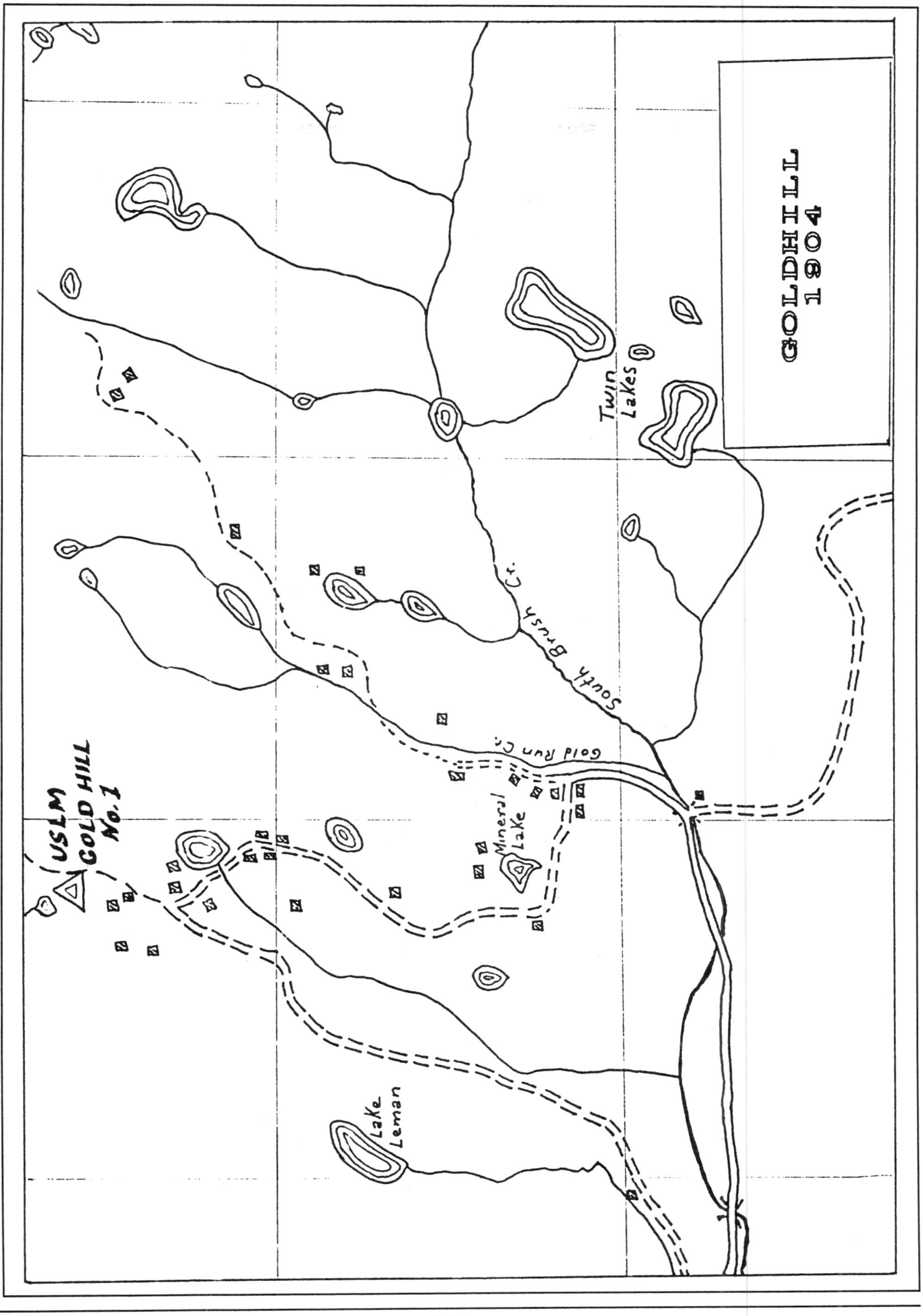
GOLDHILL
1904
Twin Lakes
South Brush Cr.
Gold Run Cr.
Mineral Lake
USLM
GOLD HILL
No. 1
Lake Leman

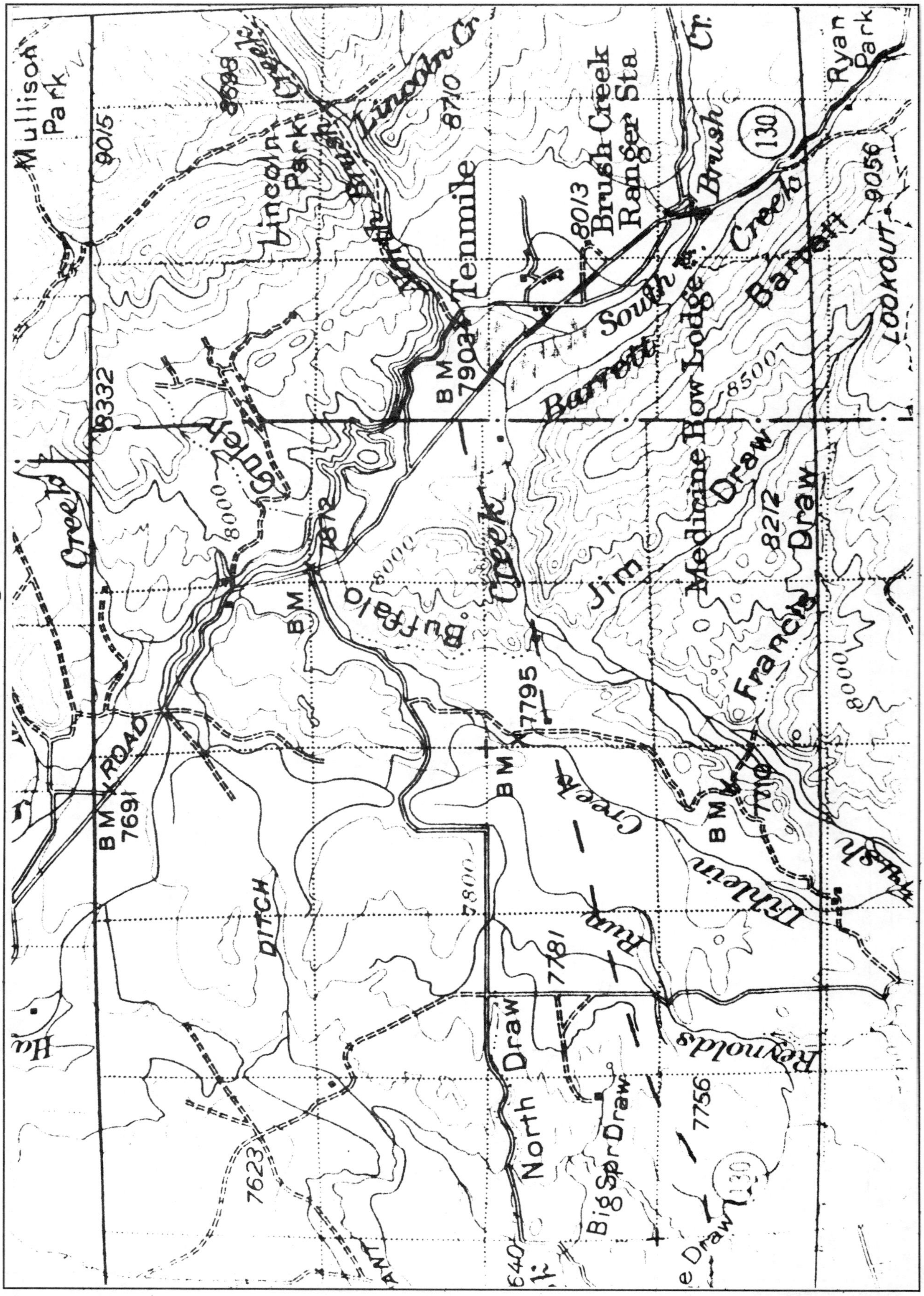

Tenmile Mining Area

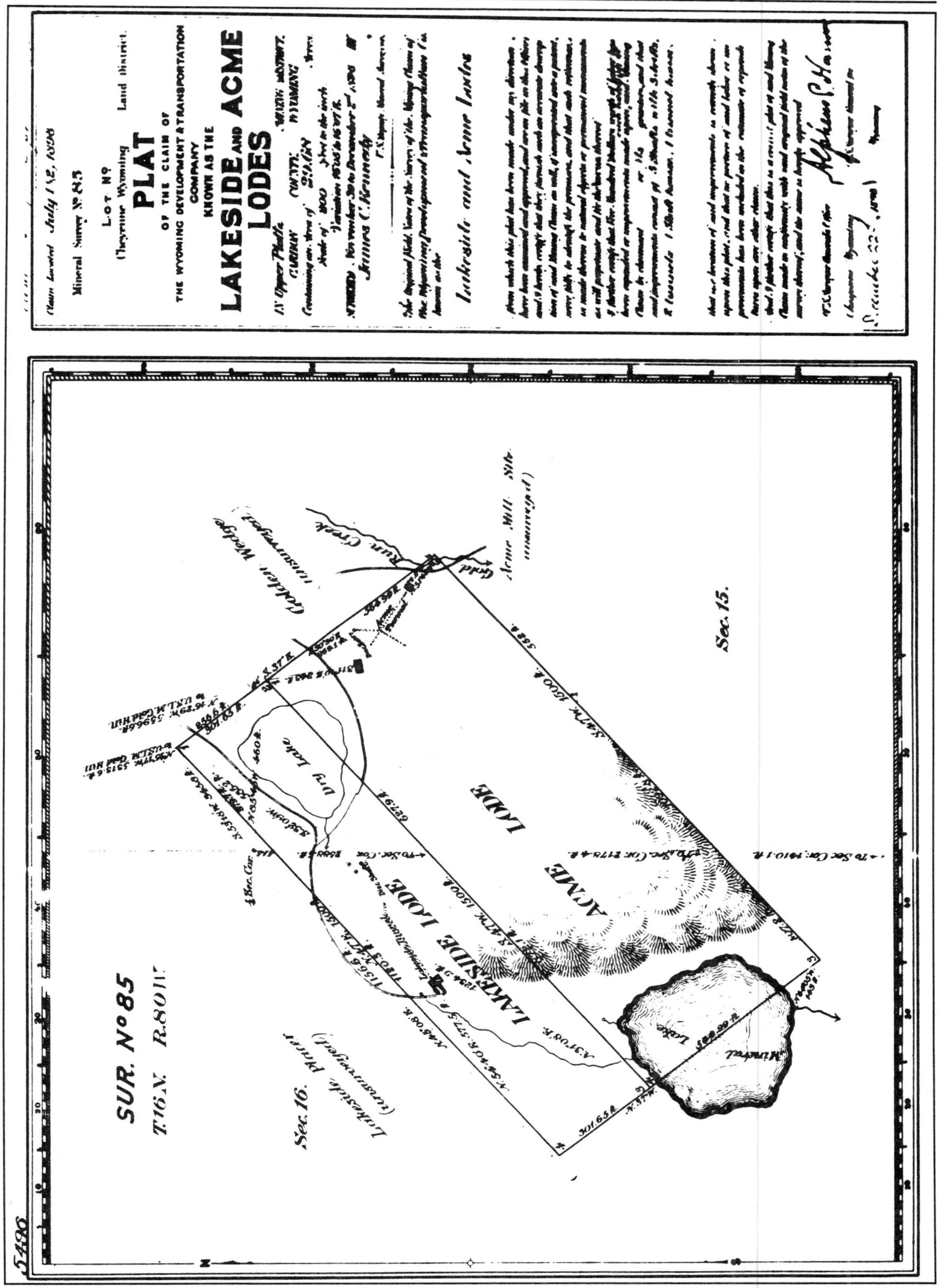
Mineral Survey No 85
Lot No
Cheyenne Wyoming Land District
PLAT
OF THE CLAIM OF
THE WYOMING DEVELOPMENT & TRANSPORTATION COMPANY
KNOWN AS THE
LAKESIDE AND ACME LODES
Lakeside and Acme Lodes
SUR. No 85
T.16 N. R.80 W.
Sec. 16.
Sec. 15.
LAKESIDE LODE
ACME LODE
Dry Lake
Lake
Mineral
Golden Wedge (unsurveyed)
Lakeside placer (unsurveyed)
Gold Run Creek
Acme Mill Site (unsurveyed)
5496

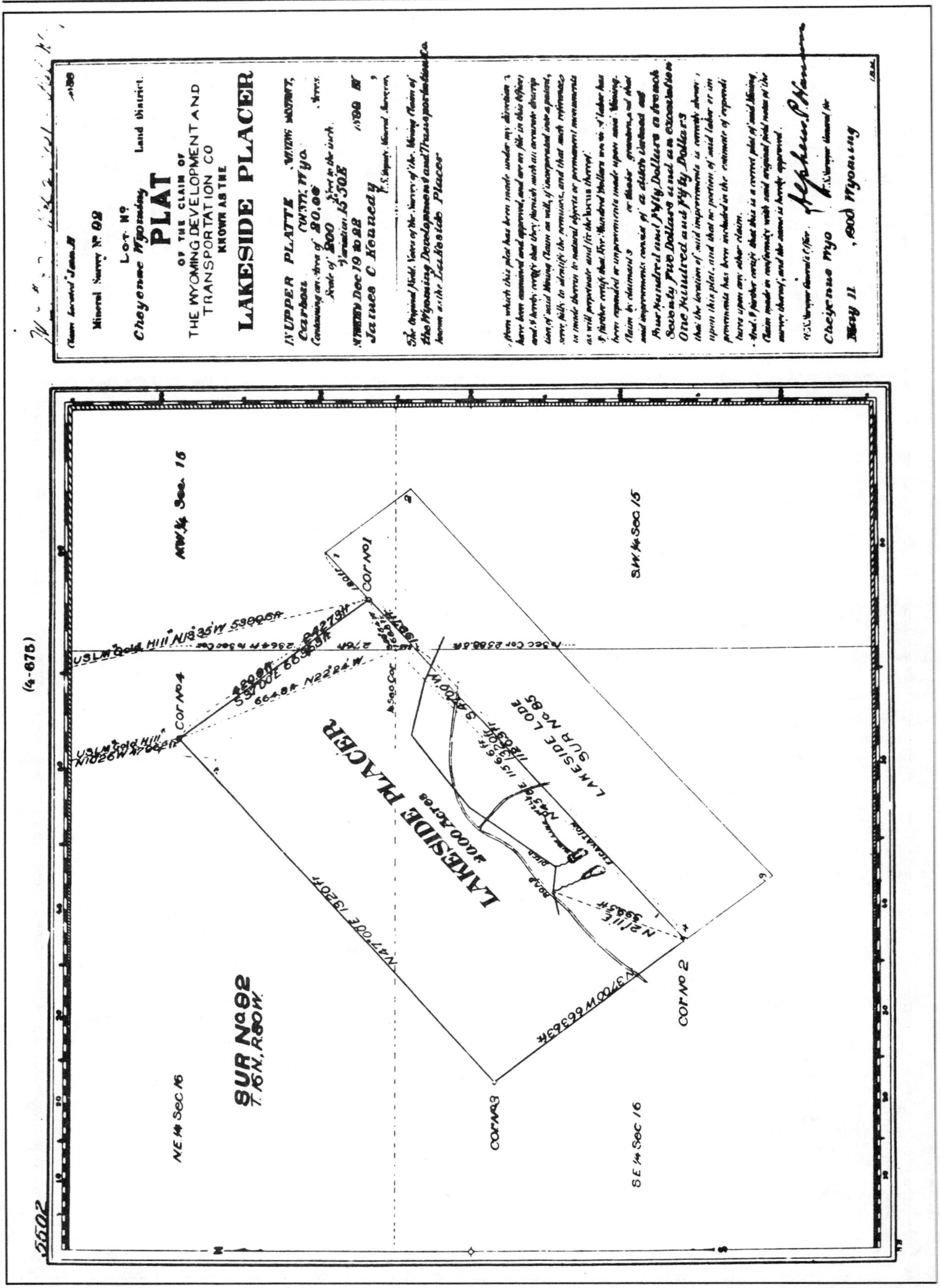
Mineral Survey No 92
Lot No
Cheyenne Wyoming Land District.
PLAT
OF THE CLAIM OF
THE WYOMING DEVELOPMENT AND TRANSPORTATION CO
KNOWN AS THE
LAKESIDE PLACER
Carbon County, Wyo.
James C Kennedy
Cheyenne Wyo
(4-675)
NW 1/4 Sec 15
SW 1/4 Sec 15
NE 1/4 Sec 16
SE 1/4 Sec 16
SUR No 92
T. 15N, R.80W.
LAKESIDE PLACER
LAKESIDE LODE
SUR No 85
Cor No 1
Cor No 2
Cor No 3
Cor No 4
N47°00'E 1320 ft
ROAD
DITCH
5502

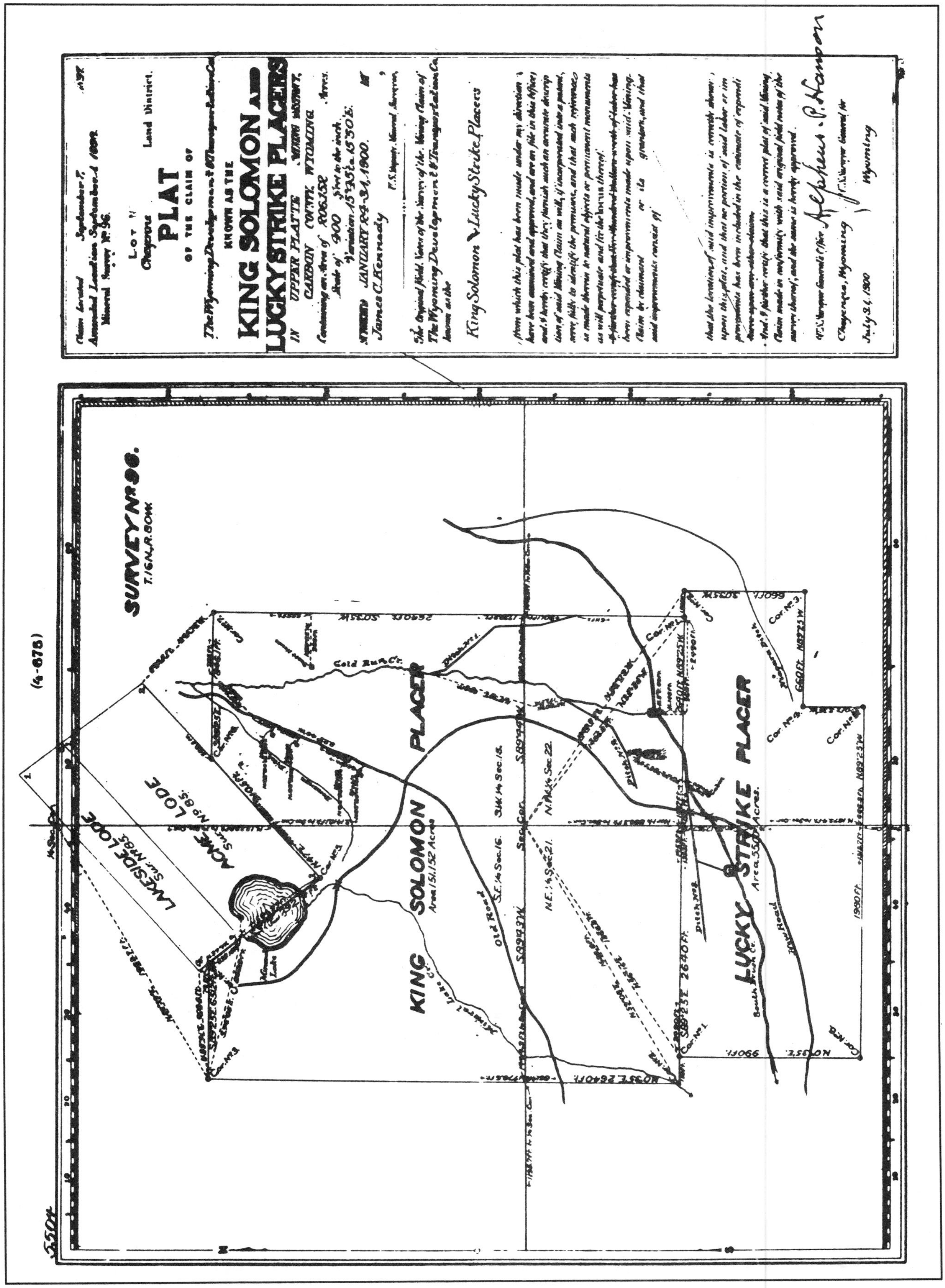
PLAT
KING SOLOMON AND
LUCKY STRIKE PLACERS
SURVEY No 96.
KING SOLOMON PLACER
LUCKY STRIKE PLACER
LAKESIDE LODE
ACME LODE

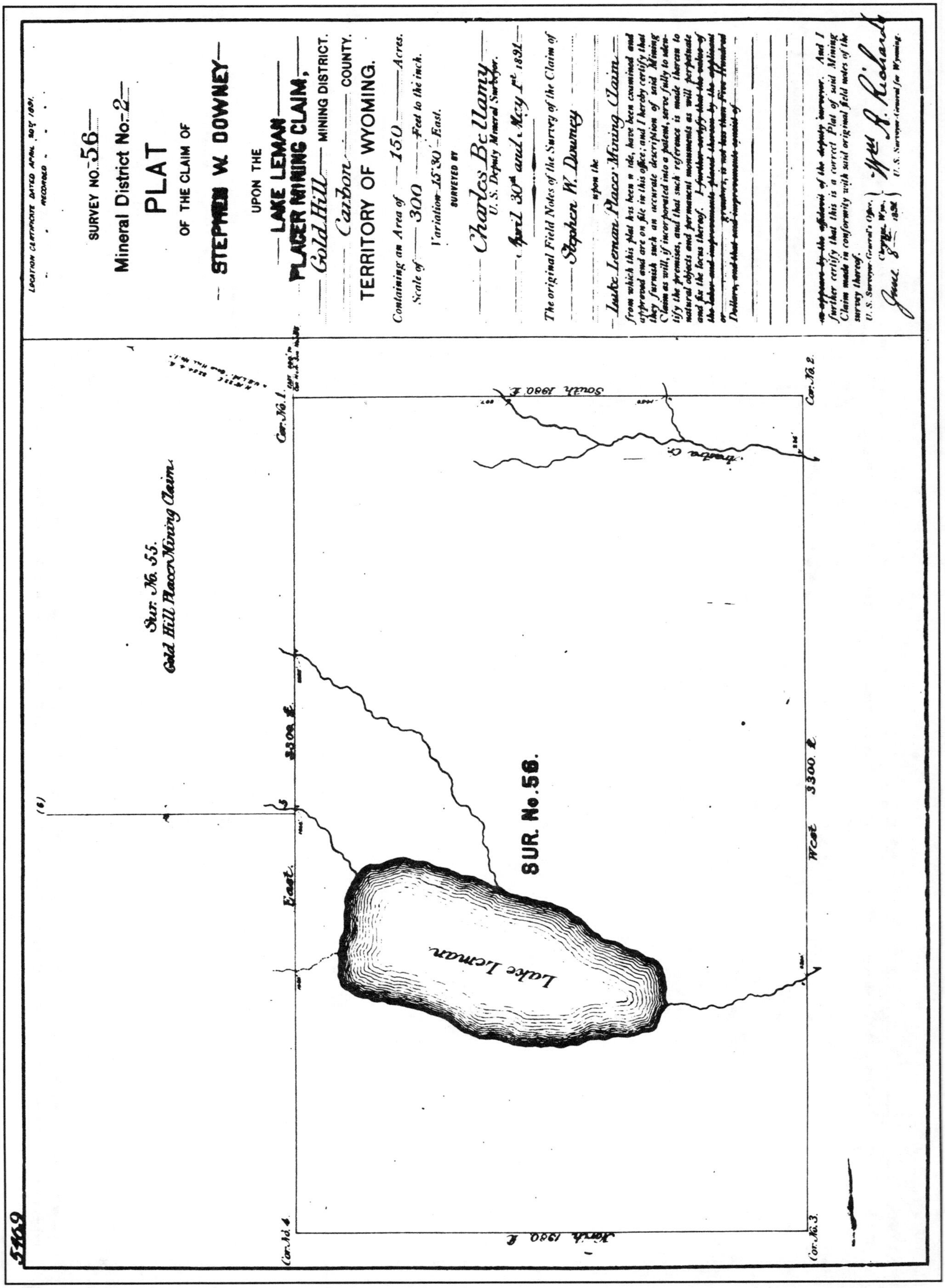
LOCATION CERTIFICATE DATED APRIL 30th 1891.
RECORDED
SURVEY NO. 56
Mineral District No. 2
PLAT
OF THE CLAIM OF
STEPHEN W. DOWNEY
UPON THE
LAKE LEMAN
PLACER MINING CLAIM,
Gold Hill MINING DISTRICT.
Carbon COUNTY.
TERRITORY OF WYOMING.
Containing an Area of 150 Acres.
Scale of 300 Feet to the inch.
Variation 15°30' East.
SURVEYED BY
Charles Bellamy
U.S. Deputy Mineral Surveyor.
April 30th and May 1st 1891
The original Field Notes of the Survey of the Claim of
Stephen W. Downey
upon the
Lake Leman Placer Mining Claim
from which this plat has been made, have been examined and approved and are on file in this office; and I hereby certify that they furnish such an accurate description of said Mining Claim as will, if incorporated into a patent, serve fully to identify the premises, and that such reference is made therein to natural objects and permanent monuments as will perpetuate and fix the locus thereof.
And I further certify that this is a correct Plat of said Mining Claim made in conformity with said original field notes of the survey thereof.
U.S. Surveyor General's Office, Cheyenne, Wyo.
June 8th 1891
Wm. A. Richards
U.S. Surveyor-General for Wyoming.
Sur. No. 55.
Gold Hill Placer Mining Claim
SUR. No. 56.
Lake Leman.
East 3300 ft.
West 3300 ft.
South 1980 ft.
North 1980 ft.
Cor. No. 1.
Cor. No. 2.
Cor. No. 3.
Cor. No. 4.
5469

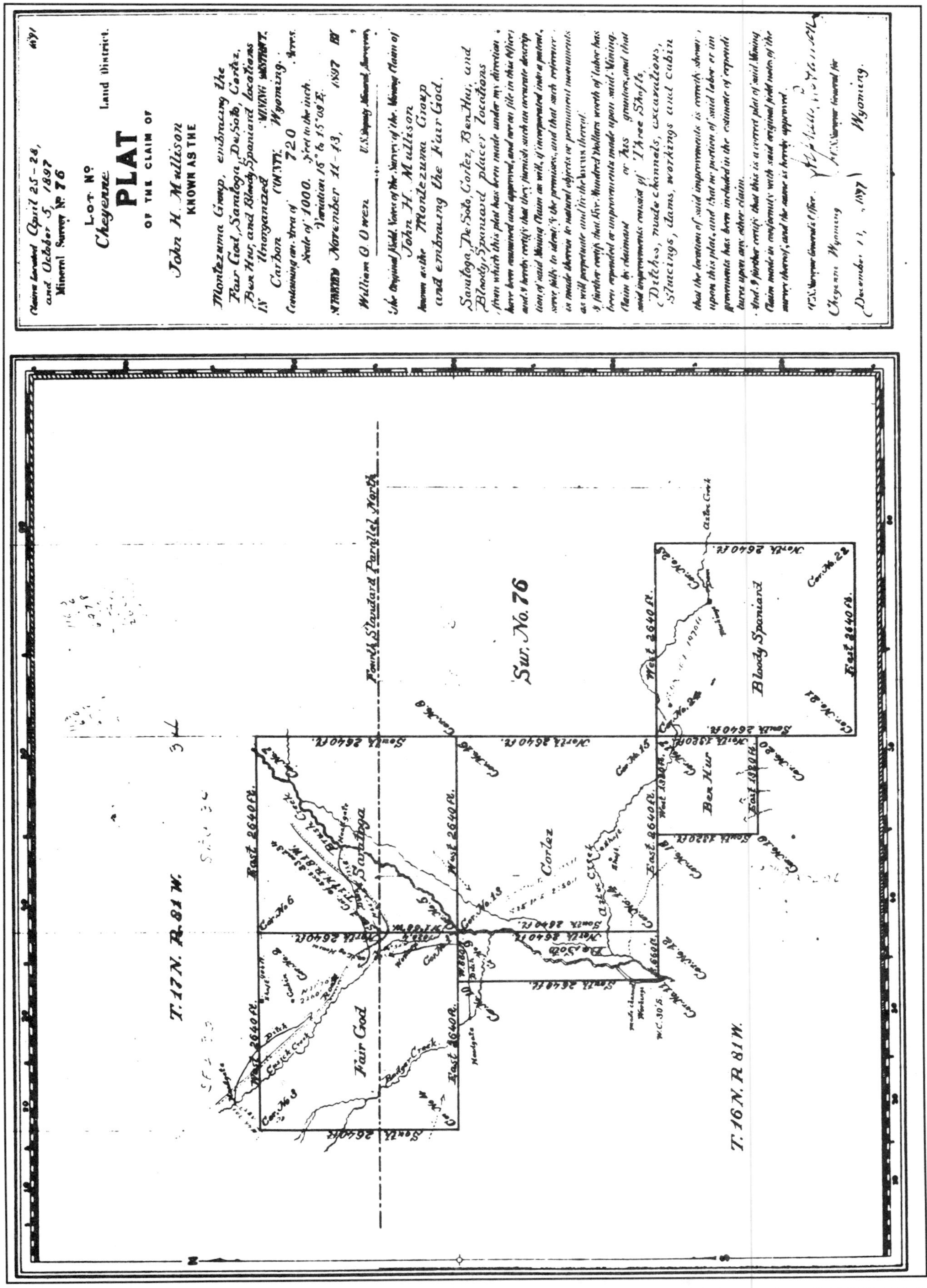
Claims located April 25-28, and October 5, 1897
Mineral Survey No. 76
Lot No.
Cheyenne
Land District.
PLAT
OF THE CLAIM OF
John H. Mullison
KNOWN AS THE
Montezuma Group, embracing the Fair God, Saratoga, DeSoto, Cortez, Ben Hur, and Bloody Spaniard locations
IN Unorganized MINING DISTRICT,
Carbon COUNTY, Wyoming.
Containing an area of 720 Acres.
Scale of 1000 feet to the inch.
Variation 15° to 15°08' E.
Surveyed November 11-13, 1897 By
William O. Owen U.S. Deputy Mineral Surveyor.
T.17N. R.81W.
T.16N. R.81W.
Fourth Standard Parallel North
Sur. No. 76
Fair God
Saratoga
Cortez
Ben Hur
Bloody Spaniard
Aztec Creek
Badger Creek

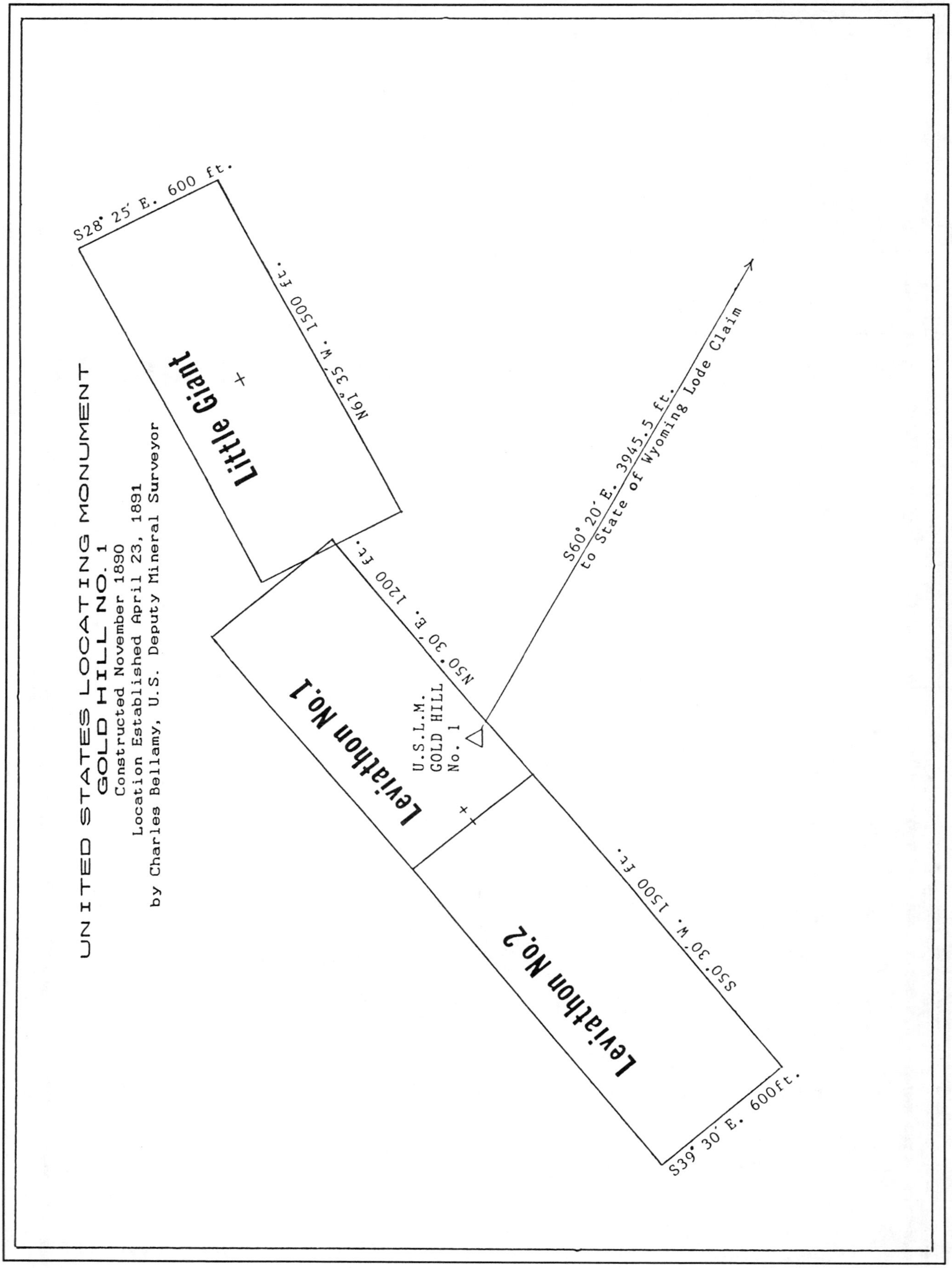
UNITED STATES LOCATING MONUMENT
GOLD HILL NO. 1
Constructed November 1890
Location Established April 23, 1891
by Charles Bellamy, U.S. Deputy Mineral Surveyor
Little Giant
S28° 25′ E. 600 ft.
N61° 35′ W. 1500 ft.
Leviathon No.1
Leviathon No.2
N50° 30′ E. 1200 ft.
U.S.L.M.
GOLD HILL
No. 1
S60° 20′ E. 3945.5 ft.
to State of Wyoming Lode Claim
S50° 30′ W. 1500 ft.
S39° 30′ E. 600ft.

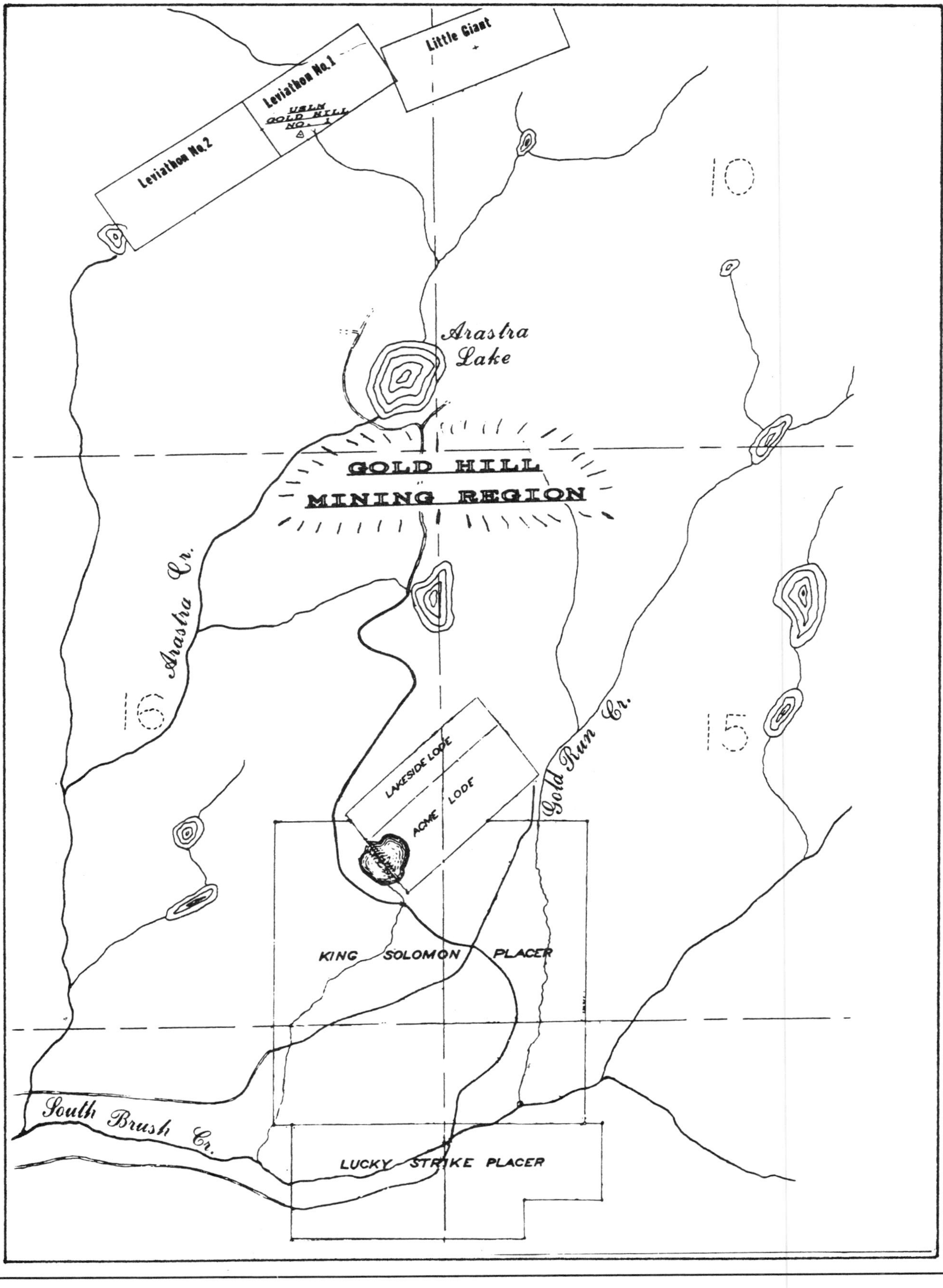
Little Giant
Leviathon No.1
USLM
GOLD HILL
NO. 1
Leviathon No.2
10
Arastra
Lake
GOLD HILL
MINING REGION
Arastra Cr.
16
15
Gold Run Cr.
LAKESIDE LODE
ACME LODE
KING SOLOMON PLACER
South Brush Cr.
LUCKY STRIKE PLACER

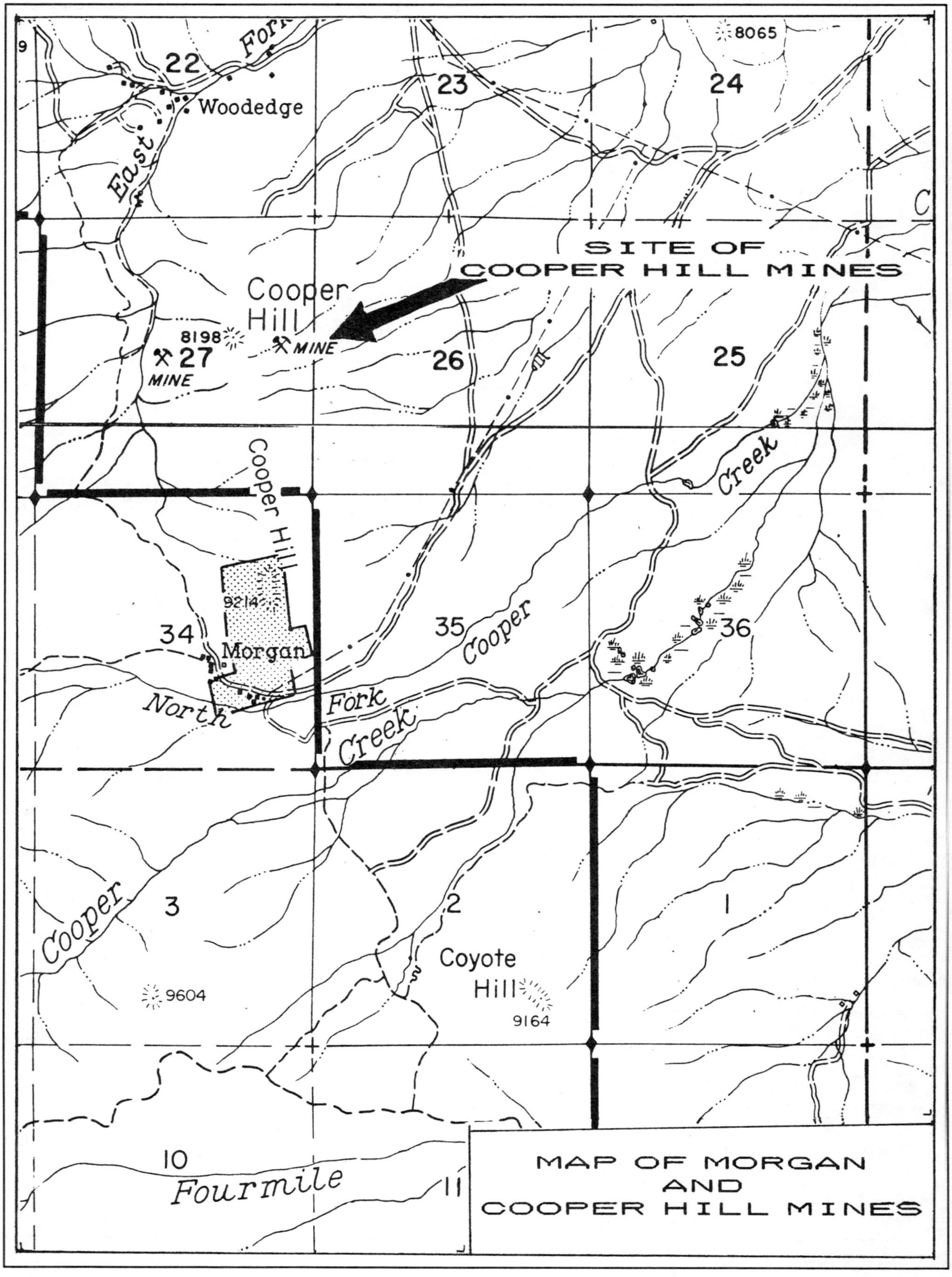
East Fork
22
Woodedge
23
24
8065
SITE OF
COOPER HILL MINES
Cooper
Hill
8198
27
MINE
MINE
26
25
Cooper Hill
Creek
9214
34
Morgan
35
Cooper
36
North Fork
Creek
Cooper
3
2
1
Coyote
Hill
9604
9164
10
Fourmile
11
MAP OF MORGAN
AND
COOPER HILL MINES

IX THE COOPER HILL MINES IX

By 1893, the Gold Hill excitement had run its course and the earlier mines of the Medicine Bow Mountains were mostly abandoned. However, a few diehard miners stayed on at the old camps, panning and sluicing the streams or running a tunnel on a suspected lode.

On the northern slopes of the Medicine Bow Mountains a mining camp was established in the early 1890s to prospect the ridges of the East Fork of the Medicine Bow River and Rock Creek. A mining district was formed on August 15,1892, in accordance with the 1872 Mining Law, and became known as the Herman Mining District. Named after tie cutter and miner, John Herman, the camp filled a dual purpose of a tie camp and a gold camp. Notable mines of the district were the "Randall," "Chicago" and "East Fork." A town site of 160 acres was laid out by Tom Morgan, Ira Carrington and S. M. Parker, but apparently only a few cabins were built. The camp, however did boast a "restaurant" advertising, on a sign over the door:

THE LOLIPOP RESTAURANT
Oysters on the Shelf
Tomatoes in the Can
Turkeys in the Barnyard
Ox Tail Soup Running in the Field
Scrambled Pig's Feet
STOP YE WEARY TRAVELERS STOP
McDonald & Sullivan
Proprietors[1]

The Lolipop Restaurant was host to the miner's Christmas banquet of 1892.

The Herman post office was authorized on March 31, 1893, and John S. Kidd was designated postmaster, but proof of actual operation is lacking.[2] Wesley F. King, the district recorder for several years, provided the Laramie newspapers with many details of life at Camp Herman and the Cooper Hill mines.

In 1893, the United States was on the brink of the "Panic of 1893" during which 600 banks would close, 15,000 businesses would fail and 74 railroads would fall into receivership. During April the National Gold Reserve would fall below $100,000,000 bringing about an even greater run on the Federal Treasury and the temporary suspension on the issuance of gold certificates. By November, the Sherman Silver Purchase Act, which called for the government to buy 4,500,000 ounces of silver each month, was repealed. Despite the repeal, by years end, the National Gold Reserve fell to $80,000,000 and the country was plunged into the depths of one of the worst depressions of its history. The successful mining states of the West were alarmed by the downward plunge of silver prices. Colorado had but one glimmer of hope in its new gold mines of Cripple Creek. In the three year old State of Wyoming, with no paying silver mines and the cattle business still reeling from the results of the disastrous blizzards of 1886-1887, the vision of successful gold mines again arose.

The winter of 1892-1893 was hard and cold. By January, 1893, nine feet of snow had fallen at Camp Herman on upper Rock Creek,and lay four to five feet deep in the trees. Not an unusual amount of snow but temperatures dropped to 32 degrees below zero at Rockdale and even colder in the mountains. A six horse team with supplies from Laramie was unable to travel closer than three miles of Camp Herman. Supplies had to be hauled in by sled pulled by men on "snowshoes" (nordic skis).

Spring brought a heavy runoff with six to seven feet of water in the creeks. As soon as the runoff began to subside, a small group of ex-cattlemen left Laramie to try their luck at gold mining. These four men, George Morgan, Tom Morgan, William Whittingham and Victor Beaumier had all been employed as foremen or managers on the Douglas Willan Sartoris ranches. In addition to mining and surveying equipment the men packed enough food for several weeks of extensive prospecting.

Their first objective was the Little Laramie River several miles below Centennial. Near the old McGill ranch they located 2,200 acres of placer ground. Proceeding north and west, they prospected all the promising streams from the Little Laramie to Rock Creek, searching for gold bearing placer gravel. Near Rock

George Morgan
U.W. American Heritage Center

Creek in the Herman Mining District, they changed from placer to lode prospecting. Here they found some leads but none they considered minable.

Towards the end of June with provisions running low, George Morgan, the more or less leader of the group, decided they should return to Laramie. He anticipated being home in Laramie by July 4, for his twenty-sixth wedding anniversary.

After a stop at the Rock Creek Stage crossing, known then as Rockdale, the four horsemen rode eastward toward Laramie. Retracing their tracks of a week before, they rode high on the slope above the old Overland Stage road. The air was cooler here and the route shorter. Besides, George wanted to further investigate a boulder he had found during previous prospecting. It was nearly noon when they reached the creek at the southeast end of a huge bald ridge. The men knew this to be Cooper Creek, having summer grazed cattle here and more recently done some placer prospecting. The grass was lush, and provided good grazing for the horses. Tom Morgan unpacked the lunch that Mrs. Williams of the stage stop had fixed for them, and George scouted around until he found the particular boulder for which he was looking. The four men studied the rock while they ate their lunch, and discussed the possible origin of the apparently good ore.

After lunch, Tom Morgan and Whittingham decided to climb the bare ridge to do some more prospecting. As they neared the top, Whittingham picked up a piece of brown ore and handing it to Tom, asked what it was. "That is what we are looking for and I will find that lead before I sleep," answered Tom and immediately began digging on the same spot. When he had gone down three feet, he struck his first piece of ore. Since it was late in the day, the four decided to make camp near the ridge and prospect further the next day.

With provisions almost gone, the next morning George Morgan went to the John Strauss ranch, some three miles distant, to replenish what he could of their food supplies. Strauss rode back to the camp with George to investigate on his own. It was Strauss who brought the first word of the new discovery to Laramie, but not until July 13.[3]

Wesley F. King, veteran miner and recorder for the Herman Mining District was also in the area and later described the antics of the prospectors when they made their discovery:

> We found Messrs. George and Tom Morgan, Wm. Whittingham and Victor Beaumier very much excited over the new find. Beaumier climbed a tree, sat down in the creek, and yelled 'lets do something.' Tom Morgan turned a back handspring and sat down on a cactus. Geo. Morgan began piling up the boulders believing them all rich ore.[4]

The four men worked three days to determine the direction and extent of the vein, and on July 7, 1893, staked out their first claim, naming it after Whittingham's daughter, "Emma G." The claim was the maximum size allowed by the Mining Law of 1872, 600 by 1,500 feet.[5]

Three days later, on July 10, they staked out another claim. This one, also of maximum size, was named the "Clara B" for George Morgan's daughter.[6] During this time, a third claim, the "Snowshoe" was also claimed by Ed Morgan, George's twenty year old son.

It was July 17 before George returned to Laramie with a first hand report of the new discoveries. On his arrival in Laramie, he hired and sent to the camp, a four horse team

John Strauss
Wyoming State Museum

to haul the 50 sacks of ore they had already sacked.

When word of the discovery circulated around Laramie, another gold rush started for the bald mountain. Again, not a gold rush as much of the west had known, but a rush nevertheless. The first people had already left for the area on the strength of John Strauss' report. Heck Tilton, Stockton Smith and W. J. Hills left on Saturday, July 15. George and Hansley Timmerman, Robert Burkhardt and Charles Richter left on the afternoon of July 16. The next day, William Frazer and Tom Dugan, and on the 20th, veteran surveyor Charles Bellamy, Mat Pennington, William Snyder, Dave Sarver, T. J. Dayton and Union Pacific Conductor A. R. Gracey joined the rush. Gracey rode his bicycle from Laramie to the mining camp. Miners from the Herman District also converged on the hill to look things over. M. S. Hampton and W. F. King visited the camp on July 14. King had been mining in the district for some years and his reports to the Laramie newspapers form the basis for many of the details in this account. It was King who reported that the first claim on the hill had been made some 12 years before by P. P. Rynders. (Actually, several claims were made at the heads of Cooper Creek and Dutton Creek as early as 1875). Rynders, King and Hampton promptly re-opened this claim, re-named it the "Producer" and began working it.[7]

A "mining expert" from Cripple Creek, Colorado, visited the mines in August and stated that he thought it could compare to the Cripple Creek area. (By 1893, Cripple Creek had a population of 12,500 and had already mined $2,783,369 in gold).

By December, 1893, only a dozen or so claims had been filed on Cooper Hill and many of the gold rushers had returned to Laramie. The Morgans and the other locators of the "Emma G" stayed on at Cooper Hill and by the end of the year had tunneled into the hill 65 feet. John Strauss and George McFadden had staked out a claim of their own, the "Cooper Hill" and were sinking a shaft. Benjamin W. Towner and T. C. VanBenthusen were working their new claim the "Rip Van Winkle" and were down ten feet with their discovery shaft. They also filed a claim on the "North Star" located at the far north end of the hill and the "Hidden Treasure" close to the south end. Ben Towner, who had prospected the length and breadth of the Medicine Bow Mountains, had come down from the La Plata District, high in the Snowy Range where he had been mining near his namesake, Towner Lake. Towner was among the first to stake a claim on Cooper Hill and was to remain longer than most of the others. Professor Wilbur C. Knight, of the University of Wyoming, was doing the assay work for the miners.[8] Knight had most recently been foreman and assayer for the "Keystone" on Douglas Creek.

• 1894

The new year of 1894 found the entire nation locked in deep depression. Fortunately the small miners, were less affected than the more affluent business men, they simply had less to lose. Gold was holding at slightly more than $20.00 an ounce and its buying power was rising. Trabing's of Laramie, the primary source of staples, listed their prices in the *Laramie Boomerang*:

Wilbur C. Knight
U.W. Western History Reasearch Center

Potatoes	$1.00 per hundred weight
Sugar	6.25 " " "
Gold Medal Flour	1.50 " " "
Kirks Moose Soap	1.00 for 25 bars
Schiltz Beer	2.25 for 12 bottles
Arbuckles Coffee	.25 per pound

The new year found the Laramie Plains sparsely covered but snow lay deep in the trees and wind sheltered pockets. During February, Wesley F. King reported that, nineteen feet of snow had fallen at Camp Herman since October.

Coe and Coe's tie cutting operation had continued throughout the winter and by spring they had 300,000 ties ready to float down the Medicine Bow River and an additional 60,000 on Rock Creek.[9] A few miners were still working at Gold Hill, La Plata and Camp Herman, but the principal point of gold mining activity was now the bald mountain between Cooper Creek and Dutton Creek where the Morgan Brothers had made their discovery the previous summer.

The Morgans and their mining associates had thus far failed to organize a new district and although located some distance from Camp Herman, the new claims were considered to be in the Herman District. The recorder for the Herman District, Wesley F. King, snowshoed (actually Nordic skied) from

Drawing by Joe Barros

Rockdale (now Arlington) to the mines, a distance of nine miles early in January. "Snowshoeing" through timber in deep powder snow is a feat demanding stamina, strength and good balance. Considering that King had only one leg, this trek seems almost unbelievable. King also visited the Four Mile prospects another two and one-half miles away. On returning to Rockdale, King sent his regular correspondence to the *Laramie Boomerang* noting in his latest observations that John Strauss, George McFadden, Ben Towner and John Watkins were on their claims. He also reported that the Morgans and Whittingham had struck another new lead on the "Emma G" and that several new placer claims had been made.[10]

By March, the snow was still four feet deep in the timber. Tie hacks were busy preparing for the spring tie drive. John R. Herman (for whom the Herman Mining District was named) was killed by a falling tree. In another accident, F. E. Skinner of the Herman District, had his leg broken by a sled and was taken to Denver for treatment.[11]

As spring struggled to arrive, the miners on

Ben Towner
Courtesy of Joe Barros

Cooper Creek began preparations for the anticipated great mining season. The owners of the "Emma G," "Clara B" and "Snowshoe" negotiated with Frederick J. Douglas Willan, Laramie Plains ranch baron, and finally signed over control of the three claims for sufficient money to continue operation of the others.[12]

W. F. King boldly declared that he fully expected a thousand men would be working the hill within sixty days. He also anticipated there would be a stage line, a mail route and most important, a smelter either at the hill or in Laramie.

Laramie merchants claimed they had sold more mining material during a three week period than they had in years.

Hickey, Naismith, Burke and Richardson, owners of the "Smuggler" mine ordered sacks for the ore they expected to ship. The "Smuggler" mine was into the hill twenty-five feet, following a vein of brown quartz. The hill was still known as "Bald Mountain" but one by one the miners began referring to it as "Cooper Hill" or "Cooper Mountain." Two and one-half miles long, 9,200 feet above sea level and rising more than a thousand feet above the surrounding prairie, the ridge seemed to the miners to be at last the great mineral deposit for which they had searched the last twenty-five years.

In May, judges Bramel and Hance visited the area. Bramel and his two sons had two claims on the hill, the "Champion" and the "Richmond." The judges noted there were twenty-six recorded claims on the hill and ten more not yet recorded. They also noted the "Emma G" had the most work accomplished. Most of the claims were for gold, they noted, but George McFadden and John Strauss were working a copper lead. All of Cooper Hill had been staked out in mineral claims, most, the maximum size allowed by mining law, 600 by 1,500 feet.[13] Prospecting, however, was not confined to Cooper Hill and numerous claims were made in the surrounding hills and canyons.

George Morgan, cattleman turned miner, was enthusiastically promoting all the mines at Cooper Hill. Morgan, it seems had supplanted Stephen Downey as the chief mentor of the mining camps. On May 28, 1894, George's wife Sarah, died at the Thornburgh Hotel in Laramie. She was 51. George was notably absent from the mining scene for several months.[14]

On August 13, 1894, the miners of the district held a meeting at the Morgan camp and elected new officials. Thomas Morgan was elected president and George McFadden was elected clerk and recorder. (Wesley King maintained that he was still recorder of the Herman District and subsequently McFadden became deputy recorder). At the meeting, Judge Bramel drew up a petition for a post office which was signed by all the miners present.[15]

During the last week of August, D. C. Abrams of the *Laramie Boomerang* paid a visit to the Cooper Hill mines. Abrams was escorted around the camp by Ben Towner. Each of the developing mines was visited and described by Abrams in the *Boomerang* on two successive days. He concluded his report with a description of the settlement. "A number of cabins have been built at the foot of Cooper Hill and these give the appearance of a little town as one approaches. . . . Cooper Hill has the promise of making one of the best mining camps in the west."[16]

• 1895

The winter of 1894-1895 was exceptionally mild and work at Cooper Hill continued without stop. The miners worked their claims

George W. Fox
University of Wyoming, American Heritage Center

C. W. Bramel
Wyoming State Museum

without covering over the shafts. The residents of Cooper Hill discussed the possibility of stringing a telephone line from Cooper Hill to Laramie. George Fox offered to furnish the wire if the miners would cut and set the poles.[17]

Four mines were working through the winter and several new claims were staked and filed. Tom Morgan and about a dozen miners lived and worked at the hill during December and January.

George McFadden kept many cats and one had a litter of nine born on New Years day. Such was the exciting news coming from the mines.

By February, 1895, only nineteen men were at work in the mining camp but in anticipation of spring, enthusiasm ran high among the miners.

In early spring more people began to arrive in camp with supplies and equipment to work during the summer. In April, one miner correspondent wrote the *Boomerang* and asked for assistance in obtaining a U.S. Post Office for the camp.[18]

Excitement regarding the ore continued to increase and a mill was proposed for treating the ore which the miners were already stockpiling in anticipation. One "mining expert," a Mr. Maltby, on his way to the South Pass Mines stopped at Laramie where he saw some of the Cooper Hill ore being assayed. He was so impressed with the apparent quality of the ore that he lost interest in his South Pass destination and went instead to Cooper Hill to examine the area.[19]

The first mining related death in the district occurred at the Richmond Mine in July. Charles W. Bramel Jr. was working at the bottom of the Richmond shaft when the ore bucket came loose from the hoist, falling 25 feet and hitting him on the head. After summoning help from nearby miners, his brother Burt rode to Laramie, leaving Cooper Hill about 2 p.m. and arriving in Laramie by 6 p.m. after two changes of horses. At 7 p.m. he left Laramie with his father, Judge C. W. Bramel Sr. and Doctor Alexander B. Hamilton. They arrived back at Cooper Hill by midnight. A remarkable ride but to no avail as his brother had died almost as soon as he had left camp. He was buried in Laramie.[20]

Towards the end of summer, mines were opening up all around Cooper Hill. The "Little Ella" north of Cooper Hill, which was considered the best developed mine in the district, was into the mountain 66 feet and was work-

Judge Bramel and C. W. Bramel Jr. at the Richmond Mine.
Laramie Plains Museum

ing an eight foot vein. Professor Knight of the University of Wyoming assayed some of the selected ore and declared, "... 32.17 ounces of gold per ton with a value of $643.40." He also found the ore assayed four to ten ounces of silver and 10 to 12 percent copper for an additional $20.00 per ton.[21] Hopes for some real producing mines ran high in Laramie and even higher at Cooper Hill. According to W. J. Huselton, who wrote to the *Laramie Daily Boomerang* on a regular basis, several of the mines were being worked even during a severe snowstorm during the last four or five days of December.[22]

By the close of the 1895 mining season, some 100 mines had been claimed on Cooper Hill. Apparently recognizing that this might be their last chance to discover a new and rich claim, prospectors from the old Medicine Bow mining camps had converged on Cooper Hill to try once again to reap the rewards of a rich mine. However, after two years of mining, the men of Cooper Hill were still looking for their first paying mine.

George Morgan spent part of the winter in Minneapolis, Kansas where he secured financial backing for the Cooper Hill mines and also courted Julia Redmond who would later become his second wife. He returned to Wyoming in March, 1896.

• 1896

As work continued on the mines without producing enough valuable ore to even pay the mining costs, some of the mine owners began to form companies. Their hope was that by combining their labor and resources, they could produce enough good ore to show a profit. Also by incorporating, they could market shares in their companies and gain the needed capital for continued mining operations.

The first Cooper Hill company to incorporate was the Carbon County Gold Mining

A. B. Hamilton
Wyoming State Museum

and Milling Company, on February 29, 1896. Capital stock was set at $500,000, at $1.00 per share. Two hundred and fifty thousand shares were retained by the trustees and 250,000 shares were offered for public sale. Trustees for the new company were Thomas Morgan, William Whittingham, Jabez Smith, William West and Joseph Gafford, all part owners of the Albion and Croesus mines. The owners divided the shares among themselves and also assigned George Morgan a trifling 100 shares. Additional trustees were George W. Fox, James Gray and C. B. Maxwell.[23]

A short time later, on April 13, 1896, the Gold Coin Mining and Milling Company was formed and incorporated. This company set their capital stock at $1,500,000, at one dollar a share. Mines owned by this company were the "Emma G," "Clara B," "Minneapolis," "King Solomon" and the "Gold Coin." Nine trustees were named: George Morgan, Tom Morgan, and William Whittingham of Laramie; B. F. Bracken, of Beloit, Kansas; William West, Samuel Blades, and Richard Goucher, of Minneapolis, Kansas; and James Gray, and Frank Siegel both from Kansas City, Missouri.[24]

Slightly more than a month later, a third

No.

THE

Morgan Mining and Improvement Company

CAPITAL STOCK, $1,000.00. 1,000 SHARES, $1.00 EACH.

This is to Certify That Geo F Morgan

is the owner of Thirty Seven Shares of the Capital Stock of

The Morgan Mining and Improvement Company,

non-assessable and transferable only on the books of the Company in person

or by attorney.

Witness the corporate seal of said company and the sig-

company was formed, the Cooper Hill Deep Mining Company. The company mines were known as the "Little Johnny" and the "Laura Bear." This company set their capital stock at $1,000,000 with a par value of $2.00. Trustees for the company were Thomas VanBenthusen, Doctor Alexander B. Hamilton, Dwight Smith and Thomas Simpson.[25]

Slightly less than a month later, on June 10, 1896, another company, the Morgan Mining and Improvement Company, was formed and incorporated. Capital stock for this company was set at a modest $1,000. One of the objectives of this company was to lay out and dedicate a town site. Nine trustees were designated, among them George Morgan.[26] On June 15, 1896, George Morgan, acting in his capacity as president of the Morgan Mining and Improvement Company, traveled to Laramie and Rawlins to file a plat of the company's proposed new town, "Morganville." The tract for the new town was located on a comparatively level area at the south-east end of Cooper Hill (the north half of Section 35 of T18N, R78W). The town was to be laid out on 49 blocks with 22 lots in each block, for a total of 1,078 lots. The streets were to be 80 feet and the alleys 16 feet wide, both devoted to public use. The company was to occupy approximately 76 lots on the west edge of the tract, leaving one thousand or more lots for sale to the public.[27] Not since John Cummins laid out Cummins City in 1880, had there been such an ambitious development proposed for the Medicine Bow Mountains. Actually two towns were beginning to grow, one on the company site and the other on Section 34 near the "Clara B" and the "Emma G" mining claims.

Although the entire hill had by 1896 been claimed, several of the mines were showing a greater potential, notably the "Albion,"

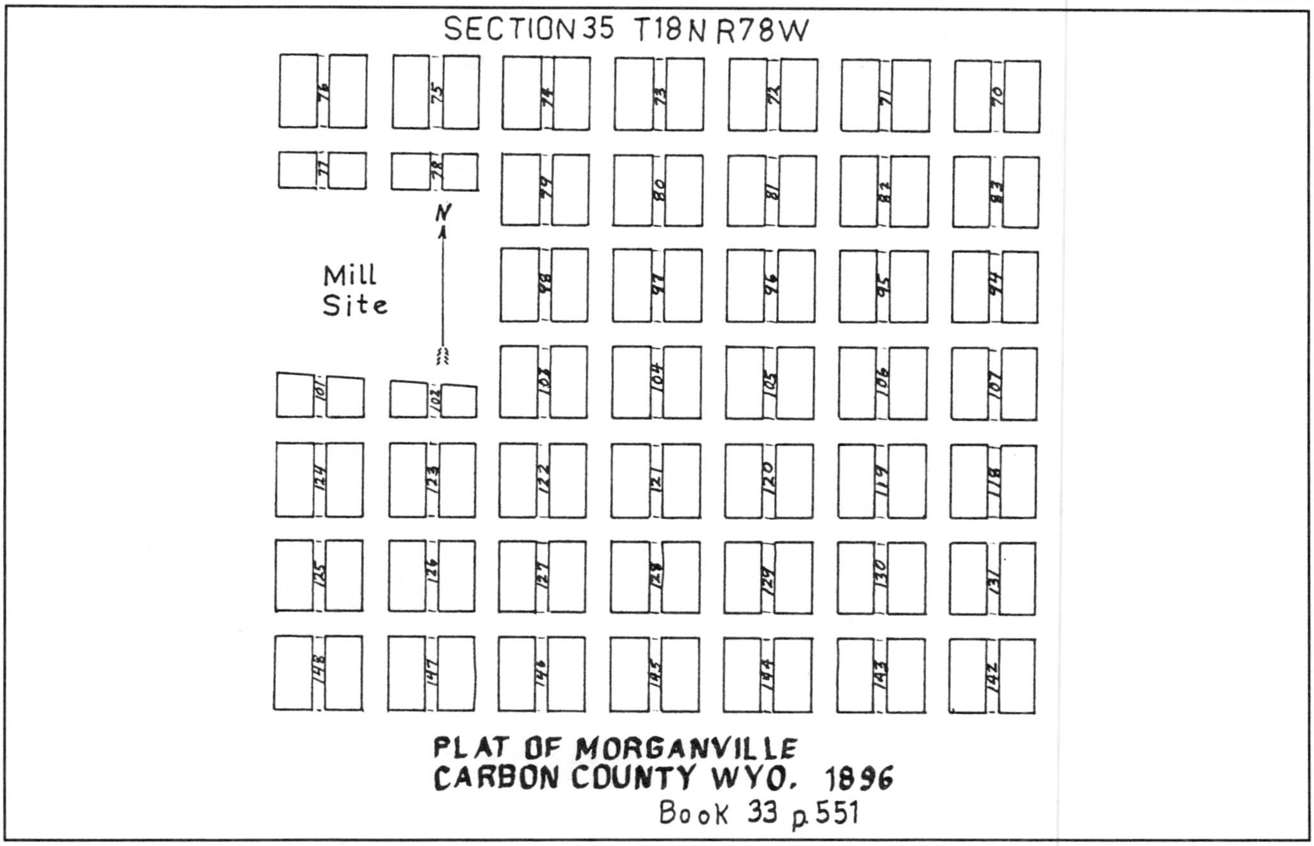

"Emma G," "Croesus," "Producer" and "Rip Van Winkle." The mining companies had gained control of these.

By June, Morganville began to take on some of the aspects of a frontier town. Ed Allen was the first to venture into business with a saloon called the "Podunk" located at the south end of the hill. His bar fixtures arrived on June 27, and he opened a few days later advertising, "Here exhilarating as well as refreshing drinks are dispensed by the insurmountable Baldy Carrington." The new hotel was also finished on June 27, and a dance was held in honor of that opening and of the new town. It was pronounced a "howling success."

On July 4, an all day program featuring horse racing and foot racing in the afternoon and dancing in the evening, provided ". . . great entertainment for all." [28]

In August, a house warming was held at William Conway's new house, the first with a shingle roof. Young people from nearby ranches attended ". . . and altogether the dance was a flattering success and a credit to Mrs. Conway." These dances usually lasted well into the night and it was not unusual for the rising sun to catch the last of the party on their way home. On this particular occasion, George Morgan's son, Ed, and Frank Brazil were the designated drivers and delivered the late dancers to their respective camps and ranches. Ed was often called upon to serve as coachman for travelers to and from the camp. On one occasion, ". . . while taking in the hill in a six-in-hand Concord coach, it is reported that his work was so well performed, that several of the party have not yet had their backbones straightened out since their memorable experience." [29]

On August 23, 1896, Morganville recorded its first birth, "a bouncing baby boy born to Mrs. Edward Willett." A short time later, on September 4, another baby boy was born to Mrs. Edward Russett.[30] Not to be outdone, Mrs. Tim McCoy presented her husband with a "bouncing baby boy" on September 22.[31]

Reports of these seemingly trivial occurrences point to the hopes of the miners for a prosperous and permanent mining town but indicate a lack of significant mining news.

Early in October, 1896, Victor Beaumier moved a sawmill into the canyon south of the mining camp. Beaumier had contracts for 200,000 feet of lumber from area ranchers. As soon as these contracts were fulfilled, he

Drawing by
Dan Swift

planned to start cutting timbers for the mines.[32]

The presidential election campaign of 1896 hinged on the issue of free coinage of silver at the ratio to gold of 16 to one, or what was known as Bi-metalism. To support their belief in Bi-metalism, the miners made a huge banner which was to be raised "if anything unusual occurs." The banner, owned by Carl Stradley and McDonald and kept at their cabin, proclaimed "16 to 1, Bryan and Sewell" in two foot high letters, indicating their choice of candidates for the impending presidential election.[33] However, when the 1896 elections were held, the mining districts as well as the rest of the state and nation did not support Bryan and the country remained on the gold standard.

As early as 1894, the miners at Cooper Hill had proposed a post office to serve the district. George Morgan had often been suggested for the duty of postmaster. However, when the petition to Washington D.C. was drawn up by Judge C. W. Bramel, on August 15, 1894, George McFadden was recommended as the first postmaster. More than two years later, on November 16, 1896, the post office was finally established. It was given the name of "Morgan" not "Morganville" as petitioned.[34] Morganville, however, continued to be used by some of the local people. McFadden was appointed postmaster. McFadden, a Civil War veteran, was one of the first miners in the area and in partnership with John Strauss, owned an interest in several mines. He had done some prospecting in the old Last Chance District and then gone into business for himself, with a small photographic salon in Laramie. Apparently business was not very good and he moved to Morgan and again took up mining. He was considered a little eccentric and was known locally for the large number of cats he kept at his cabin.

A stage to the new post office was soon established which ran twice a week, leaving Laramie on Tuesday and Friday and returning on Wednesday and Saturday, carrying mail both ways. Freight was also hauled for "fifty cents a hundred weight." The stage ran throughout the year, but not without some difficulty. In April, 1897, the stage was eleven days late due to the deep snow and rising creeks.[35]

Since the spring of 1895, the miners at Cooper Hill had been mining enough promising ore to begin hoping for a reduction mill for processing. A stamp mill with a minimum capacity of 25 tons a day was proposed. During the month of April 1895, George Morgan and an "English gentleman" had inspected the area with the intention of locating a mill site. In May, 1895, the announcement was made that a mill would indeed be constructed. It was contemplated that $5,000 would cover the cost of construction and machinery, but no action was taken to begin construction during the remainder of 1895.

In February, 1896, the encouraging announcement was made that ". . . when it is determined which method of processing the ore is best, they will put in their plant." Finally, in August 1896, the survey for the mill site was completed.

During November, 1896, the first positive move toward construction was completed with the appointment of William Stradley to take charge of construction and the movement of two large wagon loads of bricks from John Mast's brickyard in Laramie, to the mill site at Cooper Hill.

By December, actual construction had begun on the building, and it was announced that the machinery was "on the way." The mill was a ten stamp affair of the latest pattern by the Frazer and Chalmers Company, and was to have a 30 horsepower steam engine and a 40 horsepower boiler for power. Working throughout the winter, the mill was completed in April 1897, and was ready for operation "awaiting the advent of spring."[36]

• 1897

On February 6, 1897, another company was formed. The corporate name of this company was The Rip Van Winkle Gold Mining and Milling Company. Capital stock was set at $2,000,000 with a par value of $1.00 a share. The company's mines were, the "Rip Van Winkle," "Maggie T," "John L," "Missing Link," "Spar" and "Ruby Trust." Trustees were owners Benjamin W. Towner, George Fox and I. R. Swigart plus financial contributors Otto Gramm, J. C. Gafford and M. F. Swigart.[37] "Rip Van Winkle" stock was carried on the local market for some years with a trading price of between five and twelve cents but apparently little if any was sold.

THE

RIP VAN WINKLE CONSOLIDATED GOLD MINI'NG

AND MILLING CO.

Capital Stock, 2,000,000 Shares
Par Value, One Dollar per Share.
Fully Paid and Non-Assessable.
Treasury Stock, 500,000 Shares.

PROPERTY:--Seven Mining Claims--Rip Van Winkle, Spar, Missing Link, Ruby Trust, Maggie T., John L. and Little Maud--Embracing 130 acres of land, located on Cooper Hill, Carbon County, Wyoming; 35 miles N.-W. of Laramie; 9230 feet above sea level.

ORGANIZED UNDER THE LAWS OF WYOMING.

Principal Office: LARAMIE, WYO.

This Company has no Indebtedness and no Salaried Officers.

To the mining camp enthusiasts, 1897 promised to be a great year at Cooper Hill. In April it was "positively stated" that two large buildings would be erected at Morgan, a large hotel and a general store. It was further stated that a smelter had been purchased by the Carbon County Mining and Milling Company and was on the way to the camp. In addition, a new school house was planned to be built. In anticipation of a great season, William Conway built a new porch on the front of his Silver Dollar Saloon, Mr. Gray built a new blacksmith shop and Vic Beaumier and Jack Smith began setting up their sawmill to provide the anticipated demand for lumber. A club calling themselves the Silver Coin Social Club was formed and met at Mrs. Nelson's on Friday nights.

All over the district, mining was taking on a new energy and the miners were sacking their

Otto Gramm
University of Wyoming, American Heritage Center

best ore in anticipation of the mill and smelter being erected.

In May, an announcement was made in the *Laramie Boomerang* that the smelter had arrived at Wyoming Station and was awaiting transportation to Cooper Hill, where it was to be set up next to the mill. The smelter apparently never made it to Cooper Hill, and its final destination remains a mystery.[38]

During the latter part of summer 1897, the mill was finally put into operation. The Carbon County Gold Mining and Milling Company processed 300 tons of ore from the "Albion" and the "Emma G" mines, yielding an average of $17.50 a ton for a net return of $10.00 a ton. A good return if this was the bulk of the ore, but this was upgraded and sacked ore, which had been stockpiled for three years. A mere $3,000 for all their efforts for three years! A mere $3,000 for their best ore! The mill alone had cost $5,000.[39] It was a great disappointment to the miners and many packed up and left. However, there were those who still believed the next few feet of digging would reveal a rich vein if only they would stick to it.

• AFTER THE BOOM

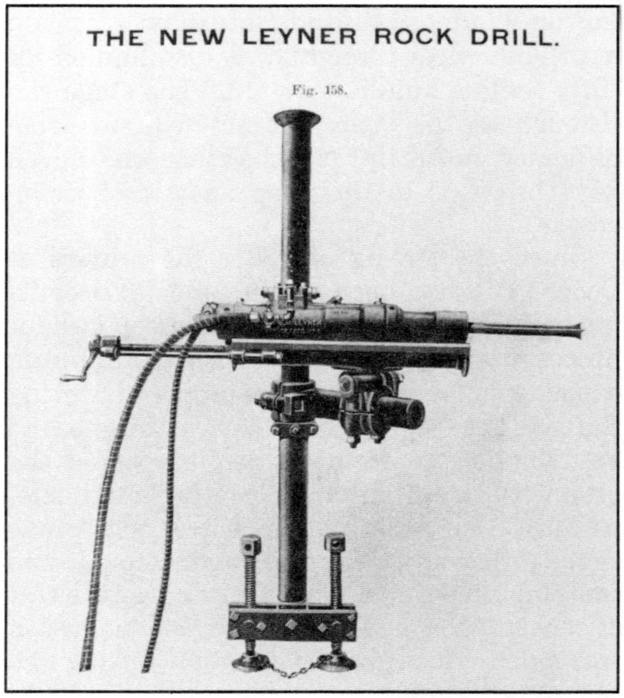

Courtesy of Bud Owens

The boom was over but some of the miners stayed on at Cooper Hill, convinced in their own mind that they would eventually strike it rich. Ben Towner continued to work the "Rip Van Winkle" although he found it necessary to occasionally find employment elsewhere to help him along. Little by little he sold off shares of his favorite mine to finance his mining operations.

In November, 1897, Terry Fee took over the stage route. Fee had a ranch about midway between Laramie and Morgan and supplemented his income by running the stage with mail, freight and passengers. Sometimes he shot jackrabbits along the way and sold them to the ranchers for fox food. Jack McCue, a Laramie plains rancher, claimed the jackrabbits were as big as baby antelope. One of Fee's more unusual cargoes was the disinterred body of Julia Lenihan and the body of her mother, Mrs. John Lenihan, which he transported from Aurora (near Rock Creek) to Laramie.

From 1890 to 1900, Wyoming's population grew from 62,555 to 92,531, seemingly a favorable increase, but generally disappointing to most of the state residents. Within the first few years of statehood, four banks failed. The cattle industry declined from its boom of the early 1880s and was only partially offset

Charles Bellamy
University of Wyoming, American Heritage Center

Al Mountford
University of Wyoming, American Heritage Center

by an expansion of the sheep raising industry. The Glass Works in Laramie closed down in 1889, the Flour Mill and cigar factory closed in 1891 and the rolling mill by the turn of the century. The Union Pacific went bankrupt in 1893. The Warren Livestock Company went into receivership in 1894.

However, Wyoming was not alone, a general depression prevailed throughout the country during the 1890s. The Panic of 1893 was the beginning of a deep depression which lasted until 1898.

The turn of the century brought a new surge of mining to the Medicine Bow Mountains. Centennial, La Plata, Cummins City, Gold Hill, Keystone and Cooper Hill all experienced a short rebirth. For a brief period of time after the turn of the century, new hope was expressed for the Cooper Hill mines by the promise of a new railroad to be built close by. The new railroad, The Denver, Laramie and Northwestern, was to begin in Colorado, cross the mountains between Fort Collins and Laramie and generally follow the old Overland Trail route across the Laramie Plains. Construction began on the most difficult portion of the grade between Fort Collins and Laramie, some track was actually laid and an extensive tunnel completed in Colorado, but the company was unable to sell enough shares to operate.

In August 1901, the Carbon County Gold Mining and Milling Company and the Gold Coin Mining and Milling Company pooled their resources and purchased a Leyner Air Drill and compressor for the "Croesus" mine. The Leyner Drill, developed in the Colorado mining camps, was at the time considered the most advanced drill available and procurement of the drill was considered an indication of a promising company and mine.

In anticipation of the proposed railroad, another company the Producer Gold and Copper Mining and Milling Company filed articles of incorporation in Cheyenne on April 15, 1904. Capital stock was listed at $3,000,000 divided into 3,000,000 shares. Principal place of business was listed as Laramie with a branch office located in Minneapolis, Kansas, where several of the trustees for the company lived.[40]

In 1907, several of the old mines were relocated and in January, 1908, a new survey was completed by Deputy Mineral Surveyor, Char-

Drawing by Dan Swift

in Chillicothe, Missouri. He was returned to Laramie for burial. Morgan will be remembered by history, not for his mining activities, not for the little town named after him, but for the introduction of Hereford cattle to the west. According to the *American Hereford Journal*, 99 percent of the Hereford cattle in the United States are descended from bulls imported by Morgan.[41]

William Whittingham spent years more, prospecting the mountains, before returning to Laramie where he died in 1919.

Benjamin Towner continued mining at Cooper Hill for years after the boom, vowing that if it took his last dime he would strike it rich. It did and he did not. He suffered a stroke at Cooper Hill in 1915 and died in Laramie a few days later.

Charles W. Bramel Sr. lost interest in the mines but apparently grubstaked an occasional prospector including his son Burt. The senior Bramel was shot to death by a deranged ex-client in 1907.

les Bellamy. Later in 1908, application for patent was made for the mines owned by The Producer Gold and Copper Mining and Milling Company; the "Clara B," "Emma G," "Gold Coin," "Olney," "Albion" and "Bi-metalist." After extensive correspondence through 1909, patent was finally granted on December 6, 1910. In the meantime, J. C. Gafford, acting-secretary of the company, had left Wyoming and the certificates of patent were mailed to him in Topeka, Kansas.

As with most all of the Medicine Bow Mining camps before, the Cooper Hill mining camp of Morgan slowly faded to virtual non-existence.

George Morgan left the mining world to return to his first love, cattle, or to be more exact, Hereford cattle. He seemed to be recovering from his business and personal problems when in 1903 he became ill and died

Thomas Falconer stayed on at Morgan and became postmaster on July 14, 1909. He was relieved of his duties in 1913 when he burned down the post office while allegedly under the influence of strong drink.

Doctor Alexander B. Hamilton, one of Laramie's premier physicians, never gave up hope on the Medicine Bow mines. Aside from his part ownership of several of the mines, he was a company officer in at least three of the companies and continued his support of the miners with an occasional grubstake. Doctor Hamilton died in Laramie in 1943.

John Strauss, never overly enthusiastic about mining, returned to his nearby ranch and cattle business. His name is perpetuated by Strauss Hill.

Al Mountford stayed on and by continuing to file proof of labor notices on the claims and eventually going to court, he made a successful claim for possession of the patented claims.

Mountford died in 1947, one of the last of the 19th century Medicine Bow miners.

The "Emma G" mine is still visible, the brown quartz for which the miners struggled, lies scattered around the old shaft. Remains of the shaft house are slowly rotting away. Towner's favorite mine, the "Rip Van Winkle" sleeps undisturbed on the westerly slopes of Cooper Hill, the shaft house reduced to a pile of crumbling logs. The ruins of the mill are still to be seen on the south of Cooper Hill. Pieces of mining equipment are occasionally picked up by the curious. The town site of Morgan, with the 1896 blacksmith shop and one or two cabins of the 19th century, is deserted except for the vacation cabins built on the old land patents. Perhaps with all the modern mining and processing technology the mines will again be opened and the dreams of the old prosp ectors will materialize.

Cooper Hill and the mining camp of Morgan became the last of the mining camps of the Medicine Bow Mountains. To be sure, prospectors and miners returned again and again to the old camps scattered around the mountains but no new camps were established, primarily because, virtually all of the mountains had, by the late 19th century, been prospected.

To this day, hopeful prospectors can be found searching the placers and even doing some lode mining on the sites, but barring an unforeseen breakthrough in technology or the expenditure to excavate for deep mining, it seems unlikely that any noteworthy discoveries will be made.

Morgan Cabin
Dan Swift Collection

Rip Van Winkle shaft, ca. 1900
Dan Swift Collection

Rip Van Winkle shaft, 1989

Rip Van Winkle tunnel, 1900
Wyoming State Museum

Rip Van Winkle tunnel, 1989

Town of Morgan, ca. 1900
Dan Swift Collection

Town of Morgan, 1989

Rip Van Winkle tailings. Top row, left to right: Al Mountford, unknown, Ben Towner, unknown, Culbertson, George Fox. Two men seated in foreground are unknown.

Dan Swift Collection

Courtesy of Bud Owens

Morgan Hotel, Boarding House, Post Office and residence, from 1896 to 1990.

Morgan 1896 Blacksmith's shop, 1990.

North Star Mine, ca. 1900.
Dan Swift Collection

Shovel and Windlass handle.

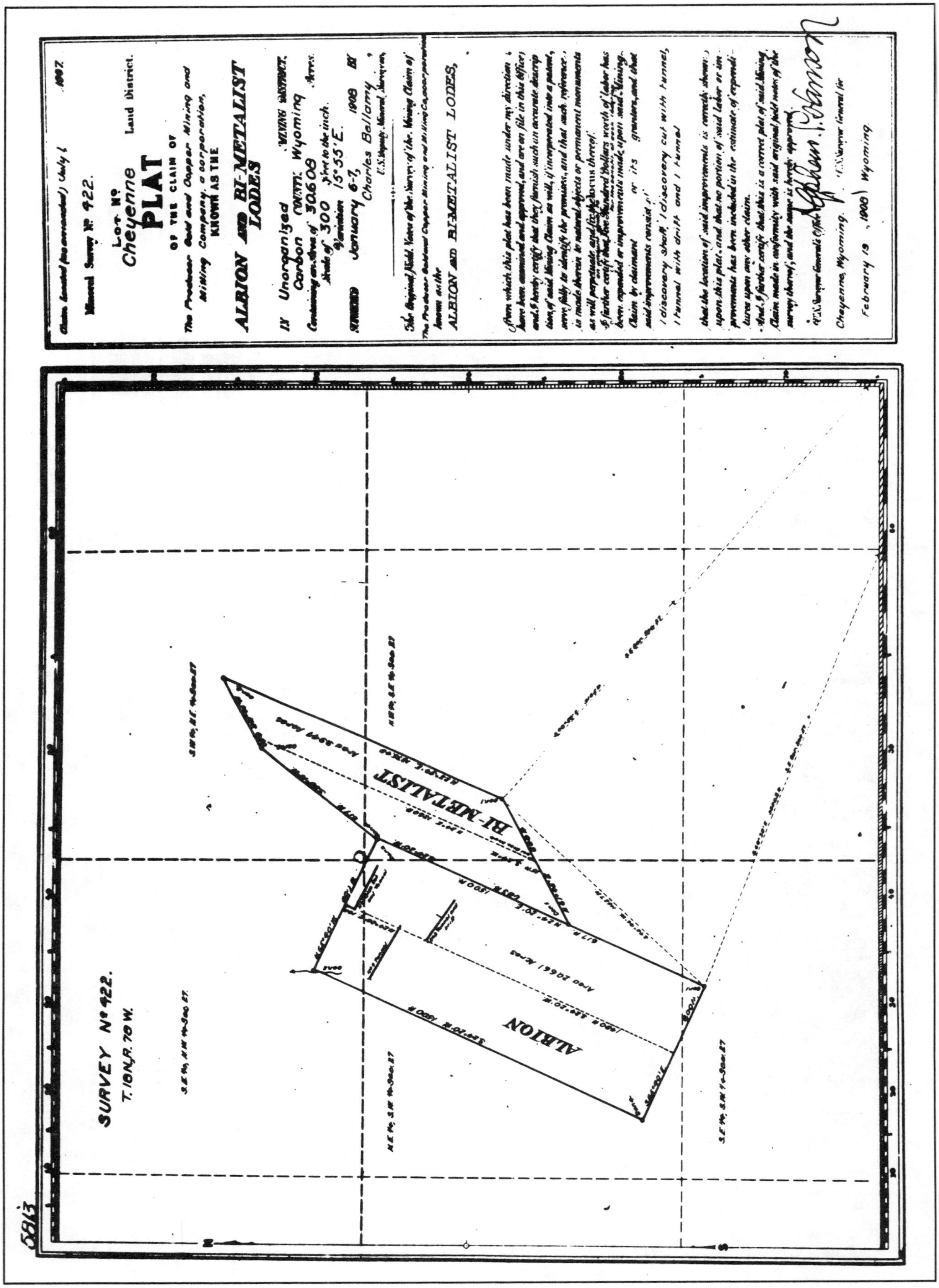
5813
SURVEY Nº 922.
T. 18N, R. 78 W.
S.E. ¼, N.W. ¼ Sec. 27
S.W. ¼, N.E. ¼ Sec. 27
N.E. ¼, S.W. ¼ Sec. 27
N.W. ¼, S.E. ¼ Sec. 27
S.E. ¼, S.W. ¼ Sec. 27
ALBION
BI-METALIST
Claim located (as amended) July 1 1907.
Mineral Survey Nº 922.
Lot Nº
Cheyenne Land District.
PLAT
OF THE CLAIM OF
The Producer Gold and Copper Mining and Milling Company, a corporation,
KNOWN AS THE
ALBION AND BI-METALIST LODES
IN Unorganized MINING DISTRICT,
Carbon COUNTY, Wyoming
Containing an Area of 30.608 Acres.
Scale of 300 feet to the inch.
Variation 15°55'E.
SURVEYED January 6-7, 1908 BY
Charles Bellamy,
U.S. Deputy Mineral Surveyor.
The Original Field Notes of the Survey of the Mining Claim of
The Producer Gold and Copper Mining and Milling Co., a corporation
known as the
ALBION AND BI-METALIST LODES,
from which this plat has been made under my direction, have been examined and approved, and are on file in this Office; and I hereby certify that they furnish such an accurate description of said Mining Claim as will, if incorporated into a patent, serve fully to identify the premises, and that such reference is made therein to natural objects or permanent monuments as will perpetuate and fix the locus thereof.
I further certify that Five Hundred Dollars worth of labor has been expended or improvements made upon said Mining Claim by claimant or its grantors, and that said improvements consist of
1 discovery shaft, 1 discovery cut with tunnel, 1 tunnel with drift and 1 tunnel
that the location of said improvements is correctly shown upon this plat, and that no portion of said labor or improvements has been included in the estimate of expenditures upon any other claim.
And I further certify that this is a correct plat of said Mining Claim made in conformity with said original field notes of the survey thereof, and the same is hereby approved.
U.S. Surveyor General's Office
Cheyenne, Wyoming.
U.S. Surveyor General for
February 19, 1908 Wyoming.

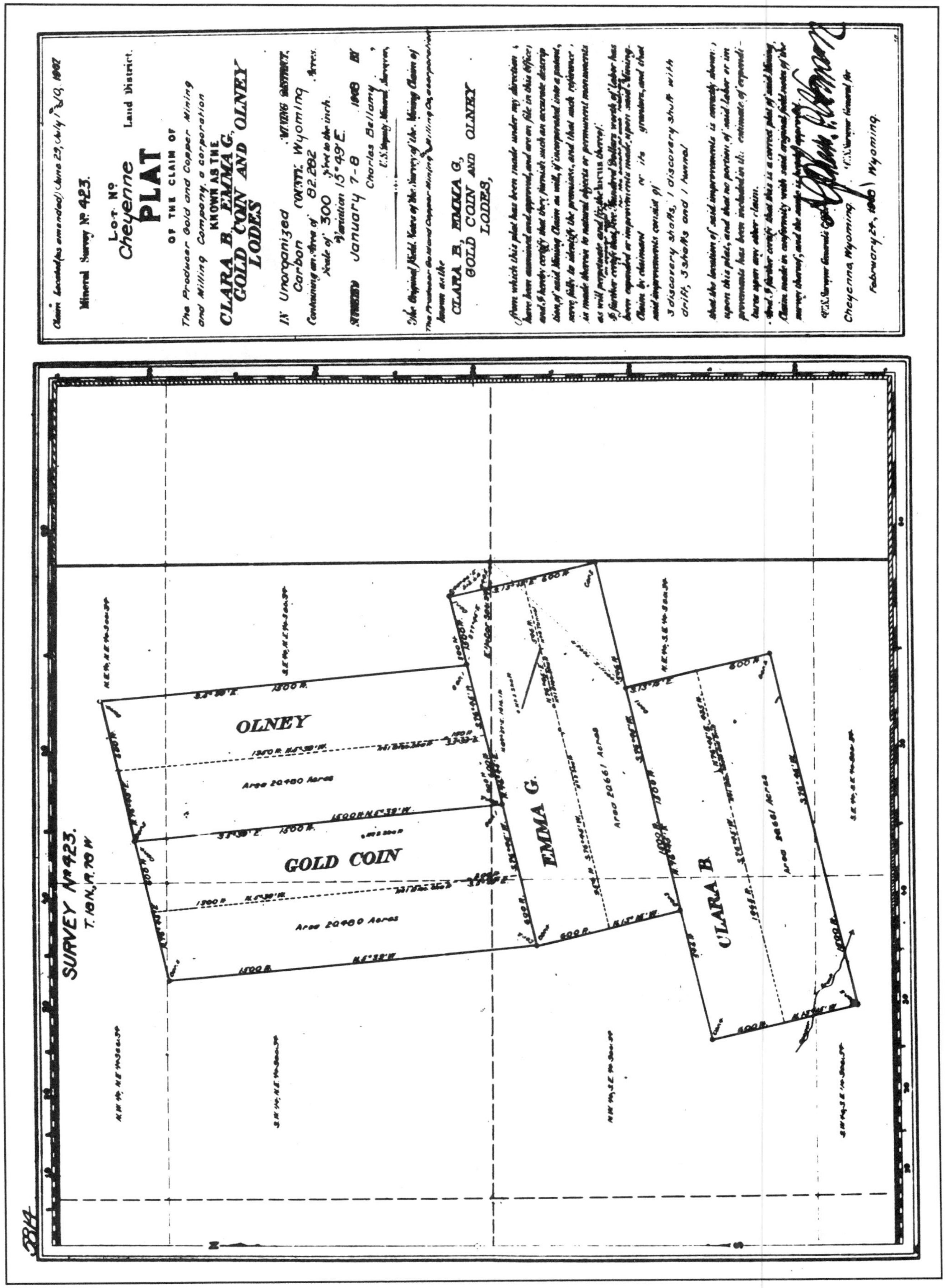
Claim Located (as amended) June 23, July 1 & 10, 1907
Mineral Survey No 423.
Lot No
Cheyenne Land District.
PLAT
OF THE CLAIM OF
The Producer Gold and Copper Mining and Milling Company, a corporation
KNOWN AS THE
CLARA B, EMMA G, GOLD COIN AND OLNEY LODES
IN Unorganized MINING DISTRICT,
Carbon COUNTY, Wyoming
Containing an Area of 82.282 Acres.
Scale of 300 feet to the inch.
Variation 15° 49' E.
SURVEYED January 7-8 1908 BY
Charles Bellamy,
U.S. Deputy Mineral Surveyor,
The Original Field Notes of the Survey of the Mining Claim of The Producer Gold and Copper Mining & Milling Co, a corporation known as the
CLARA B, EMMA G, GOLD COIN AND OLNEY LODES,
from which this plat has been made under my direction, have been examined and approved, and are on file in this office; and I hereby certify that they furnish such an accurate description of said Mining Claim as will, if incorporated into a patent, serve fully to identify the premises, and that such reference is made therein to natural objects or permanent monuments as will perpetuate and fix the locus thereof.
I further certify that Five Hundred Dollars worth of labor has been expended or improvements made upon said Mining Claim by claimant or its grantors, and that said improvements consist of
3 discovery shafts, 1 discovery shaft with drift, 3 shafts and 1 tunnel
that the location of said improvements is correctly shown upon this plat, and that no portion of said labor or improvements has been included in the estimate of expenditures upon any other claim.
And I further certify that this is a correct plat of said Mining Claim made in conformity with said original field notes of the survey thereof, and the same is hereby approved.
U.S. Surveyor General's Office
Cheyenne, Wyoming
February 24, 1908
U.S. Surveyor General for Wyoming.
SURVEY No 423.
T. 18 N., R. 78 W
OLNEY
GOLD COIN
EMMA G.
CLARA B
Area 20.480 Acres
Area 20.480 Acres
Area 20.661 Acres
Area 20.661 Acres
5814

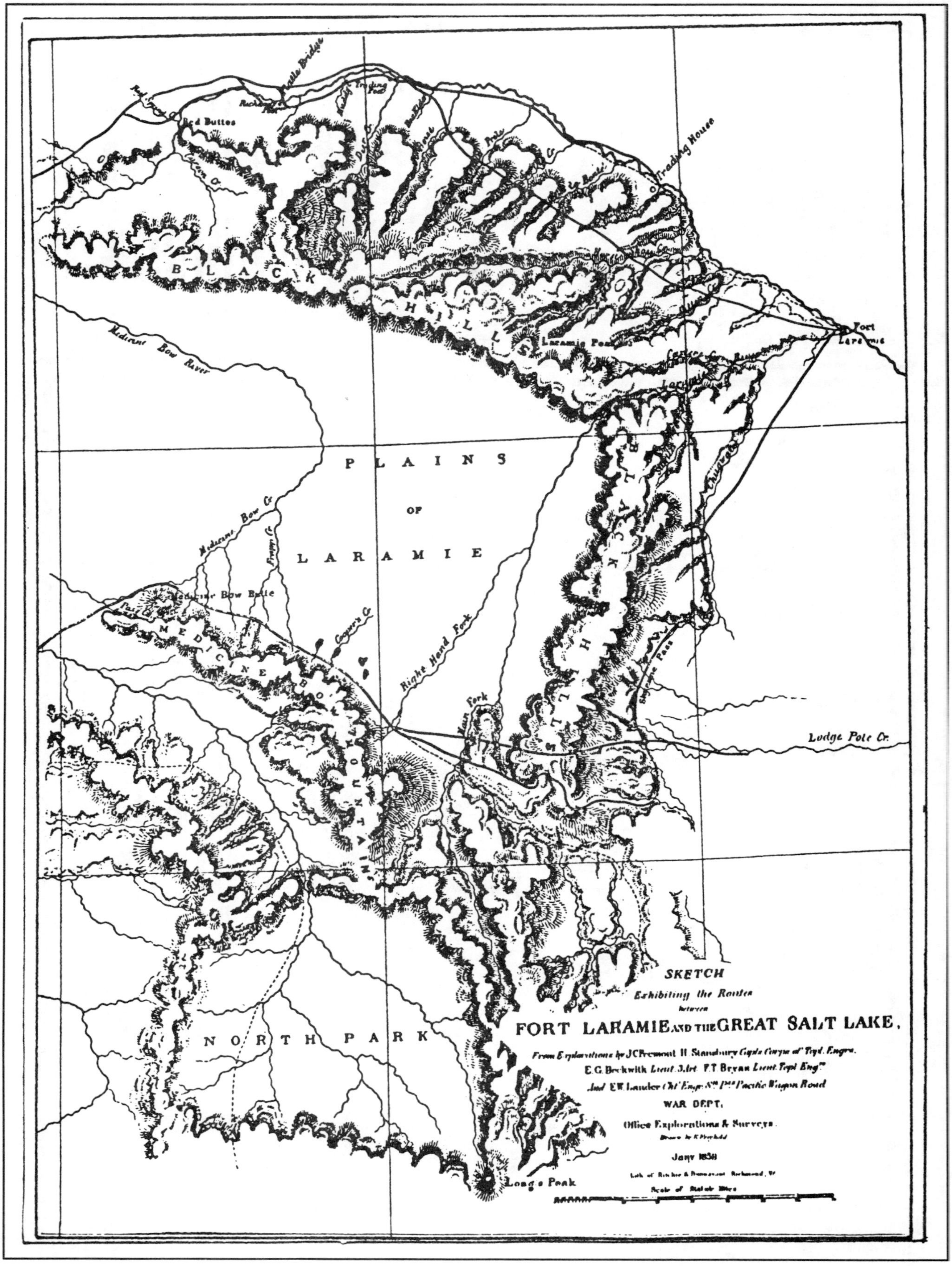
SKETCH
Exhibiting the Routes
between
FORT LARAMIE AND THE GREAT SALT LAKE.
WAR DEPT.
Office Explorations & Surveys.
Long's Peak
BLACK HILLS
PLAINS OF LARAMIE
NORTH PARK
MEDICINE BOW MOUNTAINS
Medicine Bow River
Medicine Bow Cr.
Right Hand Fork
Lodge Pole Cr.
Fort Laramie
Red Buttes
Trading House
Laramie Peak

X EARLY MINING X

The 1868 prospecting and mining at Last Chance was the earliest recorded in the Medicine Bow Mountains, but these early prospectors and those who followed invariably reported evidence of even earlier mining activities in virtually every new area they explored. Trails, campsites, excavations and even what appeared to be caved in mine shafts and tunnels were found throughout the mountains. Many of these miners believed these were just indications of their recent predecessors, but some believed this evidence pointed to much earlier mining, possibly by the Spaniards, Aztecs or Incas. However, then as well as now, no firm evidence remained to establish the date or identity of these early miners. One might presume that the realization that someone had prospected and mined before them would have led to discouragement or, at least been an incentive to prospect elsewhere. However, the discovery of an earlier excavation with little or no evidence of precious metal did nothing to discourage the miners and in fact had the opposite effect. Many marveled at the apparent skill of the miners before them, as evidenced by the lack of any precious metal remaining in the excavations.

Newspapers of the area printed many stories of early mining in the area, presumably to inspire the local miners to expand their search and, hopefully, discover enough paying mineral to boost the general economy.

• SPANISH INFERENCE

On April 25, 1875, the *Laramie Daily Sentinel*, James H. Hayford, editor, printed an article expressing his viewpoint on early mining in the Medicine Bow Mountains:

> The existence of gold in what is known now as the Medicine Bow range has been known for centuries. The aborigines of the country recognized the value of what we term the precious metals. It was the gold of Mexico and Peru that incited the cupidity of Cortez and Pizarro in the sixteenth century, and it is not at all improbable that the coffers of Montezuma, if not of the Incas, contained gold from the mountains of what is now Wyoming. Our assertions may seem wild and visionary to many, but there are unmistakable evidences that our mines were worked centuries ago, and most likely by the Aztecs, before America was discovered by Europeans. In that section where Douglas, French and Rock Creeks rise, are relics of ancient mining operations. Traces of tunnels or drifts remain, and on the mounds of earth that were removed, stately pines stand . . . .[1]

Not only the Medicine Bow Mountains but the whole of what we now call Wyoming was included in the premise that early mining had been done by the Incas, Aztecs or Spaniards. In support of this assumption, another story was related by Robert E. Strahorn in his 1877 *Handbook of Wyoming* :

> During the progress of the Mexican War, in 1846, a highly educated Mexican padre, named Ortiz was captured near El Paso. . . . While a prisoner in the hands of General Stephen W. Kearney's followers, he volunteered the statement to different American army officers that the Spanish had early in the seventeenth century obtained a footing in the mountainous region some six or seven hundred miles north of Santa Fe, but still south of the great muddy river (the Missouri); they had built stone houses and arrastras, and for nearly a quarter of a century, had sent trains south laden with gold and rich furs. About 1650, however, the savages of the region commenced a wholesale massacre of these pioneers and all were swept away as far south as Santa Fe.

Strahorn does not say where he learned these "facts" but continues on:

> These facts become especially interesting in connection with the developments of American explorers in our own time. Members of General Conners Big Horn expedition of 1865 . . . vouch for the statement . . . that ruins of stone houses, evidently more than a century old were found near the shores of the beautiful Lake DeSmet, at the base of the Big Horn Mountains. In 1866, the remains of an old Spanish arrastra - a quartz crushing implement - were found in the same region, at a point about fifty miles southwest of Fort Phil Kearney. Ruins of stone houses and fortifications were also discovered by Colonel Mills' expedition in the Big Horn country as late as the fall of 1874.[2]

Stories of what was presumed to be evidence of early activity of the Spanish, Incas, Aztecs or other early occupants continued to be aired and in 1879 the first reports of the so-called "Spanish Diggings" were printed. The diggings were discovered by pioneer rancher A. A. Spaugh in the area east of present day Glendo Reservoir. For some time these excavations were believed to be the work of Spanish Conquistadors and added to the speculation that the Spanish had indeed visited Wyoming. However, investigations by scientists later proved beyond any reasonable doubt that the quarries were the work of prehistoric stone age man.

The scientific revelation that the Spaniards had nothing to do with these excavations discouraged further speculation by Wyoming miners and subsequently fewer articles, editorials and speeches were apparent. However, not everyone was dissuaded from the Spanish inference and in later years another story which tends to lend credence to Strahorn's was told by Ed Patrick, a rancher of the Torrington area, before his death in 1916. In the early 1880s, Patrick had taken a party to the Big Horn Mountains and they camped one day on the shores of Lake DeSmet. Here they found the remains of old building foundations near the south shore of the lake. Patrick went on to say that this was a period of low water, the lake having since risen over them.

Some of the early settlers based their belief of early Spanish visits on rather fragile evidence. John P. Dodd who lived between Sheridan and Buffalo had a flat stone about two feet long with the numbers 1775 chiseled into it. The sevens had a short crossbar which the Spanish used and still use, making the number seven look something like the reverse letter F. He said that he had found the stone at the dump ground at Shoshoni, Wyoming at an old Spanish oven. According to Al Bartlett, who told the story of his 1935 conversation with Dodd, "this would indicate that the Spaniards got as far as Shoshoni in 1775, or they may even have gotten to Lake DeSmet on the expedition." Although this was based on the flimsiest of evidence, other evidence was based on more concrete findings.[3]

In northern Wyoming, another, more recent discovery, adds one more piece of evidence on the side of early Spaniards. In 1961, a Spanish sword was dug up in Dayton, Wyoming. One of the theories presented was that it was left at the site by Spanish explorers. Of course, on the other hand, it could have been carried into the area by later visitors.[4]

In his 1899 book, *History of Wyoming*, historian C. G. Coutant supported the theory of early Wyoming Spanish occupation. Another more widely recognized historian, H. H. Bancroft, however, in his *Volume XXV* maintained there was nothing true in the story of early Spanish occupation. Coutant countered that Bancroft failed to account for the iron tools found by many Wyoming people in the latter part of the century, and concluded, "I regret that I am obliged to leave the subject in an unsettled condition, but some future historian will undoubtedly solve the vexed question."

In 1891, during the peak of the Gold Hill mining, the *Saratoga Sun* published an article, "They Knew How To Mine," which said that prospectors were first attracted to the region by stories told by the Redskins. "... The Utes used to tell about the great quantities of gold that had been taken out of that section. Their legends were so misty tho, that the identity of the race that worked on these treasure fields could not be established." The article went on to tell of extensive old workings in the gulches below Gold Hill which "... prospected only a few colors and furnished further evidence of the thoroughness of the work." (Not a thought on the possibility there was little gold to begin with.) Growing out of the workings was a tree, seven inches in diameter, indicating it was 80 to 100 years old and affording the only basis for the time elapsed since work had ceased there.[5] From the description of the workings, one could conclude that someone had indeed placer mined in the area, or one could just as easily conclude that this was an act of nature. However, in a nearby gulch, Jim Cusick, had in 1888, found a square timbered shaft more than 50 feet deep. The shaft, he said, was put down a little to one side of the gulch, was five feet square and timbered with square sets. The only clue as to the age of the shaft was the fact that the timbers were rotting away. This was a solid piece of evidence for the miners and strengthened their belief that someone had mined there long before them.[6]

Taylor Pennock, an old time resident of the North Platte Valley, who came to Wyoming in

1872 was another who told of finding evidence of early mining. In the course of prospecting in Mullison Park on Brush Creek in 1886, Pennock told of finding evidence of old mine workings, including a long shallow tunnel that had collapsed in so many places that he was able to trace it on the surface and so old that trees one and one-half feet in diameter were growing in the cavings. He and his partners dug down into the collapsed tunnel and found some of the timbers which were used for support. At the time he thought that it was mining done by the Spaniards.[7]

John H. Mullison was probably one of the greatest supporters of the early mining theory. Mullison had come into the territory in 1867, and prospected, mined, hunted and ranched along the North Platte River and the Medicine Bow Mountains. Perhaps no other individual had his extensive knowledge and experience in the area. Born in 1842, raised in Pennsylvania, he served in the Union Army and headed west after the war. He spent about two years in Colorado then came to Wyoming where he spent a few more years and then went to Nevada and Utah for two more years before returning to Wyoming. He settled in the North Platte Valley and took up ranching and farming. Mullison was one of the first forest rangers and served from 1900 to shortly before his death in 1912. Perhaps no man believed more strongly, that mining had been carried out in the Medicine Bow Mountains at a much earlier date, than Mullison and he sometimes speculated as far back as one thousand years. Mullison with a party of ten had been prospecting in the Brush Creek area around 1870 when they discovered what they thought were old Aztec mines. His description said the ruins "... consists of a system of shafts and tunnels, from the bottom of one shaft to that of another, leading up from Brush Creek in the direction of the field of recent discoveries, [Gold Hill]. The series of ancient workings is some two and one-half miles in length and is still there [1891] to be seen by those curiously inclined." He also said that, "Similar but less perfectly preserved workings were found near Laramie Peak four or five years ago, and this spring [1891] a party of men engaged in opening a road from Carbon to Gold Hill accidently stumbled on a third set of workings." Mullison's discoveries were probably the basis for many of the stories of early mining near Gold Hill.[8]

The belief that the Spaniards were early visitors to the Medicine Bow Mountains was widespread among the mining camps and in the early 1880s, several Big Laramie miners went to North Park to investigate what they described as old Spanish Land Grants in the vicinity of Independence Mountain. Neither the origination of their information nor the results of their expedition was, at the time, disclosed.[9]

During the 1901 Wyoming Industrial Convention, George W. Fox read a paper about the history of minerals in Albany County. He claimed that when gold was discovered in Moore's Gulch in 1868, it was found that gulch mining had been carried out along the lower tributaries of Douglas Creek at a much earlier period. His theory was that Spaniards mined, in what is now known as Albany County, in the 16th century.[10]

Laramie remained as the mining center for the Medicine Bow Mountains and occasionally would publish promotional mining information. One such publication was the 1906 pamphlet, *Mineral and Allied Resources of Albany County*, which advanced the early mining theory:

> Mining in this region has been carried on since the first Spanish explorers worked their way northward along the Rocky Mountain Range from their landing places on the Mexican coast, as traces of these ancient workings have been found, together with old tools, weapons and other articles indicating the presence of these the earliest of pioneers.[11]

Stories that seem to support early Spanish activity in Wyoming continued to be told even in later years. In 1928, Bruce Torgny, early day forest ranger, wrote in his stories of area pioneers that, "John Allen claims that several years ago he found in a canyon on Coad Mountain the skeletons of several people and that he also found a bronze hatchet there." (John Allen was an early day Pass Creek rancher).[12]

Another staunch believer in early mining evidence was Judge Charles W. Bramel, who had done considerable prospecting and mining on his own. His thoughts were published in both the *Centennial Post* and the *Laramie Boomerang*:

> The Snowy Range of the Medicine Bow Mountains

has a history very ancient in character and little known. There is nothing mythical in the story that the inhabitants of Mexico denominated as the Aztecs, occupied in the twelfth century almost the whole of the middle and the southern portions of the now United States, and that in the middle of the fifteenth century sent explorers and the most scientific men of their nation so far north as Vancouver's Island. From the north end of Elk Mountain of the Medicine Bow Range to the City of Mexico and beyond, ancient ruins mark the pathway of these explorers and intrusive Anahuac. Old reservoirs, sluice boxes, excavations and huts are not unfamiliar to the trapper and miner of the present day and are frequently met within the Snowy Range of the Medicine Bow Mountains.

Contemporary historians might differ with Bramel's history of the range of the Aztecs but one must wonder at the discoveries that led Bramel and others to the conclusion that the area was mined long before their time.

As wide spread as the theory of Spanish in Wyoming was, still no solid proof was offered. However, many of the old miners and prospectors held faith that there were miners in the Wyoming mountains before them. If not Spaniards, Incas or Aztecs, maybe Utes or possibly early prospectors on their way west during the "49 gold rush."

• THE '49er THEORY

As the trails west opened up to the gold seekers and others headed west, stories began coming back of rich discoveries along the trails. Among these were tales from what is now known as Wyoming and included the Medicine Bow Mountains which had been known by name at least as early as Fremont's expedition of 1842.

The Overland Trail, through what is now Wyoming, was used by a large percentage of the California Gold Rushers from 1849 to 1862. In 1862, the trail was moved south to the old Cherokee Trail a portion of which skirted the northern reaches of the Medicine Bow Mountains.

At the northern end of the Medicine Bow Mountains, Fort Halleck was established in 1862, to protect travelers on the Overland Trail. In 1864, Doctor Finfrock, acting surgeon at the fort, counted 17,584 people passing that point.[13] It seems quite probable that at least a few of these men ventured into the mountains to prospect. At least one instance is recorded.

In October, 1862, Edward Bliss took a couple of days layover at Fort Halleck and prospected a "likely looking" canyon with apparently no success.[14]

Virtually all the mining regions had areas which the 19th century prospectors claimed had been worked at a much earlier time. During his prospecting along the northern edge of the Medicine Bow Mountains, in 1893, M. S. Hampton found an old shaft about a mile west of Cooper Hill in the center of which was growing a tree 14 inches in diameter. (This would indicate that the tree was approximately 150 years old). He apparently believed that much earlier mining had been done in the area and said, "Many other evidences of ancient work can be found near here." His discovery was within sight of the Overland Trail.[15]

The southern end of the Medicine Bow Mountains, in what is now Colorado, were also the subject of speculation. On the headwaters of the North Platte River tributaries in the southern reaches of North Park were found traces of precious metals. In Laramie, a story was published about one early visit:

> In '49 a party of hardy prospectors on their way to California were snow bound on the northern part of the park. During the winter they hunted and when spring came prospected along the range. In a gulch on Owl Mountain they struck good pay dirt and hewed out sluice boxes from solid spruce trees. That spring and summer they took out over 50 ounces of gold. In August work was stopped for want of water.

On descending into the park, all the members of the party, with the exception of two, one of these the writer's grandfather, were killed by the Indians. The survivors escaped into Middle Park, and after fearful hardships reached Fort Bridger.

The writer went on to say, "Within the last two years, [of 1896] the old sluice boxes which they used have been discovered and work resumed." [16]

Even the old mining records indicate an occasional reference to early mining. In 1888, Peter Neeb found what he described as "An ancient mine having been worked and abandoned, 10 miles N.E. of Rock Creek station."[17]

• THE CHERRY CREEK MINERS

The "Pikes Peak" gold rush of the 1850s brought a good many hopefuls into the Cherry Creek region of what is now Colorado. Numerous guide books were printed to guide the itinerant prospector to his goal. One such book was printed by Byers and Kellom in Omaha:

> Again, from Fort Laramie the gold seeker may strike west, up the valley of Laramie River, prospecting as he goes, traveling through a most beautiful and romantic country, and reaching the Medicine Bow in 120 miles from the Fort, and 618 miles from Omaha.
>
> The Medicine Bow is reported by some to be the richest mining region yet discovered.
>
> Passing the Medicine Bow, the North Platte is reached in 30 miles; here diverging to the left, and following up the North Platte and traveling 50 miles, we enter the North Park, or Bull Pen, 200 miles from the Fort and 700 miles from Omaha.[18]

Although the writers of the handbook had never visited the area, they apparently drew freely from reports they heard and published their own version which probably had the desired effect of selling guide books and attracting more people to the area.

Author of the guidebook, William N. Byers, moved west and set up his printing press in the new mining camp in Cherry Creek, Kansas Territory, which was later to become Denver. In his first edition of his new paper, the *Rocky Mountain News*, he reprinted the map which he had earlier used in the guide, and had originally been published in *Harpers Weekly*.

It would not be unreasonable to assume that some of the miners examined the map, read the accompanying story and possibly explored some of the ares designated as a "gold field" not knowing that the originator had little knowledge of the area.

In 1858, William Green Russell, James Pierce and about 10 others prospected north of the Cherry Creek gold finds and supposedly took a trip into present day Wyoming, primarily it was said, for hunting.[19]

Professor Wilbur C. Knight in his address before the Wyoming Mining Convention, held in Cheyenne in January, 1891, said:

> I have been told that placer gold was discovered on Douglas Creek in 1868, and since that time the gulches have been worked every year. I have no doubt that the two men who came to Denver from 1865 to 1870 from the northwest each fall, with gold to sell, washed it from these gulches. . . . In traveling over this section one finds old workings in nearly every gulch. Old pieces of rocks and sluice boxes are common and in one place I saw a sluice box that was hewn out of a log with an axe, which signified very early work.[20]

J. H. Hayford, editor of *The Laramie Sentinel*, printed this story on April 21, 1883:

> Gulch Mining; We recollect being up in this vicinity in 1863 and meeting a party of miners who had been at work on Rock Creek and had been driven out by the Indians. Some of the party we were personally acquainted with and they reported that they had found good paying mines up on Rock Creek and believed they could make fully ten dollars per day to the man if they could only hold their ground against the Indians.[21]

These diggings could have been later misidentified as even earlier diggings.

The concept of early miners in what is now Wyoming lives on to this day and some believe that proof will yet be found, but it is unlikely at this late date that physical evidence will be uncovered, although artifacts discovered during the building of Rob Roy Reservoir have yet to be fully evaluated. The contemporary historians need more than hearsay to convince them that there were indeed early visits by "others" in what we now know as Wyoming. One thing for sure, the old pioneer prospectors strongly believed that the area had not only been visited, but that the visitors had mined for gold.

• LOST MINES

Although not as prevalent as in the surrounding states, Wyoming had a small share of "Lost Mine" stories which were subject to speculation and besides the "Lost Pick and Shovel" mine story related in the Gold Hill chapter, several stories were circulated.

Among the lost mine stories is one told in the *Laramie Boomerang* August 2, 1901, apparently printed from an earlier *Denver Post*. "Away up in the Medicine Bow mountains, not far from the Wyoming line . . . there is a hidden mine for which a generation of men have searched in vain. It is owned, operated and its location kept secret by two young women who have kept their secret since one was 18 and the other 14 years." The story goes on to tell of a man named Smithers who returned to Fort Collins after a prospecting trip to find that his wife had died and left him with two young daughters, aged 7 and 3. Smithers took the girls back to the mountains where he built a cabin and did a little farming and ranching. He also showed the girls his secret vein and taught them how to mine. Some years later when the girls were 18 and 14 years, Smithers himself died leaving the girls to fend for themselves. According to the story, the girls were able to take care of themselves and only occasionally went to town to buy provisions, paying in gold dust and nuggets. Speculation was that their mine must have been fabulously rich to enable the two girls to mine such amounts of gold.[22]

Stories of other lost mines were occasionally printed by Wyoming newspapers. The *Boomerang* on December 20, 1894, published a story of a lost silver mine. "A Ledge of Silver; Discovered about six years ago near Laramie. 20,000 ounces to the ton, it was almost solid silver - its location has been forgotten - a search for it will be begun soon." The story that followed told of a sheep herder who, during the course of his wandering collected various mineral specimens. One day, close to Laramie he picked up what he thought to be a specimen of "white iron." Some years later, the mineral sample was assayed as silver. The former sheep herder had unfortunately forgotten the exact location of the ledge, but the *Boomerang* promised a search would be launched to find the fabulous silver ledge. Bernard Holtum, local seer specializing in locating gold and silver, was called in to aid in the search. "Mr. Holtum, it is believed, can locate the ledge if they can succeed in passing near it, through his magnetic power." If a search was indeed instigated, no such silver ledge was found.[23]

The *Boomerang* related another story of a lost gold mine. The story was of an Army officer who started out for Fort Steele and during the course of his journey became separated from his party. He stumbled onto a ledge studded with free gold. A later attempt to relocate the ledge was unsuccessful.[24]

There were in the course of time many stories of lost mines throughout Wyoming, among them, the Lost Cabin Mine of the Big Horns, and the "Lost Seminole" mine of central Wyoming. As with most stories of this kind, none of the mines were every "found."

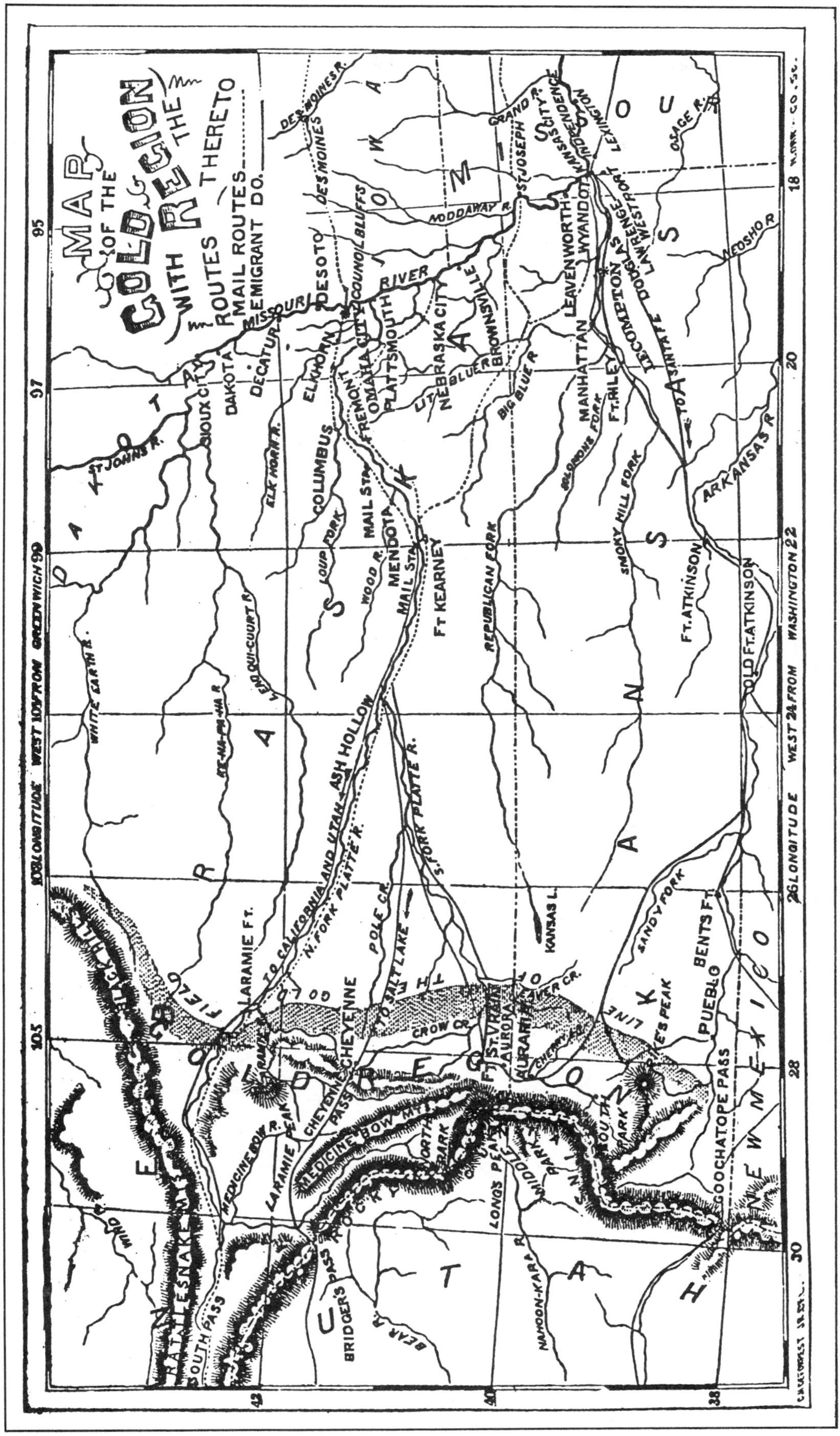
MAP OF THE GOLD REGION WITH THE ROUTES THERETO
ROUTES
MAIL ROUTES
EMIGRANT DO.
108 LONGITUDE WEST 109 FROM GREENWICH 99
WEST 24 FROM WASHINGTON 22
26 LONGITUDE
MISSOURI RIVER
DES MOINES R.
DES MOINES
COUNCIL BLUFFS
DESOTO
DECATUR
DAKOTA
SIOUX CITY
ST JOHNS R.
ELKHORN
ELK HORN R.
OMAHA CITY
FREMONT
PLATTSMOUTH
NEBRASKA CITY
BROWNSVILLE
NODDAWAY R.
ST JOSEPH
GRAND R.
LEAVENWORTH
WYANDOT
KANSAS CITY
INDEPENDENCE
LEXINGTON
LECOMPTON
LAWRENCE
DOUGLAS
WESTPORT
OSAGE R.
NEOSHO R.
MANHATTAN
FT. RILEY
BIG BLUE R.
COLUMBUS
LOUP FORK
WOOD R.
MAIL STA.
MENDOTA
FT KEARNEY
REPUBLICAN FORK
SMOKY HILL FORK
ARKANSAS R.
FT. ATKINSON
OLD FT. ATKINSON
WHITE EARTH R.
ASH HOLLOW
TO CALIFORNIA AND UTAH
N. FORK PLATTE R.
S. FORK PLATTE R.
POLE CR.
TO SALT LAKE
KANSAS L.
LARAMIE FT.
GOLD FIELD
THE GOLD LINE
CHEYENNE
CHEYENNE PASS
CROW CR.
ST. VRAIN
FT. ST. VRAIN
AURORA
AURARIA
CHERRY CR.
PIKES PEAK
PUEBLO
BENTS FT.
SANDY FORK
GOOCHATOPE PASS
SOUTH PARK
MIDDLE PARK
NORTH PARK
LONGS PEAK
MEDICINE BOW MTS.
MEDICINE BOW R.
LARAMIE PEAK
LARAMIE R.
BLACK HILLS
RATTLESNAKE MTS.
SOUTH PASS
BRIDGERS PASS
BEAR R.
NAHOON-KARA R.
ROCKY MTS.
NEW MEXICO
UTAH
NEBRASKA
KANSAS
MISSOURI
IOWA
DAKOTA
95
97
105
18
20
22
28
30
43
40
38

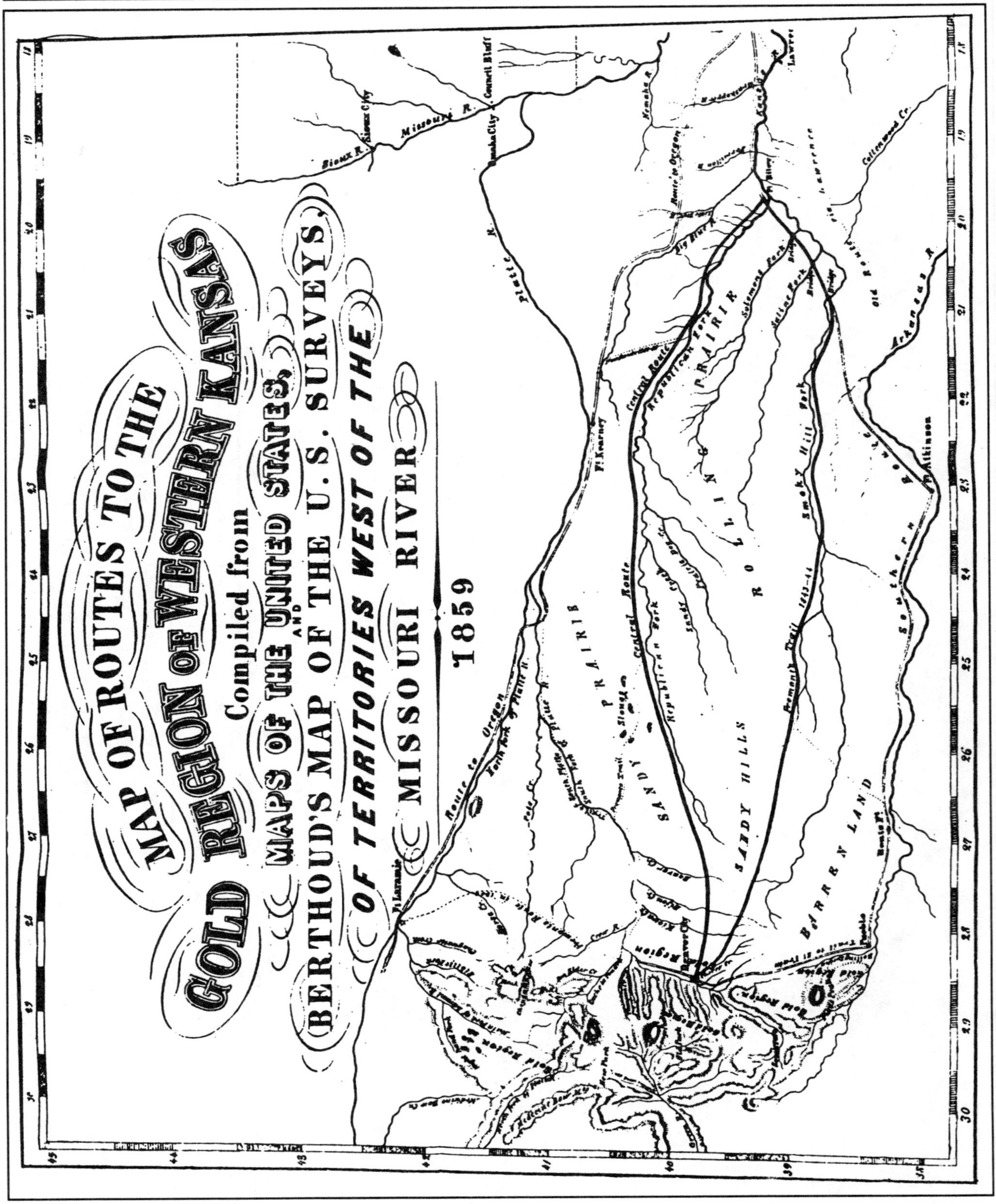
MAP OF ROUTES TO THE
GOLD REGION OF WESTERN KANSAS
Compiled from
MAPS OF THE UNITED STATES,
AND
BERTHOUD'S MAP OF THE U.S. SURVEYS,
OF TERRITORIES WEST OF THE
MISSOURI RIVER
1859
Platte R.
Ft. Kearney
Missouri R.
Route to Oregon
Central Route
Republican Fork
Smoky Hill Fork
SANDY HILLS
BARREN LAND
ROLLING
PRAIRIE
Arkansas R.

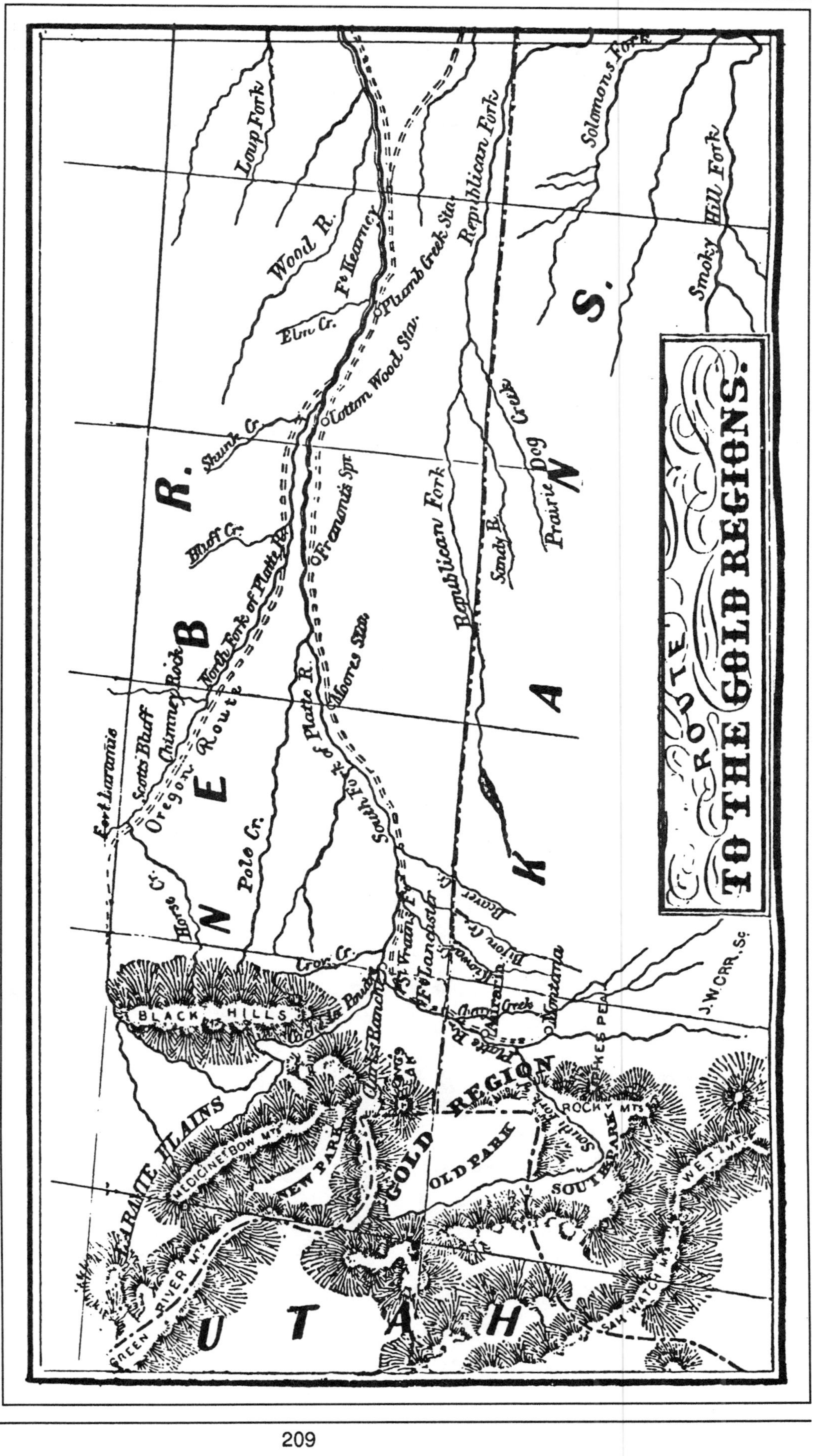
ROUTE TO THE GOLD REGIONS.
NEBRASKA
KANSAS
UTAH
Loup Fork
Wood R.
Ft Kearney
Plumb Creek Sta.
Republican Fork
Solomons Fork
Smoky Hill Fork
Elm Cr.
Cotton Wood Sta.
Skunk Cr.
Fremonts Spr.
Bluff Cr.
North Fork of Platte R.
Prairie Dog Creek
Sandy B.
Platte R.
Moores Sta.
Chimney Rock
Scotts Bluff
Fort Laramie
Oregon Route
Polo Cr.
South Fk of Platte R.
Horse Cr.
Crow Cr.
Beaver Cr.
BLACK HILLS
LARAMIE PLAINS
MEDICINE BOW MTS
NEW PARK
GOLD REGION
OLD PARK
SOUTH PARK
PIKES PEAK
ROCKY MTS
GREEN RIVER MTS
J. W. ORR, Sc.

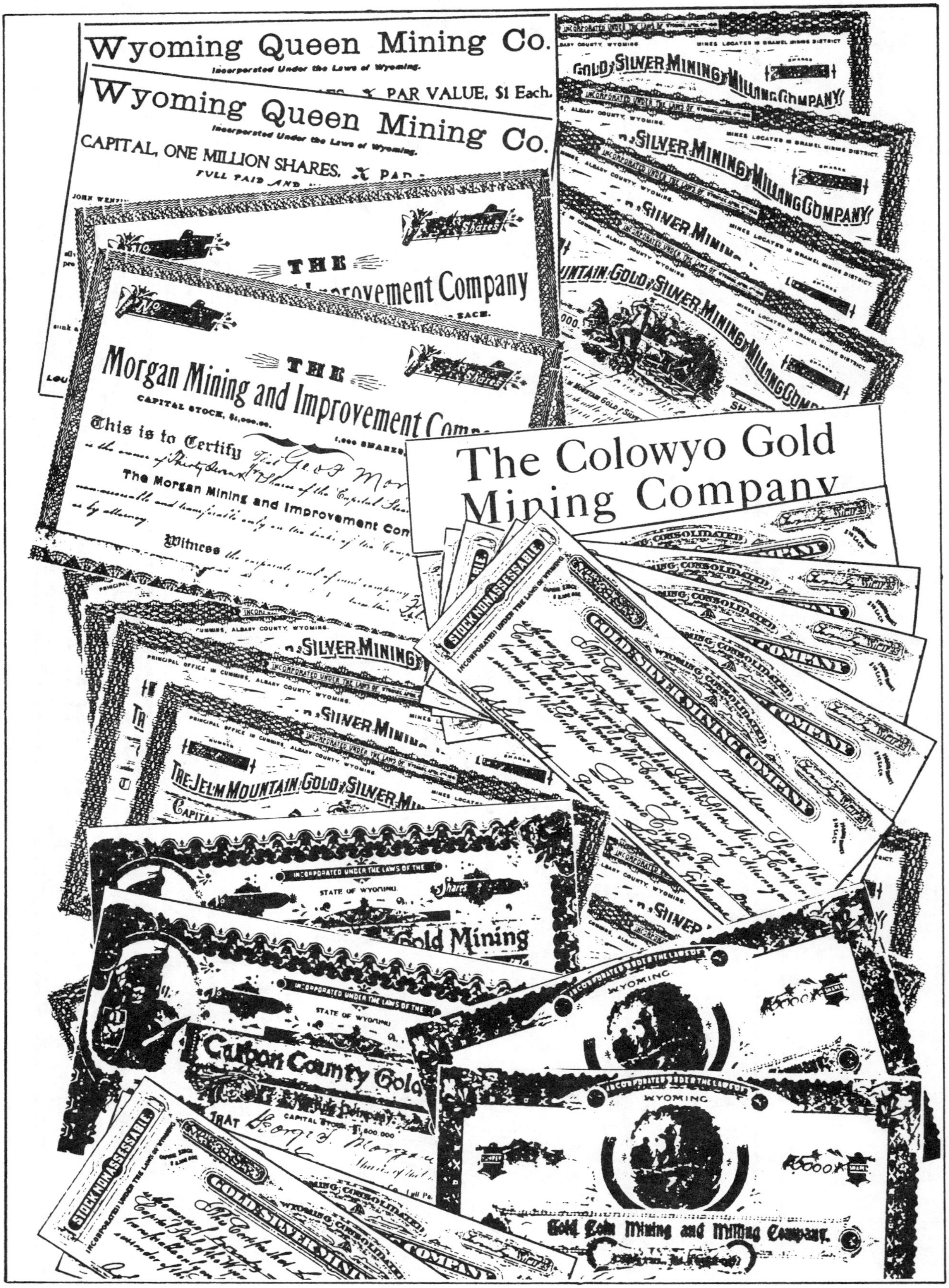
Wyoming Queen Mining Co.
Incorporated Under the Laws of Wyoming.
PAR VALUE, $1 Each.
Wyoming Queen Mining Co.
Incorporated Under the Laws of Wyoming.
CAPITAL, ONE MILLION SHARES.
THE
Improvement Company
THE
Morgan Mining and Improvement Comp
This is to Certify
The Morgan Mining and Improvement Com
Witness
SILVER MINING
MILLING COMPANY
THE JELM MOUNTAIN GOLD & SILVER MI
The Colowyo Gold
Mining Company
STOCK NON-ASSESSABLE
GOLD & SILVER MINING COMPANY
INCORPORATED UNDER THE LAWS OF THE
STATE OF WYOMING
Gold Mining
Carbon County Gold
WYOMING
Gold Coin Mining and Milling Company

XI EPILOGUE XI

• THE LEGACY OF THE MEDICINE BOW MINING CAMPS

By the closing decade of the nineteenth century, it was becoming increasingly apparent that the Medicine Bow Mountains were not going to yield the great quantities of gold and silver needed to bring Wyoming to the forefront. For a time during 1890, it appeared that the Gold Hill District might be the long awaited gold producer, but it too failed to meet the expectations. Again in 1894, the Cooper Hill mines promised briefly to be the long awaited bonanza but they too went the way of the Medicine Bow Mines before. As the Cooper Hill mines faded, so did the hopes for a gold producing region in the Medicine Bow Mountains.

By the end of the century, virtually all of the Medicine Bow Mountains had been prospected. Generally speaking the nation's gold rushes had progressed from west to east after the California Gold Rush of 1849. Although no great rush developed, Wyoming became one of the last areas to be explored and prospected. As the nineteenth century came to a close, so did the era of the Medicine Bow Mining Camps. Never again would there be a rush to the mountains and a new mining camp. In the coming years prospectors would return again and again to the mountains but there would be no more rushes, no new mining camps and no gold production of note.

• VALUE

By the turn of the century, some 5000 claims had been staked in Albany County and some 2000 in Carbon County. From 1868 through 1899, Wyoming's gold production was placed at $992,300. Of this, the Medicine Bow mines had produced less than half, or perhaps $500,000 in gold and silver combined. To do this they had expended a countless amount of time, effort and money, probably more money expended than the value of gold produced.

Some of the gold recovered may not have been reported, some was carried out of the territory and some squirrelled away. Consequently the total value of gold and silver taken from mountains may have exceeded the official total but even had it been double the official figures, the total amount is insignificant when compared to the surrounding states. Indeed, if you triple or even quadruple the official gold figures from all of Wyoming for 100 years, it does not equal the production, for example, of even one of the best ten mines of Cripple Creek, Colorado. The traditional stories of millions in gold taken from Wyoming gold mines are apparently a product of the gross exaggerations claimed during the booms.

Inflated assays and "million dollar" companies that could not afford to buy equipment or pay their laborers did nothing to improve these figures. The assays were probably accurate for the ore being tested, but only the miner knew where the ore had originated and how abundant was the deposit. A very small piece of ore could bring a very high assay.

Inflation is an elusive aspect and while it is difficult to place in perspective, the amount of money expended was indeed substantial for that time and place. Throughout the period, the price of gold was set at $20.67. While some things have stayed relatively stable, others have inflated to the point that even accomplished economists find it hard to make comparisons. A $60,000 mining and milling operation of 1890 would probably be equivalent to a million dollar operation in today's world.

Historically, in the mining camps of the West, there was a transition from the individual gold seeker to the moneyed companies who had the funds to develop the mines. This transition from individualistic to a capitalistic enterprise came to only one of the Medicine Bow Mines, and even then was limited.

The larger mines absorbed money like some giant financial sponge, taking an ever increasing amount of money to operate but producing less and less gold. The longer they operated, the more in debt they fell. Even the smaller mines took the total time and effort of its owners during the mining season.

Only two of the Medicine Bow mines (the Centennial and the Keystone) produced a notable amount of gold and those only for a brief moment in history.

• ECONOMY

In retrospect, however, even failures can produce some benefits. If an area attracts people, if the people expend time, effort and money, the region will benefit even though the endeavor was a failure. True, that if the region had prospered and become a great gold producer, the benefit would have been greater, but even the great gold rushes of the surrounding states had their heyday and then died off.

Nearly everyone in the area realized some benefit from the mining activity. For certain the local storekeeper's business was enhanced by the mining trade. Surely the local stables experienced an increase in business when one of the camps was undergoing a boom. To read

the old newspapers it becomes apparent that the businessmen expected an increase in trade for the entire region. Perhaps it was this hope that kept some from leaving the Territory and thus adding, however small, a measure to the population and economy of the struggling territory.

• STATEHOOD

Returning focus to 1889, as Wyoming looked forward to statehood, the most important item was not her mineral wealth but her population.

The year of 1889 was not a good time to look at gold mining in Wyoming. Only one gold camp post office remained in operation, (Centennial). Only 226 people were engaged in metal mining in the entire Territory, of that, only 120 were in the Medicine Bow mines and although some $80,712 had been expended in the search, only $14,512 in gold had been produced for the entire year. Although these figures were compiled during the 1890 census, they were not published until 1892, but at the time, as Wyoming prepared for statehood, few could or would doubt that the proposed state must have hidden wealth in undiscovered gold mines. After all, her neighboring states of

Montana, Colorado, Dakota, Idaho and even Utah all had rich gold mines.

Just how many people were drawn to Wyoming, looking for their personal bonanza, is impossible to say. The inflated assays, the wild claims of rich gold mines, the continued optimism of the miners and the local newspapers had done their part in attracting new people to the Territory. True that many stayed only a short time and then moved on to greener pastures, but many stayed on, some to prospect and mine, though most to return to a more stable means of income. Their contribution to the important aspect of population can never be accurately determined.

Just as the miners inflated the richness of their claims and the population of the camps, Wyoming's Delegate to Congress, Joseph M. Carey, before the House of Representatives inflated his estimates of the wealth of Wyoming and its population. Although Carey was usually cautious about giving any exact population numbers and based his estimates on the most recent election, at one time he estimated there were between 110,000 and 125,000 people in the territory awaiting statehood. (When the 1890 census was completed it enumerated 62,555 people.) At the time, admission to the Union was loosely based on an arbitrary population figure of 100,000. [1]

Although he expounded upon the resources of Wyoming to some length, telling of Wyoming's great agricultural potential and her pastoral and lumber possibilities, it was in soda, oil and coal that Carey correctly speculated would be Wyoming's future wealth.

However, in his presentation Carey was at least able to point to a gold mining industry as potential wealth.

The subject of population, however, was the primary factor for admission to the Union. There can be little doubt that the Medicine Bow mines drew people to the area, not only miners and prospectors but also merchants and workers in their supporting role.

• POPULATION

The population of the mining camps in 1889 and 1890 had little impact on the population claims for the Territory. There can be little doubt that in the preceding twenty years the mines had added to the Territory's population

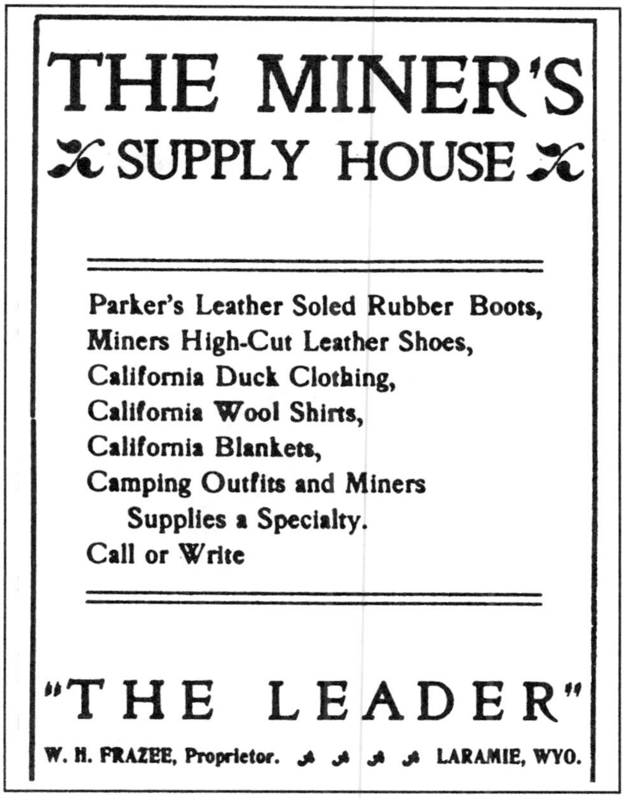

and indirectly to the economy. From 1868 to the turn of the century and beyond, the Medicine Bow Mountains lured the prospectors, not only from the nearby settlements but also from clear across the country.

There is no question that the Medicine Bow Mining Camps contributed to the building of Laramie, Saratoga, and all of Wyoming, only the degree is in doubt. Laramie had grown from 788 in 1870, to 2696 in 1880 and 6388 in 1890. Certainly at least some of this was due to the potential mineral wealth of the nearby mountains. Saratoga had doubled its size in one year and in the early 1890s, during the Gold Hill boom, was proclaimed to be the fastest growing city in Wyoming. At least a portion of this was due to the Gold Hill boom. Rawlins, not located near the mines had grown only a small amount, from 1451 people in 1880, to 2235 in 1890. The new city, no doubt, derived some benefit from the traffic through the town headed for the mines of the Medicine Bow and Sierra Madre Mountains.

• PEOPLE

Beyond the limited addition to the population, the men who worked the Medicine Bow Mines were, for the most part, an asset to the general populace. Many of Laramie's civic

Isaac P. Caldwell
University of Wyoming, American Heritage Center

John W. Blake
Wyoming State Museum

leaders had come first to Wyoming's gold fields and after failing to make their great discovery, reverted to the settlement and a more stable trade. Many of Laramie's independent businessmen had first been in business in the mining camps and when they failed to strike it rich had returned to the city.

Many of these men who were bitten by the gold bug also had a sense of responsibility to the future. The 1886 Board of Trustees, the founding fathers of the University of Wyoming, (Hoyt, Finfrock, Ivinson, Holliday, Aughey, Hayford and Brown) all had mining interests.

Two of the early Medicine Bow miners, Isaac P. Caldwell and John W. Blake,[2] who had gone on to be active in the legal profession and then the Territorial Legislature, were chosen to revise the laws and statutes of the Territory. These laws were subsequently used by the Constitutional Convention in formulating the laws for the new state of Wyoming.

Stephen Downey, the outstanding proponent for the Medicine Bow mines for thirty years, was also one of the most popular individuals of the time. Downey held numerous public offices, (at times more than one), from a member of the Territorial Council to Delegate to Congress. An outstanding lawyer with a flourishing business, both corporate and private, he poured most of his wealth back into the mines and died in debt.

Wilbur C. Knight, assayer and superintendent for several of the mines continued his support of the mining industry and apparently believed that deeper mining would eventually yield rich returns. After the Keystone "bust", Knight moved to Laramie and set up an assay office. In 1893 he was appointed Professor of Geology and Mining Engineering of the University of Wyoming. In 1897 he was appointed State Geologist (although apparently without salary). In July, 1899 Knight led the Wyoming Fossil Fields Expedition out of Laramie across the Laramie Plains to Bates Hole and return. This expedition was sponsored by the Union Pacific Railroad, the University of Wyoming and the National

Geographic Society and was the largest of its kind to that date. Wilbur Knight died rather prematurely in 1903. The Knight family name is perpetuated at the University of Wyoming.

After 1887, a disastrous year for the cattle industry, it is well documented that at least some of the cattlemen turned to the gold fields for relief rather than leave the state. One well known cattleman turned miner was George Morgan. The mines attracted Morgan for a few years but could not hold him. He returned to the cattle business.

Benjamin W. Towner, civil war veteran, drawn to the mountains by the promise of gold, also served in many public offices but his primary goal was to strike it rich in the Medicine Bow Mines.

Herman V. S. Groesbeck spent his early days in Wyoming at Cummins City tending bar. He later became a member of the more prestigious bar and entered the legal field where he eventually became Supreme Court Justice.

Only during the early days of their association with the Medicine Bow Mines did these well known men work in the mines. They were the planners, the dreamers, the speculators and the financiers. Men like Laramie's first banker, Edward Ivinson were attracted to the mines but aside from a small tentative investment here and there could not be persuaded to invest.

Otto Gramm, said to be Laramie's first millionaire, had apparently grub-staked some of the miners, often visited the mines but apparently did not engage in any actual mining or prospecting.

Men like Louis Miller, who, during the mining season was an avid miner, and during the off season returned to Laramie to be their leading jeweler. He also served as County Sheriff, Mayor of Laramie City and was a member of the First Territorial Legislature.

Perhaps it was the likes of editor James H. Hayford to whom greatest credit is due. As inflated as his editorials were, he seldom had a discouraging word to say about the mines. Truthful or not, sincere or not, there can be little doubt that his enthusiastic exaggerations fueled the miniature gold rushes to the Medicine Bow Mountains.

• LEGACY OF THE MINING CAMPS

Now, a hundred or more years later we look back at the mining activities of the Medicine Bow Mountains and in retrospect ask, What is the legacy of the mining camps? Is it merely a legacy of empty holes, abandoned shafts, piles of tailings and altered streams? For certain the Snowy Range and the entire Medicine Bow Mountains would be a more pristine and beautiful area without the several thousand discovery shafts and the several hundred mines. One of the sidelights and possibly the most enduring legacy of the miners was the lending of their names to places in the mountains. These miners were generally the first to spend any degree of time in the high Medicine Bow Mountains and it is for certain they left their names upon the land. Centennial, Gold Hill, Arrastra Lake, Stamp Mill Lake, Cummins City and Keystone were obviously named for their link to mining. Lewis Lake, Brooklyn Lake (first called Trout Lake and later Brooklyn after the nearby Brooklyn Mine) and Towner Lake are more subtly linked. After discounting the names of natural phenomenon such as, Elk Mountain, Muddy Creek, Beaver Creek, Rock Creek, and Brush Creek, most of the names in the Medicine Bow Mountains reflect the activity of the miners of the nineteenth century. Many of these names remain today although some have been changed, Lake Leman to Phantom Lake, Virginia Lake to Reservoir Lake and Mineral Lake to Stamp Mill Lake are but a few examples.

• ROADS AND RAILROADS

Throughout the Medicine Bow mining era the city fathers of Laramie had a definite concern that commerce with the mines was being impeded by the lack of adequate roads and rail service to the mines. Although the tie hacks and timbermen were also concerned with roads into the mountains, theirs was a concern of penetrating the timbered areas. It was the miners that were the instigators of the construction of roads to the high mountains.

It was not until 1927 when the road as we now know it was finally constructed across Snowy Range. Prior to that time the only road across the area was either by way of Douglas Creek or the Centennial to Gold Hill road which followed up Jim Creek, crossed the French Creek Valley and then went on to Gold Hill and down Brush Creek to Saratoga. These were roads used primarily by the miners.

The Laramie Plains railroad under its various names and management was indeed slow in building across the plains and several times through the years changed its route and destination. There can be little doubt that the Medicine Bow mines influenced the building of the railroad, only the degree is uncertain.

Perhaps as important as any worthwhile legacy, is also the legacy the miners did not leave. While the Medicine Bow Mining Camps may not have been attractive, they never reached the extent of the ugly mining town sprawling across the gulches as in the neighboring states. Not only did they miss the thrill of finding a great bonanza but they also missed the smelters belching smoke, the cyanide vats leaking into the streams and the wholesale destruction of the forest to fuel the mills.

The lasting legacy of these camps is not one of great economical impact, not of great mining camps drawing thousands of miners overnight, but the people and their pioneer spirit, the spirit to mold the West, the spirit to build Wyoming, a spirit that, although battered, lives on today.

"Mining is basic," "Gold is where you find it."

Members of the Constitutional Convention of Wyoming, 1889
Melville C. Brown, president of the convention, front row center.
Also, George W. Fox, Mortimer N. Grant, Stephen W. Downey and John W. Hoyt.
Wyoming State Museum

APPENDICES

Post Offices of the Medicine Bow Mining Camps

Wyoming Territory

Name	Opened	Closed
Last Chance	June 25, 1874	October 13, 1878
Centennial	July 23, 1875	Remains Open
Douglass	October 14, 1878	October 4, 1884
La Plata	May 16, 1879	January 6, 1881
Cummins	May 26, 1880	August 20, 1888
Keystone	November 30, 1889	February 15, 1893

State of Wyoming

Name	Opened	Closed
Gold (Gold Hill)	January 9, 1891	November 30, 1892
Headquarters	October 19, 1891	November 15,1892
Altamont	February 5, 1892	November 30, 1892
Herman	(Authorized March 31, 1893)	(No Records)
Morgan	November 16, 1896	November 29, 1924
Jelm	May 20, 1898	April 23, 1965

Colorado

Name	Opened	Closed
Teller City	July 19, 1880	December 16, 1885
Park City	November 26, 1879	April 14, 1881
Tyner	October 24, 1879	July 20, 1881
Crescent City	April 18, 1880	November 22, 1880

CENSUS FOR CINNABAR CITY, 1869

Name	Age	Occupation	Birthplace
1. William W. Healum	22	Miner	North Carolina
2. Peter Kennesaw	26	Miner	Michigan
3. A.S. Blackburn	26	Carpenter	Missouri
4. John Metcalfe	29	Carpenter	New York
5. Thomas Feast	32	Carpenter	England
6. Edwin Morgan	34	Miner	Vermont
7. Stephen W. Downey	30	Lawyer	Maryland
8. William O. Downey	20	Miner	West Virginia
9. Chs W. Hilliker	38	Carpenter	New York
10. John L. Roberts	52	Miner	England
11. Hein Rosenblade	28	Miner	England
12. Charles Tyler	54	Miner	New York
13. Wm W. Waters	26	Merchant	Ohio
14. Samuel S. Phillips	31	Miner	England
15. John F. Nunn	24	Merchant	Virginia
16. Robert H. Erwin	21	Merchant	Pennsylvania
17. David C. Rice	31	Miner	Illinois
18. D.D. Wolff	37	Miner	Ohio
19. Michael Corcoran	30	Miner	Ireland
20. Charles H. Davis	29	Miner	New York
21. Isaac Stephenson	33	Miner	Virginia
22. S.C. McKinsley	40	Miner	New York
23. James Allison	25	Miner	Missouri
24. James T. Foley	21	Miner	Ireland
25. A.J. Wilkinson	26	Miner	Virginia
26. C.C. Aldrich	26	Miner	New York
27. John Walbright	32	Miner	Germany
28. C.C. Dunn	26	Miner	Maine
29. Wm Geikey	25	Miner	Germany
30. Lewis Pelton	35	Miner	Norway
31. Porter Horten	39	Miner	???
32. Suduph Abrams	30	Miner	Hannover
33. John H. Brocker	32	Miner	Germany
34. William Fesch	24	Miner	Switzerland
35. H. Brammer	29	Miner	Holstein
36. H.C. Nichols	33	Merchant	Vermont
37. J.A. Hornback	29	Merchant	???
38. John B. Wands	55	Miner	New York
39. Smith McFall	29	Miner	Pennsylvania
40. John F. Smith	22	Miner	Illinois
41. Edward Chaplin	24	Miner	Virginia
42. David VanEman	32	Miner	Pennsylvania
43. H.R. Rather	33	Miner	Kentucky
44. Wm Hill	40	Miner	Vermont
45. John Hormed	24	Blacksmith	New York
46. Ths L. Roberts	35	Blacksmith	Pennsylvania
47. S.P. Watkins	28	Saloon Keeper	New York
48. S.J. Mills	26	Teamster	North Carolina
49. J.M. Jones	26	Teamster	Vermont
50. Noah Wallace	30	Teamster	Iowa

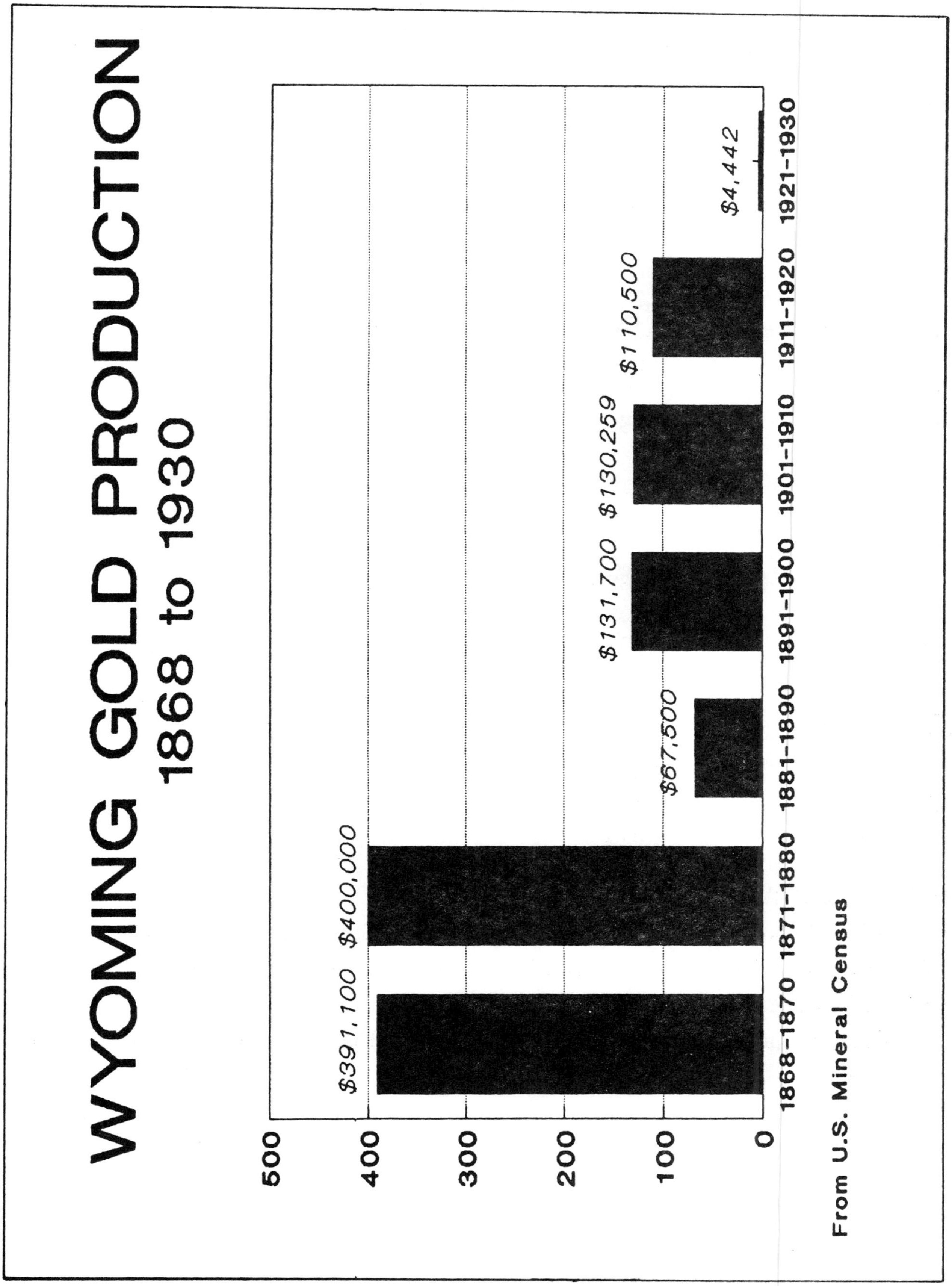
WYOMING GOLD PRODUCTION
1868 to 1930
500
400
300
200
100
0
$391,100
$400,000
$67,500
$131,700
$130,259
$110,500
$4,442
1868-1870
1871-1880
1881-1890
1891-1900
1901-1910
1911-1920
1921-1930
From U.S. Mineral Census

WYOMING GOLD PRODUCTION

(From statistics furnished by United States Bureau of Mines)

1842-1866	(Production unknown but very small)
1867-1870	$391,100
1871-1880	400,000
1881-1890	67,500
1891-1900	131,700
1901-1910	130,259
1911	19,200
1912	22,235
1913	24,184
1914	5,006
1915	14,592
1916	20,566
1917	3,762
1918	871
1919	84
1920	...
1921	62
1922	1,207
1923	296
1924	...
1925	...
1926	...
1927	1,200
1928	677
1929	1,000
1930	6,578

*OTHER STATISTICS:

1867 to 1931: $1.250,819; Ore Deposits of the Western States, Lindgren, 1933.

1867 to 1924: $1,237,624; Bulletin XXI, Wyoming Geological Survey, Marzel, 1929.

1867 to 1929: $1,240,501; U.S.G.S.

*Note: Although not in exact agreement, these figures represent a general consensus of gold production.

STATISTICS OF GOLD AND SILVER MINES OF WYOMING 1889

From Report on Mineral Industries in the United States Eleventh Census: 1890

Gold Production		**$14,512**
Expenditures		
Wages		45,945
Contractors		1,305
Office Force	(5 Males)	3,525
	(1 Female)	60
Total Wages		50,835
Supplies		24,193
Other Expenditures		5,864
Grand Total		$80,892
Value Of Mining Property		
Buildings		$19,745
Railroads		1,600
Underground Improvements		335,982
Machinery		121,550
Mine Supplies		13,515
Mine Proper		1,868,444
Cash		181
Total		$2,361,017
Value Of Mill Or Reduction Works		
Buildings		$27,425
Machinery		102,600
Supplies		550
Total		$130,575
Total Capital Invested		$2,491,592

Amount Of Bullion Produced Per Dollar Of Capital	**.01**
Value Of Bullion Per Dollar Of Expense	**.18**

STATISTICS OF GOLD AND SILVER MINES OF WYOMING 1889

MINES

NUMBER OF MINES

Producing Less Than $1,000	7
$1,000 to $10,000	4
Total of Producing Mines	11
Number of non-producing Mines	41
Number of Mines Idle	4
Total Number of Mines	**56**

MOTIVE POWER

Number of Boilers	8	Total Horsepower	378
Number of Engines	12	Total Horsepower	449
Number of Motors	7	Total Horsepower	70

(Other Power, Waterwheels, Whims and Windlass)

STATISTICS OF GOLD AND SILVER MINES IN WYOMING 1889

EMPLOYEES ABOVE GROUND

Average number employed

	Foremen	14
	Mechanics	7
	Laborers	51
	Total	72

Average wages per day

	Foremen	$3.86
	Mechanics	3.07
	Laborers	2.36

Average number of days worked

	Foremen	105
	Mechanics	56
	Laborers	81

EMPLOYEES BELOW GROUND

Average number employed

	Foremen	13
	Miners	130
	Laborers	11
	Total	154

Average Wages per Day

	Foremen	$4.84
	Miners	3.05
	Laborers	2.36

Average Number of Days Worked

	Foremen	80
	Miners	58
	Laborers	108

GOLD PRODUCTION STATISTICS BY COUNTY 1889

TOTAL NUMBER OF MINES

Albany County	15
Carbon "	4
Crook "	3
Fremont "	21
Laramie "	13

NUMBER OF MINES PRODUCING

Albany County	6
Fremont "	5

TOTAL AMOUNT OF ORE

Albany County	525	tons
Carbon "	10	"
Fremont "	701	"
Laramie "	100	"

TOTAL ORE TREATED

Albany County	250	tons
Fremont "	640	"

TOTAL BULLION VALUE

Albany County	$6,172
Fremont "	8,340

TOTAL EXPENDITURES

Albany County	$42,469
Carbon "	2,200
Crook "	250
Fremont "	25,637
Laramie "	10,156

TOTAL VALUE OF MINING PROPERTY

Albany County	$409,670
Carbon "	80,200
Crook "	11,000
Fremont "	1,070,147
Laramie "	790,000

GOLD PRODUCTION STATISTICS BY COUNTY 1889

EMPLOYEES ABOVE GROUND

Albany County		52
Crook	"	1
Fremont	"	17
Laramie	"	2

EMPLOYEES BELOW GROUND

Albany County		68
Carbon	"	10
Crook	"	1
Fremont	"	49
Laramie	"	26

MOTIVE POWER, BOILERS

Albany County		3
Fremont	"	4
Laramie	"	1

MOTIVE POWER, ENGINES

Albany County		6	(Total Horsepower...207)
Fremont	"	4	(Total Horsepower...177)
Laramie	"	1	(Total Horsepower...65)

NUMBER OF ANIMALS FOR POWER

Albany County		18
Carbon	"	6
Fremont	"	13
Laramie	"	2

GLOSSARY OF MINING TERMS

Adit; Entrance to a mine, usually a tunnel.

Amalgam; The combination of gold and mercury after the ore has and passed over the plates.

Argentiferous; Ore containing silver.

Arrastra; (Also spelled Arrastre, Arastra, etc.) A device to crush or grind ore (usually gold ore).

Assay; A test of ore to determine the content of precious metals.

Auriferous; Ore that contains gold.

Battery; A set of stamps for crushing ore. A stamp mill.

Bed Rock Flume; A water ditch dug down to bedrock to wash gold from its base material.

Bedrock; Solid rock underneath the earth and gravel.

Blossom Rock; Ore from nearby outcroppings found on the surface.

Chlorination; A chemical process of removing gold from its base material by the injection of chlorine gas.

Claim; To file a claim with either the district or the county recorder.

Color; A visible speck of gold.

Concentrator; Anyone of several devices for separating ore, usually a table that vibrates causing the heavier ore to separate.

Cradle; A device to improve the panning process. Usually a wooden trough with a cleated bottom that was continually rocked as ore was passed through it.

Cyanide Process; A method of separating gold from base material by dissolving it in Potassium Cyanide and then depositing it on zinc plates or strips.

Double Jack; The drilling of blasting holes by two men, one holding the rock drill while the second man strikes it with a eight pound sledge.

Drift; A mine tunnel, usually running from the main shaft.

Flume; A device (usually wooden) to carry water to the gold separating machinery, whether it be cradle, long tom or stamp mill.

Free Milling; Ore from which gold can be separated by mechanical means.

Frue Vanner; A variant of the concentrator, a table with a rubber top that both vibrates and tilts.

Giant Powder; An early name for dynamite.

Ground sluice; A ditch for washing gold from the base material.

Headframe; The framework over a mine shaft used to hoist ore from below.

Hydraulic mining; The working of placer deposits by the use of powerful streams of water to wash away alluvial deposits.

Lead; The visible course of the ore vein. (Pronounced to rhyme with greed.)

Ledge; A visible section of rock containing ore, used interchangeably with "lode".

Lode; A fissure in a rock formation which is filled with metalliferous ore.

Long Tom; A long wooden trough to separate gold from the ore by passing it over riffles or cleats on the floor of the trough.

Mucking; The process of loading the loose ore into a rail car after it has been blasted from the head of the shaft.

Patent; A grant by the government to an individual (or group) conveying to him fee simple title to public lands.

Placer; An alluvial deposit, usually gravel of various sizes.

Porphyry; Rock containing large red or purple crystals. Any very hard rock.

Quartz; Ore composed of crystallized silicon dioxide. Commonly used to denote hard rock gold ore.

Riffle; A groove or raised strip to catch gold in a sluice.

Rocker; Same as cradle.

Shaft; A mine sunk either vertically or on an incline as opposed to a tunnel which is near horizontal.

Single Jack; The drilling of blasting holes by one man who holds the rock drill with one hand while striking it with a three or four pound sledge with the other.

Sink; To dig a mine shaft.

Sluice; A man made channel for conducting water for washing gold ore.

Smelter; A furnace for extracting precious metal from the ore by the application of heat.

Stamp Mill; A machine to crush ore by using a series of heavy stamps which are raised by a camshaft and released to fall on the ore.

Tailings; The refuse material left after the ore has been removed.

Vein; A clearly defined body of ore lying between rocks that are said to be "in place."

Whim; A machine for hoisting that converts rotary motion to vertical lift.

End Notes

SOURCES FOR LAST CHANCE

The Frontier Index provides an early insight into the prospecting carried on in 1868. The *Laramie Sentinel* was established in 1869, however, early editions are not available for research. The first available are 1870 editions. The *Cheyenne Daily Leader* occasionally re-printed a newsworthy mining item from the early *Sentinel* and sometimes an original article of their own. The Albany County Mining Records provide the corroborating evidence for much of the mining activity. The records and newspapers are all available on microfilm in the State Archives or the University of Wyoming.

• Last Chance Endnotes

1. *Cheyenne Leader*, June 25, 1868, p. 1.
2. *Cheyenne Leader*, February 4, 1869.
3. Albany County, Book A, p. 6.
4. *Cheyenne Leader*, September 29, 1869.
5. F. V. Hayden, *Preliminary Report of the United States Geological Survey of Wyoming and Portions of Contiguous Territories*, Washington D. C. : Government Printing Office, 1872, p, 128.
6. J. H. Triggs, *History and Directory of Laramie City*, Wyoming Territory, (Laramie City, 1875).
7. Albany County, Book A, p. 6.
8. C. W. Shores letter to Louis Coughlin, October 10, 1927, MSS #297, p. 18.
9. *Cheyenne Daily Leader*, February 4, 1869.
10. A small stream entering Douglas Creek just below the present site of Rob Roy Dam. See G. L. O. Map T14N, R79W, 1879.
11. Albany County, Book A, p. 10; i. e. Stephen W. Downey, William O. Downey, Owen D. Downey, A. S. Blackbourne, John Metcalf, Thomas Feast, C. M. Hillaker, J. L. Roberts, E. C. Greene, D. C. VanEman, Thomas C. Powell, Ed Morgan, W. W. Heelum and Ira B. Moore.
12. Albany County, Book A, p. 10; John Crowley, Enoch Sutphen, Charles Tyler and Samuel Phillips.
13. Erwin, Marie, *Wyoming Blue Book*, Vol. 1, p. 68.
14. Erwin, Marie, *Wyoming Blue Book*, Vol. 1, p. 292.
15. Albany County, Book A, p. 13; i. e. Thomas Beard, Alexander Shaw, W. E. Henshaw, W. E. Porter.
16. Albany County, Book A, p. 14; i. e. Thomas Augenbach, James Duffield, John Augenbach, Edward Chaplin, William Warren, D. D. Wolff, Thomas Paxton and William Combs.
17. Reprinted in the *Cheyenne Daily Leader*, July 5, 1869.
18. Now known as Rush Creek. See G. L. O. Map T14N, R79W, 1879 and U. S. G. S. Keystone Quadrangle, 1969.
19. Now known as Bear Creek, see note 18.
20. W. C. Knight before the Mining Convention, Cheyenne, 1890.
21. *Laramie Daily Sentinel*, June 11, 1870.
22. *Laramie Daily Sentinel*, June 20, 1870.
23. Arrastra (Arrastre, Arraster), a primitive device for crushing ore.
24. The story of the killings was carried in several newspapers; the *Cheyenne Daily Leader*, July 7, July 20; the *Laramie Daily Sentinel*, July 12, July 14, August 12, 1870.
25. The *Laramie Daily Sentinel*, July 12, 1870.
26. Kuykendall, W. L., *Frontier Days*, privately printed, 1917, p. 145.
27. Albany County, Book A, pp. 55, 59.
28. Albany County, Book A, p. 60.
29. Raymond, R. W., *Mines and Mining West of the Rocky Mountains*, U. S. Government Printing Office, 1873.
30. Territory of Wyoming, Article of Incorporation #32, September 3, 1873.
31. *Laramie Daily Sentinel*, September 4, 1873.
32. Albany County Mining Records, 1869-1879, pp. 75, 101.
33. *Laramie Daily Sentinel*, June 23, 1874.
34. *Laramie Daily Sentinel*, June 30, 1874.

CENTENNIAL SOURCES

The *Laramie Daily Sentinel* is the outstanding source of material for the Centennial Mining District. The *Centennial Post* was not published until 1902 and early editions are not available for research. The *Post*, however, does on occasion make reference to the earlier days of Centennial and events which were certainly fresher in the minds of those who participated in the early boom days. The county mining records of Carbon and Albany provide verification of most of the claims. The General Land Office maps and records now held by the Bureau of Land Management are an excellent source of maps and land descriptions. The census and the voting record provide an insight to the fluctuations of population.

The fading days of the Centennial Mine are well described in the *Superintendents Report of the Centennial Mining Company* which is available at the State Archives and the University of Wyoming Library.

A visit to the area , however, remains the best way to get a feeling for the activities of a time gone past.

• Centennial Endnotes

1. Albany County Record of Mining Claims, 1869-1879, p. 112.
2. Albany County Record of Mining Claims, 1869-1879, p. 107 and Carbon County Book B, p. 1.
3. *Laramie Boomerang*, July 24, 1908.
4. Albany County Record of Mining Claims, 1869-1879, p. 110.
5. Albany County Record of Mining Claims, 1869-1879, p. 111.
6. Albany County Record of Mining Claims, 1869-1879, p. 108.
7. *Laramie Daily Sentinel*, August 3, 1875.
8. Microfilm Records of the Post Office Department, MA #2501.
9. Erwin, Marie, *Wyoming Blue Book*, Vol 1, p. 207a.
10. *Laramie Daily Sentinel*, November 23, 1875.
11. Albany County Mining Records, 1869-1879, p. 144.
12. *Laramie Daily Sentinel*, August 3, 1875.
13. Albany County Mining Records, 1869-1879, p. 155.
14. Albany County Mining Records, 1869-1879, p. 154.
15. *Laramie Daily Sentinel*, September 1, 1875.
16. *Laramie Daily Sentinel*, January 16, 1876.
17. *New York Time*, January 16, 1876.
18. *Laramie Daily Sentinel*, January 16, 1876.
19. *Cheyenne Daily Leader*, January 10, 1876.
20. Microfilm Records of the Secretary of State, Wyoming.
21. *Laramie Daily Sentinel*, August 16, 1876.
22. *Laramie Daily Sentinel*, August 5, 1876.
23. Carbon County Mining Record, B, p. 212.
24. Erwin, Marie, *Wyoming Blue Book*, Vol. 1, p. 298.
25. *Laramie Daily Sentinel*, January 27, 1877.
26. *The Centennial Gold Mining Company, Superintendents Report*, October 10, 1877.
27. Erwin, Marie, *Wyoming Blue Book*, Vol. 1, p. 216.
28. *Laramie Weekly Sentinel*, April 22, 1878.
29. United States General Land Office Survey, T15N, R78W, June 15, 1878.
30. *Laramie Sentinel*, July 12 and August 2, 1879.
31. Erwin, Marie, Wyoming Blue Book, Vol. 1, p. 223.
32. Erwin, Marie, Wyoming Blue Book, Vol. 1, p. 229.
33. *Laramie Sentinel*, December 25, 1880.
34. Erwin, Marie, Wyoming Blue Book, Vol. 1, p. 235, 243.

LA PLATA SOURCES

Although one of the more intriguing and certainly the most picturesque, the La Plata District has a rather meager list of sources. Through the years, many prospectors were involved here, but at any point in time there was a lesser concentration of people than the other camps. The harsh weather conditions, the short mining season and the difficulty in mining in a region that spewed icy water into every shaft that penetrated the rocky terrain was surely a deterrent. The La Plata mining era spanned several decades and information is scattered through the years. The Forest Service records hold little on the early mining but do contain an occasional reference to some of the activity around the turn of the century just prior to the creation of the National Forest. The Bureau of Land Management holds many of the General Land Office surveys and field notes that present many clues to the lay of the land and the location of the mines and mining camp. The county records, both Carbon and Albany, contain many claim files and transactions of the area. However, without the news papers I would have been unable to piece together the puzzle of land records. The *Laramie Sentinel* is the best single source of information, although pertinent articles are scattered through many issues requiring many hours of research and probably resulted in some information being overlooked. The *Centennial Post*, although not printed until 1902, contains references to, not only the La Plata activities of the time, but also an occasional note on earlier events. Not to be overlooked are the court records, including probate court which contain many details of the final days of some of the miners. As with all the old mining districts, a visit to the area is an excellent way to gain insight into this mining camp of the past.

• La Plata Endnotes

1. Albany County Mining Records, 1969-1879, pp. 64 - 66.
2. Albany County Mining Records, 1869-1879, p. 68.
3. Albany County Mining Records, 1869-1879, p. 69.
4. Carbon County Book B, p. 120.
5. *Laramie Sentinel*, May 21, 1878.
6. *Laramie Sentinel*, August 13, 1878.

7. *Laramie Sentinel*, August 10, 1878.
8. *Laramie Sentinel*, August 14, 1878.
9. *Laramie Sentinel*, August 20, 1878.
10. Carbon County Mining Record B, p. 177.
11. *Laramie Sentinel*, August 5, 1878.
12. Carbon County Mining Deed Record B, pp. 1,2.
13. Carbon County Book B, p. 175.
14. *Laramie Sentinel*, September 19, 20, 1878.
15. Carbon County Book B, p. 173.
16. Albany County Book 3, p. 287.
17. *Laramie Sentinel*, November 9, 1878.
18. *Laramie Sentinel*, November 23, 1878.
19. *Laramie Sentinel*, November 23, 1878.
20. Carbon County Commissioners Records, 1879.
21. Carbon County Book B, p. 176.
22. Carbon County Book B, p. 176.
23. Microfilm Records of the Post Office Department, MA #2501.
24. Carbon County Book B, p. 178.
25. Carbon County Mining Record B, p. 3.
26. Microfilm Records of the Post Office Department, MA #2501.
27. Albany County Mining Record 3, pp. 79, 99, 108, 115, 134.
28. Albany County Mining Record 3, p. 136.
29. Albany County Mining Record B, p. 402.
30. Albany County Mining Record B, p. 496.
31. Microfilm Records of the Post Office Department, MA #2501.
32. *Laramie Boomerang*, November 12, 1892.
33. Field Notes of re-survey of T16N, R79W, June 8, 1898.
34. State of Wyoming Article of Incorporation #6748.
35. *Laramie Republican*, July 6, 1904; *Centennial Post*, July 7, 1904.
36. Albany County Book 301, p. 380.

SOURCES FOR THE BIG LARAMIE MINES

As might be expected, the Laramie newspapers and the county mining records provide most of the information on Cummins City and the Big Laramie Mines. As a supplemental source, the other various county records includes the County Commissioneres proceedings, Coroners Reports, Probate Court Records and a cursory search of the background of the principle people involved. A detailed but yet unpublished history of Cummins City and the surrounding area has been done by Keith Jones, a copy of which is in the State Archives.

• Big Laramie Mines Endnotes

1. Variously spelled; McGreevy, McGreavey, McGreavy, McGreevey. When the Survey of 1873 passed along the territory's southern border, McGreavey's sawmill was near Diamond Peak. Apparently he moved his camp several times to take advantage of available timber. *The Laramie City Directory, 1875,* lists "McGreevy" as a contractor living at the corner of South A and and Street.
2. Erwin, Marie, *Wyoming Blue Book*, Vol. 1, p. 294.
3. *Laramie Daily Sentinel*, May 11, 1874. Albany County Record of Mining Claims 1869-1879, p. 76 also list W. S. Bramel and J. L. Murphy.
4. Albany County Book A, p. 59.
5. Albany County Record of Mining Claims, 1869-1879, p. 76.
6. Albany County Record of Mining Claims, 1869-1879, p. 112.
7. Erwin, Marie, *Wyoming Blue Book*, Vol. 1, p. 296.
8. Albany County Mining Record, 1869-1879, p. 129.
9. Albany County Mining Record, 1869-1879, pp. 135, 136.
10. *Laramie Daily Sentinel*, August 4, 1875.
11. Albany County Mining Record, 1869-1879, p. 165.
12. Albany County Mining Record, 1869-1879, p. 166.
13. Albany County Book A, p. 168; i. e. James T. Clark, Athen Clark, W. P. Davis, Day K. Smith, J. Ward Boyles, N. Shelton, James R. Scott, James McCorvan, John McCorvan, Ianil Smith, Wm. Woodruff and Milton Woods.
14. *Laramie Daily Sentinel*, August 16, December 12, 1876.
15. Albany County Mining Record, 1869-1879, p. 169.
16. Albany County Mining Record, 1869-1879, p. 176.
17. *Laramie Daily Sentinel*, December 12, 1876.
18. Erwin, Marie, *Wyoming Blue Book*, Vol. 1, p. 298.
19. Albany County Mining Record, 1869-1879, p. 174.
20. Whim; a vertical hoist, sometimes horse powered.
21. *Laramie Daily Sentinel*, April 20, 1878.
22. Albany County Probate Court Records, 1878, #32.
23. GLO Survey, T13, R77W, June 3, 1878.
24. Albany County Book B, p. 498.
25. Albany County Book C, p. 131.

26. Proceedings of the Albany County Commissioners, pp. 438, 479-489.
27. Albany County Mining Records, 1869-1879, 176.
28. Albany County Mining Records, 1869-1879, 176.
29. *Laramie Daily Sentinel*, April 24, 1880, p. 3, Col. 3.
30. *Laramie Daily Sentinel*, April 24, 1880, p. 3, Col. 2.
31. *Laramie Daily Sentinel*, April 24, 1880, p. 3, Col. 3.
32. Albany County Book C, p. 132.
33. *Laramie Daily Sentinel*, May 1, 1880.
34. *Laramie Daily Sentinel*, May 8, 1880.
35. *Laramie Daily Sentinel*, May 22, 1880.
36. Albany County Clerk of Court, Judgement Docket, #1003 and #1004.
37. Cummins City Plat, on file in Albany County Courthouse; also *Laramie Daily Sentinel*, May 15, 1880.
38. *Laramie Sentinel*, May 29, 1880.
39. *Laramie Sentinel*, May 15, 1880.
40. Albany County Book B, p. 7.
41. *Laramie Sentinel*, May 29, 1880.
42. Albany County Book C, p. 210.
43. Albany County Book C, 212, 213.
44. WPA #1912, Biography of H. V. S. Groesbeck.
45. *Laramie Sentinel*, August 7, 1880.
46. The *Cheyenne Daily Leader* of October 3, 1880 quoting the *Laramie Times*, October 2, 1880.
47. *Laramie Sentinel*, August 7, 1880.
48. *Laramie Sentinel*, October 23, 1880.
49. Erwin, Marie, *Wyoming Blue Book*, Vol. 1, pp. 303, 404, 406.
50. *Report on Mineral Industries in the United States*, U. S. Government Printing Office, 1892, p. 41.
51. *Annals of Wyoming* 60:2:36.
52. Cummins City Court Docket, p. 1.
53. *Laramie Sentinel*, April 3, 1881.
54. *Laramie Sentinel*, April 16, 1881.
55. *Laramie Sentinel*, May 28, 1881.
56. Territory of Wyoming Article of Incorporation #121, April 4, 1881.
57. Territory of Wyoming Article of Incorporation #124, April 30, 1881.
58. Albany County Liens Book B, p. 5.
59. *Laramie Sentinel*, June 18, 1881.
60. *Laramie Sentinel*, July 2, 1881.
61. *Laramie Sentinel*, September 3, 1881.
62. *Laramie Sentinel*, August 27, 1881.
63. Proceedings of the Albany County Commissioners, pp. 524, October 22, 1881.
64. *Cheyenne Daily Sun*, January 20, 1882.
65. *Cheyenne Daily Sun*, January 20, 1882.
66. GLO Survey No. 1 District No. 9, p. 572.
67. *Cheyenne Daily Sun*, August 2, 1882.
68. Albany County District Court, Civil Action No. 1208.
69. Proceedings of the Albany County Commissioners, pp. 574, 575, October 3, 1882.
70. Erwin, Marie, *Wyoming Blue Book*, Vol. 1, pp. 234, 438, 411, 413.
71. *Cheyenne Daily Sun*, December 15, 1882.
72. *Laramie Boomerang*, September 22, 1881.

SOURCES FOR JOHN CUMMINS

Aside from the same sources as for the Big Laramie Mines, the Colorado newspapers of Trinidad and Pueblo provide information on the escapades of John Cummins. However, some incidents have not been verified and may be erroneous. Additional research on John Cummins is in order and should be completed by an accomplished historian / geneologist who is willing to take the time and effort to do an in-depth search.

• John Cummins Endnotes

1. Not everyone shared Thompson's high opinion of Frank Smith. Herman Groesbeck called Smith mean, vulgar, stingy, a wife beater, a child neglector, an animal abuser and a heavy drinker during Smith's divorce proceedings. H. V. S. Groesbeck Collection #596, University of Wyoming Coe Library.
2. Albany County Book E. p. 59.
3. Letter from California State Archives to Author, October 21, 1988. Letter from California Historical Society to Author, December 14, 1988.
4. The *Laramie Boomerang*, June 12, 1893.
5. Albany County Mining Deed Book O, p. 623.
6. Albany County Book C, p. 163.
7. The *Inter-Ocean Water Power and Mining Company*, Pamphlet.
8. Biography of H. V. S. Groesbeck, WPA #1912.
9. Wyoming Industrial Convention Pamphlet, 1901.

SOURCES FOR NORTH PARK AND TELLER

The best source for this Colorado mining camp is probably the Laramie newspapers, primarily the *Laramie Sentinel*. A good secondary source is the *Larimer County Express* and the *Fort Collins Express*. Since many Cheyenne citizens were involved, the *Cheyenne Daily Leader* is also a good source of activity in the area. The few remaining papers that were printed in the camp are excellent sources, however they are very limited. These are located in the Colorado Archives.

The original county records seem to have been lost, however, Jackson County has transcriptions of some, or possibly all, of the original claim filings. Larimer County has records of transactions after 1880. The first maps of the area were apparently the General Land Office surveys which took place in 1881 and 1882. Later maps are also useful to familiarize one with the mining area. As with all the camps, A visit to the old camp is still the best way to get a feeling of what transpired here in the 1880s.

• North Park And Teller Endnotes

1. Medicine Bow National Forest Collection, MSS #297.
2. F. Trench Townshend, *Ten Thousand Miles*, London, 1869.
3. F. V. Hayden, *U. S. G. S. of Wyoming*, Washington D. C., 1872.
4. *Annals of Wyoming*, 60:2:36.
5. *Annals of Wyoming*, 7:4:476,478 & 479.
6. *Colorado Magazine*, XIV, p. 230.
7. *Laramie Sentinel*, October 9, 1876.
8. *Colorado Magazine*, XLVIII, No. 4. p. 273.
9. G. L. O. maps of Wyoming and Colorado. T6, 7, 8, 11 & 12 N, R77, 78, 79 & 80 W.
10. *Laramie Sentinel*, May 28, June 16, 1881.
11. Territory of Wyoming Article of Incorporation #182.
12. *Larimer County Express*, March 19, May 28 and June 11, 1880.
13. *Saratoga Sun*, April 11, 1912.
14. Jackson County Book A-1, p. 69.
15. Jackson County Book A-1, p. 81, #7265.
16. Jackson County Book A-1, p. 73, #7160.
17. Jackson County Book A-1, p. 80, #7090.
18. *Cheyenne Daily Sun*, July 22, 1879.
19. Jackson County Book A-1, p. 74, #7203.
20. Jackson County Book A-1, p. 75, #7207.
21. Jackson County Book A-1, p. 80, #7262.
22. *Larimer County Express*, June 4, 1880.
23. *Larimer County Express*, June 18, 1880.
24. *Cheyenne Daily Leader*, March 24, 1880.
25. *Laramie Sentinel*, March 27, 1880.
26. *Laramie Sentinel*, March 27, 1880.
27. *Cheyenne Daily Leader*, April 18, 1880.
28. *Larimer County Express*, June 4, 1880.
29. *Cheyenne Daily Leader*, June 24, 1880.
30. Bartlett, I. S., *History of Wyoming*, p. 136, 137. Bartlett refers to this trip as taking place in the summer of 1879. However, neither Teller City nor Cummins City were located at that time. It appears his trip was during the summer of 1880.
31. *Laramie Sentinel*, October 9, 1880.
32. *Larimer County Express*, November 11, 1880.
33. *Cheyenne Daily Leader*, November 27, 1880.
34. *Grand County Times*, May 12, 1881.
35. Miss Clark was most likely the daughter of W. H. Clark editor and proprietor of the *Grand County Times*.
36. *Grand County Times*, May 12, 1881.
37. G. L. O. Survey T5N, R77W, July 11, 1882.
38. *Laramie Sentinel*, June 9, 1883.
39. *The North Park Miner*, September 6, 1883.
40. *The Laramie Republican-Boomerang*, December 17, 1931.
41. Larimer County Book 27, p. 54.

DOUGLAS CREEK SOURCES

The *Laramie Sentinel* is the primary source for the early saga of the Douglas Creek mines. The *Laramie Boomerang* is a good source after the *Sentinel* ceased publication in 1895. Whether Albany County transcribed records from Carbon County or whether Carbon transcribed records from Albany County, most of the mining transactions can be verified in one or the other. The voting record and the census provide an insight to the people of the mining camp. The records of the Post Office Department also are useful. The "Keystone Ledgers" are of limited use but do provide verification of the names of those who had transactions with the company. Earliest maps and plats of the mines appear to be those of the General Land Office with later maps providing reference to the area. Again, a visit to the area, although revealing very few remnants of the early camp, gives one a feeling of the region and possibly some idea of the lifestyle of the miners.

• Douglas Creek Endnotes

1. Carbon County, Book B, p. 85 - 87, 103.
2. *Laramie Daily Sentinel*, September 26, 1876.
3. Although there is no specific reference to Foley's Stamp Mill being transported and set up, reference to its purchase was made in the *Laramie Daily Sentinel* of September 2, 1876, and reference to its location in Carbon County Book B, p. 86.
4. Albany County Mining Records, Book 3, p. 536.
5. Albany County Mining Records, Book 3, p. 45 and Carbon County Book B, p. 237.
6. Albany County Mining Records, Book 3, p. 100.
7. Albany County Mining Records, Book 3, p. 33.

8. Microfilm Records of the Post Office Department, MA #2501.
9. *Laramie Daily Sentinel*, January 14, 1878.
10. *Laramie Weekly Sentinel*, February 26, 1878.
11. *Laramie Weekly Sentinel*, March 4, 1878.
12. Albany County Book A, p. 139 and the Keystone Ledger, 1878.
13. *Laramie Daily Sentinel*, March 25, 1878.
14. *Laramie Weekly Sentinel*, March 11, 1878.
15. *Laramie Weekly Sentinel*, April 27, 1878.
16. *Laramie Daily Sentinel*, June 10, 1878.
17. *Laramie Daily Sentinel*, July 1, 1878.
18. *Laramie Weekly Sentinel*, July 6, 1878.
19. *Laramie Daily Sentinel*, July 13, 1878.
20. *Laramie Weekly Sentinel*, July 6, 1878.
21. *Laramie Weekly Sentinel*, July 13, 1878.
22. *Laramie Weekly Sentinel*, July 13, 1878.
23. *Laramie Daily Sentinel*, July 16, 1878.
24. GLO Survey of T14N, R79W.
25. *Laramie Daily Sentinel*, July 22, 1878.
26. *Laramie Daily Sentinel*, August 24, 1878.
27. *Laramie Daily Sentinel*, September 11, 1878.
28. *Laramie Daily Sentinel*, September 23, 1878.
29. *Laramie Weekly Sentinel*, September 21, 1878.
30. *Laramie Daily Sentinel*, September 24, 1878.
31. *Laramie Weekly Sentinel*, September 28, 1878.
32. *Laramie Weekly Sentinel*, October 11, 1878.
33. Microfilm Records of the Post Office Department, MA #2501.
34. *Laramie Daily Sentinel*, October 9, 1878.
35. *Laramie Daily Sentinel*, November 1, 1878.
36. Erwin, Marie, *Wyoming Blue Book*, Vol. 1, p. 301.
37. *Laramie Daily Sentinel*, November 20, 1878.
38. *Laramie Daily Sentinel*, December 20, 1878.
39. *Laramie Daily Sentinel*, December 30, 1878.
40. *Laramie Sentinel*, January 3, 1879.
41. *Laramie Sentinel*, January 31, 1879.
42. Reprinted in the *Laramie Sentinel*, February 21, 1879.
43. *Laramie Sentinel*, February 14, 1879.
44. Microfilm Records of the Post Office Department, MA #2501.
45. *Laramie Sentinel*, February 14, 1879.
46. *Laramie Sentinel*, February 28, 1879.
47. *Laramie Sentinel*, March 7, 1879.
48. *Laramie Sentinel*, March 7, 1879.
49. *Laramie Sentinel*, April 5, 1879.
50. *Laramie Sentinel*, April 25, 1879.
51. *Laramie Sentinel*, May 16, 1879.
52. Erwin, Marie, Wyoming Blue Book, Vol. 1, p. 223.
53. Erwin, Marie, Wyoming Blue Book, Vol. 1, p. 229.
54. *Report on Mineral Industries in the United States*, U. S. Government Printing Office, 1892.
55. Carbon County Book B, p. 188.
56. Microfilm Records of the Post Office Department, MA #2501.
57. Territory of Wyoming Article of Incorporation #147.
58. *Annals of Wyoming*, 20:1:5.
59. Albany County Probate Court Records, Index #43.
60. *Laramie Sentinel*, December 8, 1883.
61. Territory of Wyoming Article of Incorporation #182.
62. Erwin, Marie, Wyoming Blue Book, Vol. 1, p. 411, 413.
63. *U. S. G. S. Mineral Resources Report*, 1891, p. 76.
64. Microfilm Records of the Post Office Department, MA #2501.
65. Carbon County Mining Deeds, Volume B, p. 404.
66. *Laramie Sentinel*, December 8, 1883.
67. *U. S. G. S. Mineral Resources Report*, 1891, p. 76.
68. Carbon County Mining Record B, p. 401.
69. *Laramie Sentinel*, August 2, 1884.
70. *U. S. G. S. Mineral Resources Report*, 1891.
71. *Laramie Sentinel*, November 5, 1887.
72. *Laramie Boomerang*, September 29, 1901.
73. *Laramie Boomerang*, September 29, 1901.
74. Microfilm Records of the Post Office Department, MA #2501.
75. *Laramie Boomerang*, January 13, 1890.
76. Albany County Book M, p. 288.
77. *Laramie Weekly Sentinel*, March 21, 1891; U. S. Mineral Application No. 79.
78. Microfilm Records of the Post Office Department, MA #2501.
79. *Laramie Boomerang*, November 12, 1892.
80. *Laramie Boomerang*, October 1, 1892.
81. State of Wyoming Article of Incorporation #1560.
82. Microfilm Records of the Post Office Department MA #2501. Holmes was, years later, said to be named for A. T. Holmes. This, however, is questionable as John E. Holmes was the first postmaster and A. T. Holmes was apparently not in the area until later.
83. State of Wyoming Articles of Incorporation #1560.
84. Albany County Book O, p. 383.
85. *Laramie Boomerang*, December 16, 1895.
86. *Albany County, Wyoming, Mineral Resources*, The Laramie Mining and Stock Exchange, April, 1896.

87. The *Laramie Boomerang*, May 27, 1896, reported the purchase but the county records do not support the transaction, this was probably a lease or option to buy.
88. Microfilm Records of the Post Office, MA #2501.
89. *Laramie Republican*, November 21, 1900.
90. *Laramie Boomerang*, September 29, 1901.
91. Microfilm Records of the Post Office Department MA #2501.
92. State of Wyoming Articles of Incorporation #1966
93. Albany County Book F, p. 440.
94. *Laramie Boomerang*, September 17, 1904.
95. *Laramie Boomerang*, September 18, 1904; Albany County Mining Record B, p. 304. The *Saratoga Sun*, of September 13, 1888 had also noted that "Kennesaw" was an alias or sobriquet, assumed after the Civil War Battle for Kennesaw Mountain.
96. Microfilm Records of the Post Office Department MA #2501.
97. Albany County Book G, p. 205.
98. Albany County Book G, p. 206.
99. *Laramie Boomerang*, April 2, 1908.
100. Albany County Book 119, p. 199.
101. *Wyoming Eagle*, July 13, 1939.

GOLD HILL SOURCES

Gold Hill probably has more information available than any of the other Medicine Bow Mining camps. The *Platta Valley Lyre* and the *Saratoga Sun* were printed close to the camp and were directly involved with the boom. The *Wyoming Commonwealth* began printing about the same time that Gold Hill began its little boom and not only reprinted articles from other newspapers but occasionally originated their own stories. The *Laramie Sentinel* and *Laramie Boomerang* continued to provide some of the best information to tie all of the activity together. The county land and mining records verify many of the claims and locations. The General Land Office records are the best source for early maps and plats. The census provides little or no information but the Post Office records give some information on the camp. The County Commissioners left some record of their concern for the developement of Gold Hill and Saratoga.

• Gold Hill Endnotes

1. *Ten Thousand Miles of Travel Sport and Adventure*, F. Trench Townshend, London, 1869.
2. *Laramie Daily Sentinel*, June 19, 1875.
3. Carbon County Book B, p. 90.
4. Microfilm Records of the Post Office Department, MA #2501.
5. Carbon County Book B, p. 372.
6. *Platte Valley Lyre*, September 13, 1888.
7. *Laramie Boomerang*, April 6, 1912, and *Saratoga Sun*, April 11, 1912.
8. Carbon County Book K, p. 184.
9. Carbon County Book K, p. 186.
10. *Platte Valley Lyre*, September 13, 1888.
11. Carbon County Book K, p. 187.
12. Carbon County Book K, p. 194.
13. *Platte Valley Lyre*, March 28, 1889.
14. *Platte Valley Lyre*, April 11, 1889.
15. *Platte Valley Lyre*, April 11, 1889, and Territory of Wyoming Articles of Incorporation #787.
16. *Platte Valley Lyre*, May 2, 1889.
17. Carbon County Book K, p. 312.
18. Carbon County Book K, p. 313.
19. Carbon County Book K, p. 316.
20. Carbon County Book K, p. 314.
21. Carbon County Book K, p. 315.
22. Carbon County Book K, p. 323.
23. *Wyoming Commonwealth*, September 14, 1890.
24. *Cheyenne Daily Leader*, September 23, 1890.
25. *Wyoming Commonwealth*, October 12, 1890.
26. *Platte Valley Lyre*, October 23, 1890.
27. *Platte Valley Lyre*, October 23, 1890.
28. *Platte Valley Lyre*, October 23, 1880.
29. *Wyoming Commonwealth*, October 12, 1890.
30. Carbon County Book K, pp. 577-582.
31. Carbon County Book K, p. 583.
32. *Platte Valley Lyre*, November 6, 1890.
33. Carbon County Book K, p. 587.
34. Carbon County Book K, p. 584.
35. G. L. O. Mineral Survey No. 55, p. 573.
36. *Wyoming Commonwealth*, October 5, 1890.
37. *Wyoming Commonwealth*, November 25, 1890.
38. *Wyoming Commonwealth*, November 30, 1890.
39. *Wyoming Commonwealth*, November 30, 1890.
40. *Platte Valley Lyre*, December 25, 1890.
41. *Wyoming Commonwealth*, November 30, 1890.
42. *Wyoming Commonwealth*, January 4, 1891.
43. *Platte Valley Lyre*, January 15, 1891.
44. *Platte Valley Lyre*, January 30, 1891.
45. *Cheyenne Daily Leader*, March 8, 1891.
46. *Laramie Republican*, January 2, 1891.
47. *Platte Valley Lyre*, April 12, 1891.
48. Microfilm Records of the Post Office Department, MA #2501.
49. *Platte Valley Lyre*, January 29, 1891.
50. *Laramie Republican*, February 7, 1891.
51. *Platte Valley Lyre*, April 16, 1891.
52. *Laramie Republican*, March 6, 1891.
53. *Cheyenne Daily Leader*, March 1, 1891.

54. *Platte Valley Lyre*, April 19, 1891.
55. *Wyoming Commonwealth*, March 15, 1891.
56. *Wyoming Commonwealth*, March 15, 1891.
57. *Platte Valley Lyre*, March 19, 1891.
58. *Platte Valley Lyre*, March 5, 1891.
59. *Platte Valley Lyre*, March 19, 1891.
60. *Platte Valley Lyre*, March 19 and April 16, 1891.
61. *Wyoming Commonwealth*, April 19, 1891.
62. *Wyoming Commonwealth*, March 15, 1891 and State of Wyoming Article of Incorporation #1074.
63. *Platte Valley Lyre*, April 30, 1891.
64. *Wyoming Commonwealth*, April 19, 1891.
65. *Platte Valley Lyre*, April 30, 1891.
66. *Platte Valley Lyre*, May 14, 1891.
67. *Wyoming Commonwealth*, May 10, 1891.
68. *Carbon County Journal*, May 9, 1891.
69. *Carbon County Journal*, May 16, 1891.
70. *Platte Valley Lyre*, May 7, 1891.
71. *Carbon County Journal*, May 9, 1891.
72. *Platte Valley Lyre*, June 11, 1891.
73. *Wyoming Commonwealth*, July 19, 1891.
74. *Platte Valley Lyre*, July 19, 1891.
75. *Wyoming Commonwealth*, July 26, 1891.
76. *Platte Valley Lyre*, July 23, 1891.
77. *Laramie Republican*, April 29, 1891.
78. *Saratoga Sun*, September 8, and *Platte Valley Lyre*, September 10, 1891.
79. Carbon County Book 27, p. 408.
80. *Saratoga Sun*, October 8, 1891.
81. *Wyoming Commonwealth*, October 3, 1891.
82. *Saratoga Sun*, October 15, 1891.
83. Microfilm Records of the Post Office Department, MA #2501.
84. *Wyoming Commonwealth*, October 17, 1891.
85. State of Wyoming Article of Incorporation #1290.
86. State of Wyoming Article of Incorporation #1283.
87. *Saratoga Sun*, December 17, 1891.
88. *Saratoga Sun*, January 7, 1892.
89. *Saratoga Sun*, January 14, 1892.
90. *Saratoga Sun*, January 21, 1892.
91. State of Wyoming Article of Incorporation #1351.
92. *Saratoga Sun*, January 28, 1892.
93. *Saratoga Sun*, January 28, and February 11, 1892.
94. *Saratoga Sun*, February 11, 1892, Advertisement.
95. *Saratoga Sun*, February 4, 1892.
96. *Saratoga Sun*, February 11, 1892.
97. *Saratoga Sun*, February 11, 1892.
98. *Saratoga Sun*, February 11, 1892.
99. *Saratoga Sun*, February 4, 1892.
100. Microfilm Records of the Post Office Department, MA #2501, and the *Saratoga Sun*, January 21, 1892.
101. *Saratoga Sun*, February 25, 1892.
102. *Saratoga Sun*, March 10, 1892.
103. *Saratoga Sun*, March 24, 1892.
104. *Saratoga Sun*, March 24, 31, 1892.
105. *Saratoga Sun*, April 21, 1892.
106. *Saratoga Sun*, April 21, 1892.
107. Carbon County Sheriff's Report, January 4, 1892.
108. *Saratoga Sun*, August 11, 1892.
109. *Rawlins Republican*, August 19, 1892.
110. *Saratoga Sun*, September 1, 1892.
111. Microfilm Records of the Post Office Department, MA #2501.
112. *Laramie Daily Boomerang*, January 4, 1893.
113. *Saratoga Sun*, January 12, 1893.
114. *Saratoga Sun*, January 5, 12, 1893.
115. *Saratoga Sun*, March 9, 1893.
116. *Saratoga Sun*, August 31, 1893.
117. *Satatoga Sun*, November 9, 1893.
118. *Saratoga Sun*, January 11, 1894.
119. Microfilm Records of the Post Office Department, MA #2501.
120. *Laramie Republican-Boomerang*, March 21, 1931.

SOURCES FOR COOPER HILL

The county mining records provide a good basic source of background information but it takes the newspapers of the time, primarily the *Laramie Boomerang*, to bring all this information together in some semblence of order. The State records of incorporation provide some insight into the great claims and expectations of the miners. A little delving into the life history of those involved brings out a bit of the more private aspects of the miners and their families.

• Cooper Hill Endnotes

1. *Laramie Boomerang*, January 9, 1893.
2. Microfilm records of the Post Office Department, MA #2501.
3. *Laramie Boomerang*, January 14, 1893.
4. *Laramie Boomerang*, January 24, 1893.
5. Carbon County Book 34, p. 191.
6. Carbon County Book 34, p. 206
7. *Laramie Boomerang*, January 24, 1893.
8. *Laramie Boomerang*, December 20, 1893.
9. *Laramie Boomerang*, March 8, 1894.
10. *Laramie Boomerang*, January 5, 1894.
11. *Laramie Boomerang*, March 8, 1894.
12. Carbon County, Book 34, p. 210.

13. *Laramie Boomerang*, May 14, 1894.
14. *Laramie Boomerang*, May 28, 1894.
15. *Laramie Boomerang*, August 15, 1894.
16. *Laramie Boomerang*, September 6, 7, 1894.
17. *Laramie Boomerang*, January 25, 1895.
18. *Laramie Boomerang*, April 24, 1895.
19. *Laramie Boomerang*, May 24, 1895.
20. *Laramie Boomerang*, July 30, 1895.
21. *Laramie Boomerang*, September 13, 1895.
22. *Laramie Boomerang*, January 3, 1896.
23. State of Wyoming Article of Incorporation #1801.
24. State of Wyoming Article of Incorporation #1838.
25. State of Wyoming Article of Incorporation #1858.
26. State of Wyoming Article of Incorporation #1874.
27. Plat on file in Carbon County Assessors Office.
28. *Laramie Boomerang*, July 3, 1896.
29. *Laramie Boomerang*, August 13, 1896.
30. *Laramie Boomerang*, September, 18, 1896.
31. *Laramie Boomerang*, October 3, 1896.
32. *Laramie Boomerang*, October 3, 1896.
33. *Laramie Boomerang*, August 10, 1896.
34. Microfilm Records of the Post Office Department MA #2501.
35. *Laramie Boomerang*, April 27, 1896.
36. *Laramie Boomerang*, January 8, March 2, April 10, 1897.
37. State of Wyoming Article of Incorporation, #4459.
38. *Laramie Boomerang*, April 10, May 12, 1897.
39. *Laramie Boomerang*, October 15, 1897.
40. State of Wyoming Article of Incorporation #6072.
41. *Annals of Wyoming*, 42:1:57.

EARLY MINING SOURCES

The sources for the Early Mining Chapter are many and varied. The mines and incidents described in this chapter were found scattered in the many newspapers and documents used in researching the entire book. Undoubtedly there are other stories of early mining and "lost" mines which are not included but enough are included to convey the impressions of the Medicine Bow Miners.

• Early Mining Endnotes

1. *Laramie Sentinel*, April 24, 1875.
2. Strahorn, Robert E. , *Hand-book of Wyoming*, 1877.
3. *Annals of Wyoming*, 36:1:59.
4. *Annals of Wyoming*, 35:2:207.
5. *Saratoga Sun*, July 28, 1891.
6. *Saratoga Sun*, July 28, 1891.
7. *Annals of Wyoming*, 6:1&2:207.
8. *Laramie Republican*, April 25, 1891.
9. *Laramie Sentinel*, August 27, 1881.
10. Wyoming Industrial Convention Pamphlet, 1901.
11. *Mineral and Allied Resources*, 1906, The Laramie Republican Printers.
12. WPA #225.
13. Homsher, Lola M. , *History of Albany County, Wyoming to 1880.*
14. *Colorado Magazine*, September 1931.
15. *Laramie Sentinel*, July 24, 1893.
16. *Mineral Resources*, The Laramie Mining and Stock Exchange, April 1906.
17. Albany County Mining Record B, p. 178.
18. Byers and Kellom, *Goldfields of Nebraska and Kansas.*
19. Greever, William S. , *Bonanza West*, University of Idaho Press, 1963, p. 158.
20. *Wyoming Commonwealth*, January 4, 1901.
21. *Laramie Sentinel*, April 21, 1883.
22. *Laramie Boomerang*, August 2, 1901.
23. *Laramie Boomerang*, December 20, 1894.
24. *Laramie Boomerang*, December 20, 1894.
25. The story of the Lost Cabin Mine appears many times over the years, one which is the *Casper Star Tribune*, December 4, 1966. The "Lost Seminole" mine story was printed in the *Cheyenne Daily Leader*, September 21, 1890 and has since been repeated several times.

EPILOGUE ENDNOTES

1. Some years earlier the a population figure of 60,000 had been set as a requirement for Statehood, but as the Union grew, Congress was reluctant to allow entry of territories with small populations, viz; Colorado, 150,000 in 1876; the combined Dakotas, 500,000 in 1889; Montana, 140,000 in 1889; Washington, 350,000 in 1889.
2. John W. Blake came to Wyoming Territory in 1874 and spent his early days engaged in mining in the Centennial District. He also had claims near Cummins City, the La Plata District and Cooper Hill. In 1880, the early days of the Teller, Colorado mining camp, he was the Lead Mountain District Recorder. He was a member of the Territorial Legislature in 1886 and 1888. In 1892, he was involved in judging the participants of the Johnson County War.

—◆—

INDEX